AF803342

Aquatic Ecosyst
Microbial Bio

Aquatic Ecosystems and Microbial Biofilms: Signi ***and Control*** provides a systematic introduction and mation on microbial biofilms in aquatic ecosystems designed and developed to understand the microbial their role, and the control strategies. The contents o get state-of-art knowledge on various topics such a ecosystems, microbial biofilms, and drinking water and biocorrosion, and beneficial aspects of biofilms wastewater treatment, and in biodeterioration of or provides comprehensive knowledge and in-depth sci biofilms and their contribution to antibiotic resistan gies to understand biofilms such as metagenomics.

The book offers comprehensive coverage of the n

- Microbial biofilms in aquatic ecosystems.
- New horizons to understand the role of biofilm their control measures.
- Beneficial role of aquatic biofilms such wastewater treatment, and biodeterioration of
- Various strategies to collaborate interdisciplina and develop new methods for cleaner drink advanced techniques such as metagenomics functional role of aquatic biofilms.

This book serves as a reference book for scientific study biofilms in aquatic ecosystems, as well as rese in this field to study biofilm formation in aquatic e disadvantages, and control strategies.

Aquatic Ecosystems and Microbial Biofilms

Significance, Dynamics, Prevention, and Control

Edited by
Siddhardha Busi and Ram Prasad

CRC Press
Taylor & Francis Group
Boca Raton London New York

CRC Press is an imprint of the
Taylor & Francis Group, an informa business

First edition published 2025
by CRC Press
2385 NW Executive Center Drive, Suite 320, Boca Raton FL 33431

and by CRC Press
4 Park Square, Milton Park, Abingdon, Oxon, OX14 4RN

CRC Press is an imprint of Taylor & Francis Group, LLC

© 2025 selection and editorial matter, Siddhardha Busi and Ram Prasad; individual chapters, the contributors

Reasonable efforts have been made to publish reliable data and information, but the author and publisher cannot assume responsibility for the validity of all materials or the consequences of their use. The authors and publishers have attempted to trace the copyright holders of all material reproduced in this publication and apologize to copyright holders if permission to publish in this form has not been obtained. If any copyright material has not been acknowledged please write and let us know so we may rectify in any future reprint.

Except as permitted under U.S. Copyright Law, no part of this book may be reprinted, reproduced, transmitted, or utilized in any form by any electronic, mechanical, or other means, now known or hereafter invented, including photocopying, microfilming, and recording, or in any information storage or retrieval system, without written permission from the publishers.

For permission to photocopy or use material electronically from this work, access www.copyright.com or contact the Copyright Clearance Center, Inc. (CCC), 222 Rosewood Drive, Danvers, MA 01923, 978-750-8400. For works that are not available on CCC please contact mpkbookspermissions@tandf.co.uk

Trademark notice: Product or corporate names may be trademarks or registered trademarks and are used only for identification and explanation without intent to infringe.

ISBN: 9781032782416 (hbk)
ISBN: 9781032782904 (pbk)
ISBN: 9781003487203 (ebk)

DOI: 10.1201/9781003487203

Typeset in Times New Roman
by Newgen Publishing UK

Contents

Contributors

V.T. Anju
Department of Biochemistry and Molecular Biology, School of Life Sciences, Pondicherry University, Pondicherry, India

Megha Barot
Department of Applied Science and Humanities, Sankalchand Patel University, Visnagar, Gujarat, India

Siddhardha Busi
Department of Microbiology, School of Life Sciences, Pondicherry University, Pondicherry, India

Natarajan Chandrakala
PG & Research, Department of Zoology, Kunthavai Naachiyaar Government Arts College (Women) Autonomous, Thanjavur, Tamilnadu, India

Indranil Chattopadhyay
Department of Biotechnology, School of Life Sciences, Central University of Tamil Nadu, Neelakudi, Thiruvarur, India

Sudha Charita Chillarige
Department of Biochemistry and Bioinformatics, GITAM School of Science, GITAM (Deemed to be) University, Visakhapatnam, India

Gutta Rishitha Chowdhary
Department of Biochemistry and Bioinformatics, GITAM School of Science, GITAM (Deemed to be) University, Visakhapatnam, India

Hiral Desai
Department of Biotechnology, Sankalchand Patel University, Visnagar, Gujarat, India

Madhu Dyavaiah
Department of Biochemistry and Molecular Biology, School of Life Sciences, Pondicherry University, Pondicherry, India

Jayaraman Eswari
PG & Research, Department of Zoology, Kunthavai Naachiyaar Government Arts College (Women) Autonomous, Thanjavur, Tamilnadu, India

Ganga G.
Department of Microbiology, Sree Ayyappa College, Eramallikara, Chengannur Kerala, India

Pitchaipillai Sankar Ganesh
Department of Microbiology, Centre for Infectious Diseases, Saveetha Dental College and Hospitals, Saveetha Institute of Medical and Technical Sciences (SIMATS), Saveetha University (Deemed to be University), Chennai, Tamilnadu, India

R.V. Geetha
Department of Microbiology, Centre for Infectious Diseases, Saveetha Dental College and Hospitals, Saveetha Institute of Medical and Technical Sciences (SIMATS), Saveetha University (Deemed to be University), Chennai, Tamilnadu, India

Rajesh Kanna Gopal
Department of Microbiology, Centre for Infectious Diseases, Saveetha Dental College and Hospitals, Saveetha Institute of Medical and Technical Sciences (SIMATS), Saveetha University (Deemed to be University), Chennai, Tamilnadu, India

Anitha K.
Department of Biosciences and Biotechnology, Krishna University, Rudravaram, Machilipatnam, India

Mounika K.
Department of Biosciences and Biotechnology, Krishna University, Rudravaram, Machilipatnam, India

Srujana Kathi
UGC-Malviya Mission Teacher Training Centre, Pondicherry University, Puducherry, India

Ashok Kota
Department of Biosciences and Biotechnology, Krishna University, Rudravaram, Machilipatnam, India

Pradeep Kumar Naik
Centre of Excellence in Natural Products and Therapeutics, Department of Biotechnology and Bioinformatics, Sambalpur University, Jyoti Vihar, Sambalpur, Odisha, India

Suseela Lanka
Department of Biosciences and Biotechnology, Krishna University, Rudravaram, Machilipatnam, India

Monika Mishra
Centre of Excellence in Natural Products and Therapeutics, Department of Biotechnology and Bioinformatics, Sambalpur University, Jyoti Vihar, Sambalpur, Odisha, India

Arpan Modi
Axenic Agritech Pvt Ltd, Ahmedabad, Gujarat, India

Mahima S. Mohan
Department of Microbiology, School of Life Sciences, Pondicherry University, Puducherry, India

Naji Naseef Pathoor
Department of Microbiology, Centre for Infectious Diseases, Saveetha Dental College and Hospitals, Saveetha Institute of Medical and Technical Sciences (SIMATS), Saveetha University (Deemed to be University), Chennai, Tamilnadu, India

Paramita Pakhira
Department of Microbiology, School of Life Sciences, Pondicherry University, Puducherry, India

Santhilatha Pandrangi
DBT BioCARe Scientist, EMBO Fellow, Group Leader, Onco-Stem Cell Research Laboratory, Department of Biochemistry and Bioinformatics, GITAM University, India

Subhaswaraj Pattnaik
Centre of Excellence in Natural Products and Therapeutics, Department of Biotechnology and Bioinformatics, Sambalpur University, Jyoti Vihar, Sambalpur, Odisha, India

Kaveti Sahithi
Department of Biosciences and Biotechnology, Krishna University, Rudravaram, Machilipatnam, India

Simi Asma Salim
Department of Microbiology, School of Life Sciences, Pondicherry University, Puducherry, India

Saravanan Shanmugam
Department of Microbiology, Centre for Infectious Diseases, Saveetha Dental College and Hospitals, Saveetha Institute of Medical and Technical Sciences (SIMATS), Saveetha University (Deemed to be University), Chennai, Tamilnadu, India

Sellamuthu Shanmugapriya
PG & Research, Department of Zoology, Kunthavai Naachiyaar Government Arts College (Women) Autonomous, Thanjavur, Tamilnadu, India

Sachidanand Singh
Department of Biotechnology, Pandit Dindayal Energy University, Gandhinagar, Gujarat, India

Zarin Taj
Department of Biotechnology, School of Life Sciences, Central University of Tamil Nadu, Neelakudi, Thiruvarur, India

Akshaya Viswanathan
Department of Microbiology, Centre for Infectious Diseases, Saveetha Dental College and Hospitals, Saveetha Institute of Medical and Technical Sciences (SIMATS), Saveetha University (Deemed to be University), Chennai, Tamilnadu, India

Gulshan Wadhwa
Department of Biotechnology, Ministry of Science and Technology, New Delhi, India

Kerodawala Zainab
Department of Biosciences and Biotechnology, Krishna University, Rudravaram, Machilipatnam, India

Editor Biography

Siddhardha Busi is Assistant Professor in the Department of Microbiology, Pondicherry University, Puducherry, India. He received his Ph.D. from CSIR-IICT, Hyderabad, India. Dr. Siddhardha has more than 15 years of research experience and his research is majorly focused on microbial pathogenesis, virulence, microbial biofilms, and natural compound-based antimicrobial drug discovery. His research area also expands to nanobiotechnology, drug delivery, and antimicrobial photodynamic therapy. Dr. Siddhardha's research group is currently working extensively on phytochemicals having anti-quorum sensing and anti-biofilm potential and their synergistic/adjuvant properties in combination with antibiotics. He completed two research projects on anti-quorum sensing and anti-biofilm activities of natural products funded by the Government of India. Dr. Siddhardha is serving as editorial member of several journals and is a member of many national and international scientific societies. He has published more than 80 research articles in the peer-reviewed international journal and authored or co-authored numerous book chapters. He also edited five books from his areas of expertise.

Ram Prasad, Ph.D. is Associate Professor, Department of Botany, Mahatma Gandhi Central University, Motihari, Bihar, India. His research interests include applied & environmental microbiology, plant-microbe interactions, sustainable agriculture, and nanobiotechnology. Dr. Prasad has more than 300 publications to his credit, including research papers, review articles, and book chapters; has seven patents issued or pending; and has edited or authored more than 50 books. He has guided one postdoc, two M. Phil, nine Ph.D., and several master dissertation students. Dr. Prasad has fourteen years of teaching/research experience and has been awarded the Sir CV Raman Scientist Award, Young Scientist Award & Prof. J.S. Datta Munshi Gold Medal by the International Society for Ecological Communications; Fellow of Agricultural Technology Development Society (FATDS); Fellow of Biotechnology Research society of India (FBRS); Fellow of the Society for Applied Biotechnology (FSAB); the American Cancer Society UICC International Fellowship for Beginning Investigators, USA; Outstanding Scientist Award in the field of Microbiology; BRICPL Science Investigator Award, Research Excellence Award, etc. He served as an editorial board members for Nanotechnology Reviews; BMC Microbiology; BMC Biotechnology; Current Microbiology; Archives of Microbiology; Annals of Microbiology; IET Nanobiotechnology; Green Processing and Synthesis; Discoverer Nano, SN Applied Sciences; Journal of Agriculture and Food Research, Agriculture, etc. including Series editor of Nanotechnology in the Life Sciences, Springer Nature, USA. Previously, Dr. Prasad served as Assistant Professor at Amity University Uttar Pradesh, India; Visiting Assistant Professor, Whiting School of Engineering, Department of Mechanical Engineering at Johns Hopkins University, Baltimore, USA; and Research Associate Professor at School of Environmental Science and Engineering, Sun Yat-sen University, Guangzhou, China.

1 Introduction to Microbial Biofilms

Zarin Taj and Indranil Chattopadhyay

1.1 INTRODUCTION

Bacterial biofilms originated roughly 32.5 million decades years ago, functioning as a progress strategy that safeguards cells and makes them persist in challenging environments (Hall-Stoodley et al., 2004). Biofilms are found in several natural habitats, such as rainfall, thermal waters, marine waters, and seas. They can also be present on diverse natural and synthetic surfaces, including stone, wood, soil, pipelines, and medical catheters (Goller & Romeo, 2008). Biofilms exert substantial impacts on human health, as well as ecological and industrial systems (A. Pandit et al., 2020).

A biofilm is a sophisticated structure composed of a network of microorganisms that come together to form a unified community. This community is surrounded by a matrix that could be secreted by the bacteria themselves(Flemming & Wingender, 2010). This matrix serves to anchor the biofilm to a variety of surfaces, whether they be inert or biotic. The cells within this matrix have a transformed phenotype, specifically in gene transcription and rate of growth, and typically have a diverse composition (López et al., 2010). The composition is regulated by different signaling mechanisms that rely on the density of cells, known as quorum sensing (QS) systems (Schluter et al., 2016). Chemical signal micromolecules of various structures. For example, l-homoserin-lactones enables the biofilm components to effectively communicate with each other through QS (López et al., 2010). Processes such as biofilm development, bioluminescence, antibiotic synthesis, sporulation, mobility, genetic competence, and the virulence potentiality are all influenced by the regulatory system known as QS. Strains lacking in QS have a reduced capacity to form thick, homogenous structures, as opposed to mature biofilms (Eick, 2021). When the concentration of signaling molecules reaches a certain level, it causes certain changes in the cells that make up the biofilm. These changes are related to the density of microbes within the biofilm, rather than the individual planktonic cells. This alteration has a significant impact on gene expression throughout a substantial portion of the bacterial genome (Tolker-Nielsen, 2015).

Microbial biofilms exhibit resilience to adverse conditions such as UV radiation, pH fluctuations, harsh temperatures, nutritional deprivation, hunger, high salinity, and pressured situations (Goller & Romeo, 2008). The extensive spectrum of environmental tolerance presents a difficulty in addressing the treatment of biofilms and

DOI: 10.1201/9781003487203-1

building habitats that are more resilient to biocontamination (Jiang et al., 2021). Nevertheless, some biofilms have demonstrated remarkable efficacy in the domain of processing and engineering. An exemplary instance of this phenomenon can be witnessed in terrestrial sewage systems, where biofilms play a crucial part in the process of breaking down organic substances. However, these biofilms can also contain and transmit genes that confer resistance to antimicrobial agents, which could possibly affect the functioning of microbial processes in wastewater treatments (Saratale et al., 2017). Biofilms that have practical applications can also be controlled and employed in microbial fuel cells, bioremediation, and specialized food product bioprocesses (Costerton et al., 1995).

1.2 BIOFILM FORMATION

Biofilms are complicated assemblies of microbes that develop on surfaces, typically enclosed by a matrix, and display a unique modified physical appearance in a three-dimensional arrangement. These biofilms can be composed either single kind or numerous types of cells (York, 2017).

1.2.1 Colonization

The creation of biofilms is a complex process involving bacterial transition from their motile stage to a more stationary form that builds biofilms (Tolker-Nielsen, 2015). The formation process is induced by extrinsic elements like pH levels, temperature, gravitational forces, Brownian motions, hydrodynamic forces, surface features, secondary messengers, and QS elements (Dickschat, 2010).

1.2.2 Attachment

Biofilm production begins with the attachment of planktonic microbes to surfaces, which is a crucial stage in the progression of these mobile microorganisms into a well-organized community. The conditioning layer is composed of organic or inorganic particles that provide a foundation for the development of a biofilm (Limoli et al., 2015). This layer provides numerous advantages to biofilm by altering the substrata to make it more accessible to bacteria. It also allows for the adjustment of surface charge, potential, and tensions as needed. The substrate supports nourishment, enhancing the development of the bacterial community (Webster et al., 2021). At the onset of biofilm formation, bacteria adhere to surfaces in an open and reversible manner. The movement of microbial cells is a result of the transfer of layer of liquid ffrom one surface to another. This process is reversible and is aided by physical factors or bacterial structures such as flagella, pili, or adhesins.

This stage is characterized by the existence of microorganisms that are firmly adhered to the surfaces in a polar manner (Dula et al., 2021). Subsequently, bacteria adjust their position to become horizontally aligned on surfaces and establish a lasting connection, enhancing their ability to withstand different physical barriers that may hinder the development of biofilms (Donlan, 2002). C-di-GMP, also known as bis-(3′-5′)-cyclic dimeric guanosine monophosphate, plays a vital role as an intracellular signalling molecule in the primitive biofilm formation phases (Andersen et al., 2021). It controls the movement of flagella and enhances the formation of the biofilm structure. Because of the Pil-Chp surface-sensing system found on microbial surfaces, the concentration of c-di-GMP raises with each adherence/detachment event. During the early phases of biofilm development, there is a transition from planktonic cells that are not familiar with surfaces (bacteria with low levels of c-di-GMP and no previous contact with surfaces) to planktonic cells that are aware of surfaces (bacteria with elevated levels of c-di-GMP and prior interaction with surfaces). Subsequently, cells adhere permanently to surfaces, leading to the development of a biofilm (Andersen et al., 2021)

1.2.3 Growth

Once microorganisms have successfully adhered to surfaces, they begin to multiply and aggregate, forming microcolonies through the production of extracellular polymeric substance (EPS) with a significant amount of c-di-GMP (Díaz et al., 2018). Flagella and type IV pili-mediated motilities play crucial roles in the interactions between microorganisms and surfaces, as well as in the formation of cell-cell aggregations to create microcolonies (Flemming & Wingender, 2010; Rather et al., 2021). Microbial cells undergo cell division and accumulate, resulting in the formation of a microcolony. During the process of binary division of progeny, cells translocate towards the periphery of the matrix and raise themselves from the place of adherence prior to dividing and forming clusters. The biofilm's creation of a mushroom-like structure enables the delivery of nutrients (G. Wang et al., 2020).

After an initial delay, a phase of fast growth commences, characterized by exponential population growth that leads to the nutrient depletion from the surrounding fluid and substratum. Polysaccharide intercellular adhesion polymers and divalent cations play a role in creating robust cell connections (Jefferson & Cerca, 2006). The gene expression is significantly impacted by the population adhesiveness. These components are crucial for upholding structural integrity and are necessary for the adhesion, organization, and survival of biofilms in demanding and constantly changing environmental conditions that construct biofilm architecture (Gan & Gong, 2017).

1.2.4 Maturation

The stationary phase is reached when the cell division and cell death percentage are equilibrated (York, 2017). QS is a biological mechanism that initiates the activation of genes, leading to the creation of the extracellular matrix through diverse biological processes. The EPS plays a vital role in the biofilm formation by assisting in the adhesion of microbes to surfaces, enhancing the stability of the three-dimensional biofilm architecture, promoting cell clustering, and providing protection against various stresses including host defense, antimicrobial agents, oxidative damage, and metallic cations (Saxena et al., 2019). Furthermore, it functions as a safeguarding barrier for signalling molecules engaged in metabolic byproducts, QS, and enzymes. A fully developed biofilm exhibits a discernible architecture resembling a "mushroom" or "tower," where microorganisms are arranged according to their oxygen tolerance and metabolic activity (Schuster & Markx, 2014).

1.2.5 Dispersion

The death phase signifies the termination of the biofilm, induced by the enzymes it produces. The process entails the enzymatic degradation of the polysaccharides that adhere to the biofilm, resulting in the vigorous liberation of the surface bacterial cells for colonization of new substrates (Limoli et al., 2015). The duration of the biofilm's existence is constrained by various reasons, including the depletion of nutrients, adverse environmental conditions, alterations in gene expression impacting cell movement and EPS breakdown, as well as the production of polysaccharides and fimbriae. Evidently, biofilms possess the capacity to disassemble their own structure (Toyofuku et al., 2015). Seeding dispersal is a phenomenon that has been extensively documented in numerous laboratories. During this process, the biofilm's interior undergoes liquefaction while the exterior is broken, leading to the liberation of planktonic bacteria exhibiting distinct features. Consequently, the colony remains unoccupied (Chua et al., 2014).

1.3 CONSTITUENTS OF BIOFILM

The assemblage of microorganisms, encompassing bacteria, resides inside a spontaneously formed matrix comprising polysaccharides, proteins, minerals, and other constituents. The EPS forms a cohesive matrix that networks the bacteria together and leads to surface adherence (Flemming & Wingender, 2010). The arrangement of EPS is dependent on the EPS constituents' interactions, which are vital in determining the mechanical characteristics of the matrix and the physiological functions of the organisms within the biofilm. The development of the biofilm structure is an intricate and dynamic procedure, propelled by the EPS molecules (Jayathilake et al., 2017). The outcome of this process is the formation of a three-dimensional arrangement in which cells inside the biofilm aggregate into microcolonies. A recent study conducted on *Escherichia coli* biofilms revealed the complex microbial clusters present in them. The study highlighted the detailed structure of these clusters, which contributes to specific physiological differences in various geographical areas.

The EPS play a crucial role in determining the three-dimensional structure of a biofilm. They develop during the attachment stage of a biofilm. Typically, the EPS matrix has a thickness ranging from 0.2 to 1.0 μm (S. Pandit et al., 2020). This has a direct effect on the circumstances and cell living conditions, while also offering crucial mechanical stability to the biofilm. The matrix contains several components that contribute to its structure and function. These substances consist of soluble polysaccharides that can form gels, proteins, and extracellular DNA (eDNA) (Kavanaugh et al., 2019). In addition, there are insoluble constituents like amyloids, cellulose, fimbriae, pili, and flagella. The existence of pores and channels inside the matrix has been determined to be essential in aiding the movement of liquids. This finding has sparked the fascinating concept of a "rudimentary circulatory system" for biofilm (Kaplan, 2014).

Several polysaccharides display a wide range of composition, while others have a more consistent structure, such as cellulose, fructans generated from sucrose, and glucans. The interaction between polysaccharides and proteins and ions is influenced by various factors, including vander Waals forces, electrostatic forces, ionic forces, and hydrogen bonds. These interactions are vital for preserving the organization and biofilm stability. Polysaccharides serve as a molecular adhesive, ensuring bacteria attach to different surfaces, both living and non-living, which aids in colonization. Moreover, they offer a safeguarding shield against the immune system and various external stresses. Secreted proteins, cellular appendages such as pili and flagella, outer membrane vesicles, and cell surface adhesions make up the biofilm matrix proteins (Branda et al., 2005). They have a vital function in interacting with EPS and nucleic acid components, which helps to stabilize the biofilm matrix, establish surface colonisation, and maintain the overall integrity and structure of the biofilm (Fong & Yildiz, 2015).

Specific proteins are essential for the degradation and dispersion of the biofilm matrix. Proteases are accountable for the breakdown of matrix proteins, whereas glycosyl hydrolase dispersin B decomposes polysaccharides (Kaplan, 2014). Furthermore, DNases are enzymes that participate in the destruction of nucleic acids outside of cells.There is a significant difference in the protein composition between EPS and particular cellular fractions (Fong & Yildiz, 2015). The EPS matrix comprises a substantial amount of protein peptidases, enzymes involved in the cell wall metabolism, chaperones, and disulfide-isomerases, such as cold shock and DNA binding proteins. Approximately one-third of the EPS matrix proteins were membrane proteins located within outer membrane vesicles (OMVs), while another fraction of proteins were from lysed cells and released proteins in *P. aeruginosa* (Seike et al., 2020).

Extracellular DNA (eDNA) is essential in the EPS matrix, aiding in the formation of bacteria clusters inside a biofilm. The creation of extracellular DNA (eDNA) is a complex process that can occur by several mechanisms, such as bacterial secretion systems, cell death induced by phages, autolysis, controlled release of DNA through QS, or even in conjunction with DNA-containing OMVs (Paulsson et al., 2021). DNABII binding proteins are thought to be involved in maintaining the secondary structure of extracellular DNA (eDNA) and aiding in the formation of a strong filamentous network structure in biofilms of aquatic bacteria (Devaraj et al., 2015). Water is the main component of the matrix, accounting for up to 97% of its makeup. The EPS matrix also contains several lipids that exhibit surface-active properties, such as surfactin, emulsan, and viscosin (Goodman et al., 2011). They improve the availability of hydrophobic compounds by distributing them outward. Rhamnolipids, a prominent category of surface-active agents studied in *P. aeruginosa*, have a vital function in beginning the formation of small colonies, determining the shape of biofilms, and aiding in the dispersion of biofilms (Kim & Lee, 2016; Harmsen et al., 2010; Yang et al., 2012). Water is regarded as an essential component in the EPS matrix. It preserves the moisture and protects it from desiccation, even amidst fluctuations in environmental water levels. The nutrient maintenance and dispersion inside a biofilm are affected by water availability (Rumbaugh & Sauer, 2020).

1.4 BIOFILMS ON AQUATIC ENVIRONMENTS

Aquatic biofilms are frequently referred to as periphyton and mycrophytobenthons. The term "periphyton" was used by Behnin in 1924 to refer to organisms capable of adhering to man-made surfaces in aquatic environments (Cunliffe et al., 2011; Cooke, 1956). Periphyton is a heterogeneous community comprising a wide range of creatures, including algae, bacteria, fungi, protozoans, metazoans, and viruses. The composition of the system encompasses both inorganic and organic detritus, which are affixed to various types of benthic substrates employed in freshwater ecosystems (Domozych & Domozych, 2008). Microphytobenthos is a term used in marine ecosystems comprised of cyanobacteria and algae that attach themselves to sediments (MacIntyre et al., 1996).

Biofilms are essential for preserving crucial ecosystem processes in aquatic environments, as they effectively capture a substantial amount of the complex biological activity occurring in the vicinity (Battin et al., 2003). The wide variety of microorganisms present in biofilms, combined with their vulnerability to changes in the environment, makes them an excellent model for investigating the effects of pollutants on complex biological systems (Besemer, 2015). Marine biofilms possess an exceptional capacity to inhabit artificial surfaces, resulting in an expedited occurrence of corrosion, biofouling, and potentially impacting the buoyancy of polyethylene plastic. Bacteria, along with diatoms and other microorganisms, have a significant impact on microfouling by aiding in the attachment of higher organisms such as algae, mussels, and barnacles, resulting in macrofouling (Chen et al., 2020).

1.5 FACTORS INFLUENCING THE AQUATIC BIOFILM

Biofilms are multi-microbial communities that flourish in certain water systems. There are several environmental elements that impact both their genesis and sustainability.

1.5.1 Fluid Mechanics

An in-depth analysis of the movement and behavior of fluids in water systems is fundamental for comprehending biofilm development. Biofilm production is likely to occur in locations where flow is impeded, either due to impediments, design flaws, or complex system structures (Stoodley et al., 1994; Di Pippo et al., 2018). Stagnation can lead to the formation of differential nutrient gradients, in which regions with low flow see an increase in important microbial nutrients. Furthermore, the absence of shear stresses in these regions guarantees that emerging biofilms can thrive without any mechanical disruptions, allowing them to proliferate undisturbed (Karande et al., 2014; X. Li et al., 2015). Complex systems, especially those equipped with supplementary pipes, extra storage capacity, or larger reservoirs, may inadvertently store water for prolonged durations, so producing a conducive environment for the proliferation of microorganisms (Voglauer et al., 2022).

1.5.2 Nature of Aquatic Medium

The microbial attachment rate to a substratum can be influenced by various factors in the aqueous medium, including nutrient levels, pH, ionic strength, and temperature (Yin et al., 2019). Multiple studies have demonstrated a correlation between seasonal variations and the attachment of bacteria as well as the formation of biofilms in various aqueous systems (Donlan, 2002). It is possible that this effect is influenced by the temperature of the water or by other factors that vary with the seasons but were not measured (Goller & Romeo, 2008). Fletcher discovered that cation concentration alteration (sodium, calcium, lanthanum, ferric iron) had an impact on the adherence of *Pseudomonas fluorescens* to glass surfaces (Fletcher, 1988). This effect is believed to be due to a repulsive forces reduction between the anionic charged bacterial cells and the glass surfaces. There is a direct relationship between a raise in nutrient concentration and the number of bacterial cells that become attached (Palmer et al., 2007).

1.5.3 Cellular Properties

The microbial cell attachment and extension can be influenced by factors such as cell surface hydrophobicity, the presence of fimbriae and flagella, and the production of EPS. The cell surface's hydrophobicity involves adhesion as it promotes stronger hydrophobic interactions between the cell surface and the substratum (Leonov et al., 2014).The pili facilitate the transfer of nucleic acids in order to increase the hydrophobicity of the cell surface.

(Krasowska & Sigler, 2014; Jauffred et al., 2017). Many fimbriae consist of a significant amount of hydrophobic amino acid residues. Fimbriae also contribute to the cell's surface hydrophobicity and attachment, likely by overcoming the initial electrostatic repulsion barrier between the cell and substratum. Several aquatic bacteria exhibit fimbriae, which have been found to play a role in bacterial attachment to animal cells (Siqueira & Lima, 2012).

1.5.4 Nutritional Components

Microorganisms within biofilms necessitate a continual influx of nutrients to sustain their existence. Carbon, an essential energy source, is typically derived from organic substances present in the system (Scribani Rossi et al., 2022; Amato & Brynildsen, 2014). Nitrogen, an essential constituent of amino acids and proteins, can enter ecosystems via agricultural runoffs, leachates, or the decomposition of dead organisms. Phosphorus, an essential ingredient for the generation of ATP and synthesis of nucleic acids, can be acquired from detergents or spontaneously emitted from geological substrates (Moller et al., 1997). The process of corrosion can result in the liberation of metallic elements such as iron and manganese, which can yield both advantageous and disadvantageous consequences (Bowden & Li, 1997; Butterfield et al., 2002).

1.5.5 Disinfection Dichotomy

The implementation of disinfectants, including chlorine, chloramine, and advanced oxidants, is intended to eradicate microorganisms. However, biofilms, with their dense EPS matrix, act as barriers, preventing effective disinfectant penetration (Mi et al., 2015; Xue & Seo, 2013). Due to the occurrence of disinfectant decay, the active concentrations that reach the deepest biofilm layers are often insignificant. Long-term exposure to lower levels of disinfectants can encourage the resistance development in the strains (Chowdhury et al., 2019) Microorganisms such as *Pseudomonas aeruginosa*, with their inherent resistance mechanisms, can display increased tolerance in biofilms, making it even more challenging to eliminate them (Bridier et al., 2011; Ciofu & Tolker-Nielsen, 2019).

1.5.6 Corrosion

Corrosion, an intriguing electrochemical phenomenon, can inadvertently promote the development of biofilm. Metals, when oxidized, develop surfaces that exhibit irregularities, resulting in the formation of pits and crevices (Butterfield et al., 2002). These specific regions offer optimal circumstances for bacteria to adhere and proliferate. Furthermore, the residual substances produced by corrosion, such as iron or zinc oxides, might serve as nutrient sources (Nozaka et al., 2014). Leptothrix species, a type of microorganism, has an exceptional capacity to oxidize manganese. This ability enables them to thrive in such circumstances and subsequently deteriorate the quality of water (Schmidt et al., 2014).

1.5.7 Sediments

Irrespective of their source (natural silt, pipeline erosion, or external contamination), sediments have a noteworthy impact on biofilm dynamics. During their descent, these particles frequently capture both organic and inorganic substances, thereby establishing favorable environments that promote microbial proliferation (Flemming, 2020; Droppo et al., 2016). Sediments possess intrinsic qualities that designate them as stable foundations, thereby rendering them exceptionally conducive to microbial colonization. This is particularly advantageous for organisms that thrive near substrates, as opposed to drifting freely. In infrastructure systems characterized by low water exchange rates, such as storage containers, sediment accumulation can be significant. This may cause biofilms, which are concentrated communities of microorganisms, to form (Lu et al., 2016).

1.6 IMPLICATIONS OF BIOFILMS ON WATER QUALITY

Bacteria have crucial functions in aquatic ecosystems, actively participating in biogeochemical processes and serving as a source of materials and energy for higher trophic levels. The versatility of bacteria is the primary reason for their extensive distribution and success. Genotypic evolution confers an environmental stimuli adaptation, whereas phenotypic plasticity allows bacterial ability to thrive in dynamic and challenging conditions of the aquatic environment (Fernandes et al., 2020). Biofilms can significantly impact the water quality, leading to important consequences. Ensuring the safety of drinking water can pose a significant challenge, given the incidence of pathogenic microorganisms. The EPS matrix provides a defensive shield for bacteria, ensuring their survival even in the face of disinfection methods such as chlorination. Consequently, biofilms possess the capacity to cause persistent contamination, leading to the potential transmission of waterborne illnesses (Lin et al., 2017; Xue & Seo, 2013).

The enzymes and metabolic byproducts produced by biofilms can modify the pH, turbidity, and amounts of organic compounds in the water. These alterations may potentially result in adverse effects on human well-being and the aquatic ecology (Rittmann, 2018). Changes in pH levels can impact the flavor, smell, and general excellence of drinking water, while turbidity can diminish the effectiveness of filtration procedures. Microbes can utilize organic compounds present in water as a source of nourishment, resulting in the formation of disinfection byproducts that can have an adverse effect on the quality of water (Greenman et al., 2020). The presence of biofilms also exacerbates the corrosion of water distribution systems, which is a cause for concern. The EPS matrix possesses the capacity to trap and intensify corrosive substances, leading to the degradation of pipes and infrastructure (Sivadon et al., 2019). Corrosion not only affects the structural soundness of the water system, but it can also result in heavy metal discharge and other pollutants into the water, which presents substantial health hazards (Edwards & Kjellerup, 2013).

The intricate circumstances seen in aquatic ecosystems facilitate the aggregation and adhesion of microorganisms on diverse surfaces, regardless of whether they are of natural or artificial origin. Submerged surfaces accumulate different substances, whereas biogenic particles mostly comprise carbon-rich skeletal remains derived from creatures such as plankton (Donlan, 2002). Bacteria employ these chemicals for diverse functions, such as carbon and energy substrates, electron donors/acceptors in metabolic reactions, and micronutrients. The availability of resources is likely one of the main advantages for bacteria in aquatic biofilms. However, biofilms offer protection to individual cells against a range of environmental difficulties, including dehydration, temperature variations, alterations in pH, competition, predation, UV radiation, and limited availability of nutrients (Tasaki et al., 2017).

1.7 APPLICATIONS OF BIOFILM

Biofilms are present throughout and provide crucial functions in both natural and synthetic environments. Microbial biofilm exhibits enormous beneficial applications.

1.7.1 Water and Sewage Water Treatment

Water and sewage water treatment have effectively harnessed the benefits of biofilms for more than a century. Biological filters are involved in reducing the presence of biodegradable organic carbon within water distribution systems (Miller et al., 2015). Phototrophic cyanobacterial biofilms have the potential to remediate an extra nutrient present in secondary effluents in sewage facilities. The waste degradation through the biofilm process effectively treats various types of waste, including human waste, agri waste and animal waste, and industrial wastes (Wan et al., 2019). Wastewater treatment includes both dispersed growth and attached growth systems. In the dispersed growth system, biomass is cultivated using a method that involves spreading it in an aqueous medium without any surface adherence. Microbes thrive and multiply, forming microcolonies that absorb organic nutrients in their environment. Sludge is formed because of the settling of these microcolonies. It undergoes a sludge treatment process and is then resuspended for reuse in the process. In an attached growth system, the formation of a biofilm occurs when the biomass is brought into proximity with a support medium. This support medium might be either absorbed into the liquid media or gathered sporadically and unpredictably (Singh et al., 2020). Biofilms form when organic materials and nutrients are injected into the wastewater. The recurrent exposure to novel substrates improves the efficacy of bacterial cells (Andersson et al., 2008).

1.7.2 Agriculture

A pioneering method for organic agriculture involves the utilization of biofilm fertilizers (Velmourougane et al., 2017). Biofilms formed by beneficial microbes, in particular PGPR (Plant Growth-Promoting Rhizobacteria), possess the capacity to fundamentally revolutionize the concept of sustainable agriculture in biogeochemical cycles. This specific group of bacteria possesses an extraordinary ability to infiltrate and establish itself within the root system of the host plant (A. Pandit et al., 2020). It has a vital function in promoting plant growth and productivity by growth hormones, ammonia, enzymes, and other secondary metabolites thereby employed as bio-fertilizers (Yang et al., 2011). PGPR consists of a diverse array of microbes. PGPR are essential in bioremediation and function as biocontrol agents (Xiong et al., 2021). Recently, there has been an increasing fascination with multispecies biofilms, particularly ones that consist of bacteria and fungi, which are typically found in mycorrhiza. Unlike monoculture biofilms, multi-species biofilms offer supplementary benefits and improved soil ecology (Burmølle et al., 2007). Bacterial biofilms have intricate relationships with plants, encompassing a spectrum of interactions that might be mutualistic, commensal, or pathogenic.The remarkable capacity of microorganisms to colonize roots and produce biofilms is harnessed to enhance nutrient availability, reduce disease outbreaks, and protect plants. There is strong evidence that adding microorganisms from healthy soil to deteriorated soil might have a positive impact on soil health (Z. Li et al., 2022).

1.7.3 Food Processing

Biofilms can exert a constructive influence on the fermentation. The utilization of biofilm-mediated microbial fermentation is highly effective in biochemical production owing to its remarkable capacity to immobilize cells, endure harmful chemicals, and maintain long-term cell viability. The diverse microbial communities in fermented food products such as beers, wines, spirits, meats, seafood, cheeses, and breads are encapsulated inside a biofilm and enveloped by EPS. This fosters an optimal environment for the proliferation of these microbes (Ghosh et al., 2022). Lactic acid bacteria such as *L. plantarum, L. rhamnosus,* and *L. fermentum* can incorporate themselves into a biofilm structure and have important industrial uses Singh et al., 2020). According to the study conducted by Gamarra et al. (2010), biofilms of *Aspergillus niger* can be used for the synthesis of cellulase. This enzyme exhibits potential for several sectors including food processing, textile manufacturing, and pulp and paper production.

1.7.4 Biofilter

Biofilters function by channeling exhaust air through a biofilter. A well-optimized biofilter efficiently diminishes the occurrence of unpleasant smells in the atmosphere by specifically targeting ammonia and hydrogen sulphide. Biofilter media comprise wood chips and compost, providing nutrients, energy, and water to microorganisms. The absorbency of the biofilter is essential for its efficacy. The biofilter generates carbon dioxide, water, minerals, and organic molecules as secondary outcomes, rendering it ecologically sustainable (Salisbury & Percival, 2019).

1.7.5 Biofuel

There has been significant interest in the investigation of biofuel generation as a potential alternative to fossil fuels. A research investigation has been conducted to investigate the production of ethanol through the utilization of biofilms that consume cellulose obtained from plants. Cellulase produced by microorganisms such as *Cellulomonas* and *Aspergillus* is essential for biofuel production. Biofilms enable the integration of many processes such as delignification, saccharification, bioprocessing, fermentation, and separation into a single reactor, providing advantages over conventional industrial techniques (Z.-W. Wang & Chen, 2009).

Bacterial biofilms are essential for the operation of microbial fuel cells (MFCs) and aid in the breakdown of waste. Microbial cells are composed of several elements, such as an anode, cathode, electrolyte, and an external circuit (Saba et al., 2017). The biofilm generates electric current through microbial metabolic processes. However, the transition of organic waste into microbial energy leads to the emission of carbon dioxide (CO2) and the release of unbound electrons. Bacteria could exploit insoluble substances as both sources of electrons and recipients of electrons.There are well-established processes and electrically conductive pili-like structures that may transfer electrons and molecules, such as phenazines, which can act as electron shuttles even in the absence of nanowires (Malvankar & Lovley, 2012). The presence of the biofilm matrix facilitates the electrical conductivity, since it envelops nanowires that are capable of conducting electricity at a microscale level (Reguera, 2018). Usually, MFCs employ bacterial species employed for several types of liquid waste, including wastewater from potato processing, farm animal effluent, and human waste (Vishwanathan, 2021).

1.7.6 Bioremediation

Due to their provision of environmentally sustainable, economically efficient, and ecologically friendly technologies, biofilms have gained significance in breaking down process of organic pollutants. The existence of many active functional groups on the biofilm surface renders microbial bioremediation a tangible prospect (Sonawane et al., 2022). Functional groups are widely recognized for their role in enhancing bioremediation by promoting enhanced diffusion within the biofilm. Bacteria belonging to the genera *Pseudomonas, Alcanivorax, Arthrobacter,Bacillus*, *Cyclocasticus*, and *Rhodococcus* could eliminate hydrocarbons produced by petroleum-based industries in marine settings. Combining surfactants with multispecies biofilms enables efficient removal of crude oil spills. Biofilms are employed for the more efficient degradation, purification, and cleansing of oil spills, as compared to the intricate procedures of chemical and physical techniques. Microorganisms have the extraordinary capacity to eliminate even minuscule quantities of contaminants, such as heavy metals, chlorinated chemicals, and polyaromatic hydrocarbons (Edwards & Kjellerup, 2013) (Figure 1.1).

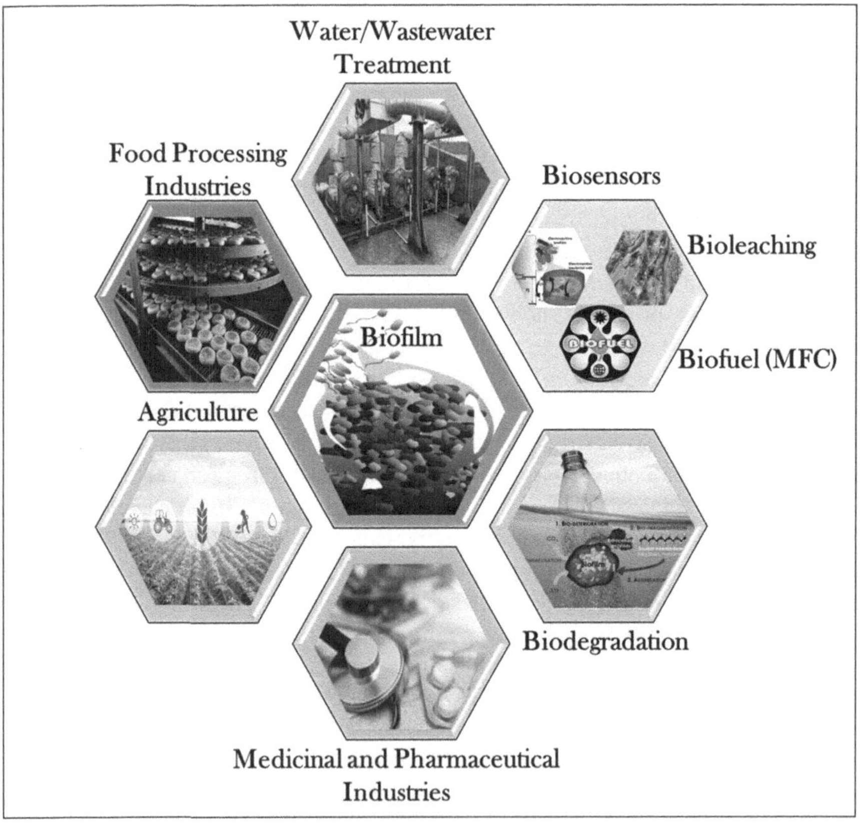

FIGURE 1.1 Applications of microbial biofilms.

1.7.7 Biodegradation

Biofilms have become significant in organic pollutant degradation due to their environmentally friendly, cost-effective, and sustainable technology (Sonawane et al., 2022). Microbial biodegradation is made possible by the existence of numerous active functional groups that have been discovered on the biofilm surface. Functional groups play a crucial role in facilitating diffusion within the biofilm, thereby supporting the process of biodegradation (Rossi, 2020). Various microbes from different species have the remarkable capability to eliminate hydrocarbons found in marine environments due to petroleum-related activities. When surfactants are used in conjunction with multi-species biofilms, they have proven to be highly effective in clearing crude oil spills (Mishra et al., 2022).

1.7.8 Biosensors

Bacteria can serve as valuable biosensors. This is due to their remarkable ability to detect pollutants and pathogens. Every year, a significant amount of waste is produced by industries and ultimately disposed of in landfills. As a result, scientists are utilizing biosensors (bacteria) capable of detecting biologically active pollutants (Eick, 2021). Bacterial biosensors possess a receptor that becomes active when exposed to pollutants, along with a reporter that documents any alterations (Mangwani et al., 2016). They utilize the lux operon from *Aliivibrio* or *Photobacterium* as a reporter. The lux operon consists of an inducer and structural genes that encode the enzyme luciferase, resulting in the production of bioluminescence that closely resembles the glow observed in fireflies (Beloin et al., 2004). In a straightforward manner, when the receptor is activated, a bacterium that possesses the lux operon will produce visible light. The lux operon has the potential to be transferred to a wide range of bacteria (Aponiene et al., 2017; Kanematsu & Barry, 2020).

1.7.9 Bioleaching

Microbial biofilms are involved in bioleaching, a process that removes metals from ores by means of bacterial microorganisms. Because the bacteria consume the nutrients found in the minerals, metals like copper, zinc, silver, and gold separate from their ores (Xing et al., 2022). Usually, bacteria that oxidize iron and sulfur dioxide carry out bioleaching. To oxidize the ore, iron ions (Fe^{+3}) are employed. The bacteria's job is to continue the ore's oxidation and regenerate the chemical oxidant Fe^{+3} from Fe^{+2}. *Thiobacillus ferrooxidans* and *Leptospirillumferrooxidans* are a few of the bacteria that are employed in this process (Yuan et al., 2020). Compared to chemical extractions, using microorganisms is less expensive and better for the environment. *Desulfitobacterium* bacteria could produce both electricity and break down contaminants like printed circuit boards (Bobadilla-Fazzini & Poblete-Castro, 2021).

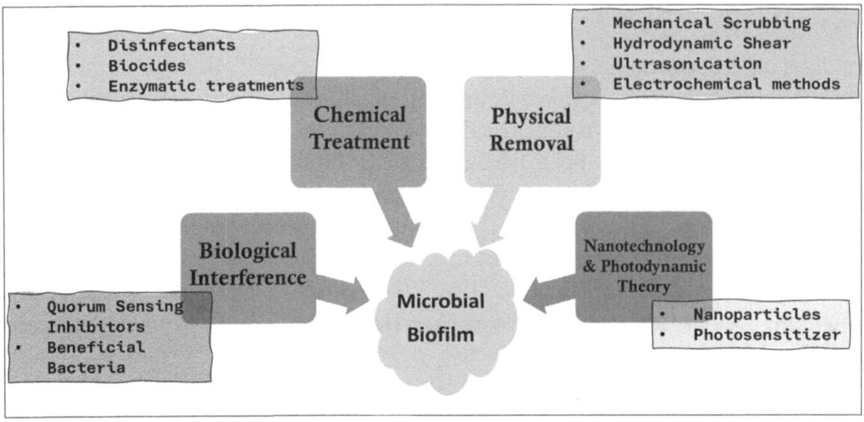

FIGURE 1.2 Eradication and management techniques involved in removal of microbial biofilm for potable water systems.

1.8 MANAGEMENT OF MICROBIAL BIOFILMS IN AQUATIC ENVIRONMENT

The existence of biofilms in usable water systems presents a substantial risk to the quality of water and the overall stability of the water distribution infrastructure. Embedded within a protective EPS matrix, biofilms consist of intricate microbial communities. This matrix not only offers protection from external dangers, but also allows biofilms to demonstrate impressive adaptive abilities, making their elimination a formidable challenge. However, utilizing a blend of traditional and cutting-edge approaches can lay the foundation for successful biofilm management (Y. Li et al., 2023) (Figure 1.2).

1.9 CONCLUSIONS AND PROSPECTIVES

Biofilms are recognized as complex ecosystems that harbor a diverse range of biodiversity across all three domains of life and form highly organized structures. Biofilms can be found on various submerged surfaces in both natural and man-made systems, creating an ideal environment for the growth, activity, and different bacterial species interaction. The profound understanding of microbial biofilms and their crucial role as key facilitators of enzymatic and metabolic activities that are vital for fundamental ecosystem processes and biogeochemical cycles. Understanding the creation and properties of biofilms is essential for developing effective measures to manage their impacts. To mitigate the potential hazards of biofilms and ensure the provision of clean and safe water to everybody, we should adopt a proactive approach by conducting routine maintenance, applying efficient disinfection techniques, and leveraging cutting-edge material technologies.

REFERENCES

Amato, S. M., & Brynildsen, M. P. (2014). Nutrient transitions are a source of persisters in Escherichia coli biofilms. *Plos One*, *9*(3), e93110. https://doi.org/10.1371/journal.pone.0093110

Andersen, J. B., Kragh, K. N., Hultqvist, L. D., Rybtke, M., Nilsson, M., Jakobsen, T. H., Givskov, M., & Tolker-Nielsen, T. (2021). Induction of native c-di-GMP Phosphodiesterases leads to dispersal of pseudomonas aeruginosa Biofilms. *Antimicrobial Agents and Chemotherapy*, *65*(4). https://doi.org/10.1128/AAC.02431-20

Andersson, S., Kuttuva Rajarao, G., Land, C. J., & Dalhammar, G. (2008). Biofilm formation and interactions of bacterial strains found in wastewater treatment systems. *FEMS Microbiology Letters*, *283*(1), 83–90. https://doi.org/10.1111/j.1574-6968.2008.01149.x

Aponiene, K., Serevičius, T., Luksiene, Z., & Juršėnas, S. (2017). Inactivation of bacterial biofilms using visible-light-activated unmodified ZnO nanorods. *Nanotechnology*, *28*(36), 365701. https://doi.org/10.1088/1361-6528/aa7a53

Battin, T. J., Kaplan, L. A., Denis Newbold, J., & Hansen, C. M. E. (2003). Contributions of microbial biofilms to ecosystem processes in stream mesocosms. *Nature*, *426*(6965), 439–442. https://doi.org/10.1038/nature02152

Beloin, C., Valle, J., Latour-Lambert, P., Faure, P., Kzreminski, M., Balestrino, D., Haagensen, J. A. J., Molin, S., Prensier, G., Arbeille, B., & Ghigo, J.-M. (2004). Global impact of mature biofilm lifestyle on Escherichia coli K-12 gene expression. *Molecular Microbiology*, *51*(3), 659–674. https://doi.org/10.1046/j.1365-2958.2003.03865.x

Besemer, K. (2015). Biodiversity, community structure and function of biofilms in stream ecosystems. *Research in Microbiology*, *166*(10), 774–781. https://doi.org/10.1016/j.resmic.2015.05.006

Bobadilla-Fazzini, R. A., & Poblete-Castro, I. (2021). Biofilm formation is crucial for efficient copper bioleaching from bornite under mesophilic conditions: Unveiling the lifestyle and catalytic role of sulfur-oxidizing bacteria. *Frontiers in Microbiology*, *12*, 761997. https://doi.org/10.3389/fmicb.2021.761997

Bowden, G. H., & Li, Y. H. (1997). Nutritional influences on biofilm development. *Advances in Dental Research*, *11*(1), 81–99. https://doi.org/10.1177/08959374970110012101

Branda, S. S., Vik, S., Friedman, L., & Kolter, R. (2005). Biofilms: The matrix revisited. *Trends in Microbiology*, *13*(1), 20–26. https://doi.org/10.1016/j.tim.2004.11.006

Bridier, A., Briandet, R., Thomas, V., & Dubois-Brissonnet, F. (2011). Resistance of bacterial biofilms to disinfectants: A review. *Biofouling*, *27*(9), 1017–1032. https://doi.org/10.1080/08927014.2011.626899

Burmølle, M., Hansen, L. H., & Sørensen, S. J. (2007). Establishment and early succession of a multispecies biofilm composed of soil bacteria. *Microbial Ecology*, *54*(2), 352–362. https://doi.org/10.1007/s00248-007-9222-5

Butterfield, P. W., Camper, A. K., Biederman, J. A., & Bargmeyer, A. M. (2002). Minimizing biofilm in the presence of iron oxides and humic substances. *Water Research*, *36*(15), 3898–3910. https://doi.org/10.1016/s0043-1354(02)00088-x

Chen, X., Chen, X., Zhao, Y., Zhou, H., Xiong, X., & Wu, C. (2020). Effects of microplastic biofilms on nutrient cycling in simulated freshwater systems. *The Science of the Total Environment*, *719*, 137276. https://doi.org/10.1016/j.scitotenv.2020.137276

Chowdhury, D., Rahman, A., Hu, H., Jensen, S. O., Deva, A. K., & Vickery, K. (2019). Effect of disinfectant formulation and organic soil on the efficacy of oxidizing disinfectants against biofilms. *The Journal of Hospital Infection*, *103*(1), e33–e41. https://doi.org/10.1016/j.jhin.2018.10.019

Chua, S. L., Liu, Y., Yam, J. K. H., Chen, Y., Vejborg, R. M., Tan, B. G. C., Kjelleberg, S., Tolker-Nielsen, T., Givskov, M., & Yang, L. (2014). Dispersed cells represent a distinct stage in the transition from bacterial biofilm to planktonic lifestyles. *Nature Communications*, *5*, 4462. https://doi.org/10.1038/ncomms5462

Ciofu, O., & Tolker-Nielsen, T. (2019). Tolerance and resistance of *Pseudomonas aeruginosa* biofilms to antimicrobial agents-How *P. aeruginosa* can escape antibiotics. *Frontiers in Microbiology*, *10*, 913. https://doi.org/10.3389/fmicb.2019.00913

Cooke, W.B. Colonization of artificial bare areas by microorganisms. (1956). *The Botanical . Review, 22*, 613–638.

Costerton, J. W., Lewandowski, Z., Caldwell, D. E., Korber, D. R., & Lappin-Scott, H. M. (1995). Microbial biofilms. *Annual Review of Microbiology*, *49*, 711–745. https://doi.org/10.1146/annurev.mi.49.100195.003431

Cunliffe, M., Upstill-Goddard, R. C., & Murrell, J. C. (2011). Microbiology of aquatic surface microlayers. *FEMS Microbiology Reviews*, *35*(2), 233–246. https://doi.org/10.1111/j.1574-6976.2010.00246.x

Devaraj, A., Justice, S. S., Bakaletz, L. O., & Goodman, S. D. (2015). DNABII proteins play a central role in UPEC biofilm structure. *Molecular Microbiology*, *96*(6), 1119–1135. https://doi.org/10.1111/mmi.12994

Di Pippo, F., Di Gregorio, L., Congestri, R., Tandoi, V., & Rossetti, S. (2018). Biofilm growth and control in cooling water industrial systems. *FEMS Microbiology Ecology*, *94*(5). https://doi.org/10.1093/femsec/fiy044

Díaz, M., Castro, M., Copaja, S., & Guiliani, N. (2018). Biofilm Formation by the Acidophile Bacterium Acidithiobacillus thiooxidans Involves c-di-GMP Pathway and Pel exopolysaccharide. *Genes*, *9*(2). https://doi.org/10.3390/genes9020113

Dickschat, J. S. (2010). Quorum sensing and bacterial biofilms. *Natural Product Reports*, *27*(3), 343–369. https://doi.org/10.1039/b804469b

Domozych, D. S., & Domozych, C. R. (2008). Desmids and biofilms of freshwater wetlands: Development and microarchitecture. *Microbial Ecology*, *55*(1), 81–93. https://doi.org/10.1007/s00248-007-9253-y

Donlan, R. M. (2002). Biofilms: microbial life on surfaces. *Emerging Infectious Diseases*, *8*(9), 881–890. https://doi.org/10.3201/eid0809.020063

Droppo, I. G., Krishnappan, B. G., & Lawrence, J. R. (2016). Microbial interactions with naturally occurring hydrophobic sediments: Influence on sediment and associated contaminant mobility. *Water Research*, *92*, 121–130. https://doi.org/10.1016/j.watres.2016.01.034

Dula, S., Ajayeoba, T. A., & Ijabadeniyi, O. A. (2021). Bacterial biofilm formation on stainless steel in the food processing environment and its health implications. *Folia Microbiologica*, *66*(3), 293–302. https://doi.org/10.1007/s12223-021-00864-2

Edwards, S. J., & Kjellerup, B. V. (2013). Applications of biofilms in bioremediation and biotransformation of persistent organic pollutants, pharmaceuticals/personal care products, and heavy metals. *Applied Microbiology and Biotechnology*, *97*(23), 9909–9921. https://doi.org/10.1007/s00253-013-5216-z

Eick, S. (2021). Biofilms. *Monographs in Oral Science*, *29*, 1–11. https://doi.org/10.1159/000510184

Fernandes, G., Bastos, M. C., de Vargas, J. P. R., Le Guet, T., Clasen, B., & Dos Santos, D. R. (2020). The use of epilithic biofilms as bioaccumulators of pesticides and pharmaceuticals in aquatic environments. *Ecotoxicology*, *29*(9), 1293–1305. https://doi.org/10.1007/s10646-020-02259-4

Flemming, H.-C. (2020). Biofouling and me: My Stockholm syndrome with biofilms. *Water Research*, *173*, 115576. https://doi.org/10.1016/j.watres.2020.115576

Flemming, H.-C., & Wingender, J. (2010). The biofilm matrix. *Nature Reviews. Microbiology*, *8*(9), 623–633. https://doi.org/10.1038/nrmicro2415

Fletcher, M. (1988). Attachment of Pseudomonas fluorescens to glass and influence of electrolytes on bacterium-substratum separation distance. *Journal of Bacteriology*, *170*(5), 2027–2030. https://doi.org/10.1128/jb.170.5.2027-2030.1988

Fong, J. N. C., & Yildiz, F. H. (2015). Biofilm matrix proteins. *Microbiology Spectrum*, *3*(2). https://doi.org/10.1128/microbiolspec.MB-0004-2014

Gamarra, N. N., Villena, G. K., & Gutiérrez-Correa, M. (2010). Cellulase production by Aspergillus niger in biofilm, solid-state, and submerged fermentations. *Applied microbiology and biotechnology*, *87*(2), 545–551. https://doi.org/10.1007/s00253-010-2540-4

Gan, T., & Gong, X. (2017). [Characterization of the physical properties of biofilms]. *Sheng Wu Gong Cheng Xue Bao = Chinese Journal of Biotechnology*, *33*(9), 1390–1398. https://doi.org/10.13345/j.cjb.170062

Ghosh, S., Nag, M., Lahiri, D., Sarkar, T., Pati, S., Kari, Z. A., Nirmal, N. P., Edinur, H. A., & Ray, R. R. (2022). Engineered biofilm: Innovative nextgen strategy for quality enhancement of fermented foods. *Frontiers in Nutrition*, *9*, 808630. https://doi.org/10.3389/fnut.2022.808630

Goller, C. C., & Romeo, T. (2008). Environmental influences on biofilm development. *Current Topics in Microbiology and Immunology*, *322*, 37–66.

Goodman, S. D., Obergfell, K. P., Jurcisek, J. A., Novotny, L. A., Downey, J. S., Ayala, E. A., Tjokro, N., Li, B., Justice, S. S., & Bakaletz, L. O. (2011). Biofilms can be dispersed by focusing the immune system on a common family of bacterial nucleoid-associated proteins. *Mucosal Immunology*, *4*(6), 625–637. https://doi.org/10.1038/mi.2011.27

Greenman, J., Hewett, K., & Saad, S. (2020). Discovery, development and exploitation of steady-state biofilms. *Journal of Breath Research*, *14*(4), 044001. https://doi.org/10.1088/1752-7163/abb765

Hall-Stoodley, L., Costerton, J. W., & Stoodley, P. (2004). Bacterial biofilms: From the natural environment to infectious diseases. *Nature Reviews. Microbiology*, *2*(2), 95–108. https://doi.org/10.1038/nrmicro821

Harmsen, M., Yang, L., Pamp, S. J., & Tolker-Nielsen, T. (2010). An update on Pseudomonas aeruginosa biofilm formation, tolerance, and dispersal. *FEMS Immunology and Medical Microbiology*, *59*(3), 253–268. https://doi.org/10.1111/j.1574-695X.2010.00690.x

https://doi.org/10.1007/BF02872373

https://doi.org/10.2307/1352224

Jauffred, L., Munk Vejborg, R., Korolev, K. S., Brown, S., & Oddershede, L. B. (2017). Chirality in microbial biofilms is mediated by close interactions between the cell surface and the substratum. *The ISME Journal*, *11*(7), 1688–1701. https://doi.org/10.1038/ismej.2017.19

Jayathilake, P. G., Jana, S., Rushton, S., Swailes, D., Bridgens, B., Curtis, T., & Chen, J. (2017). Extracellular polymeric substance production and aggregated bacteria colonization influence the competition of microbes in biofilms. *Frontiers in Microbiology*, *8*, 1865. https://doi.org/10.3389/fmicb.2017.01865

Jefferson, K. K., & Cerca, N. (2006). Bacterial-bacterial cell interactions in biofilms: detection of polysaccharide intercellular adhesins by blotting and confocal microscopy. *Methods in Molecular Biology*, *341*, 119–126. https://doi.org/10.1385/1-59745-113-4:119

Jiang, Y., Liu, Y., Zhang, X., Gao, H., Mou, L., Wu, M., Zhang, W., Xin, F., & Jiang, M. (2021). Biofilm application in the microbial biochemicals production process. *Biotechnology Advances*, *48*, 107724. https://doi.org/10.1016/j.biotechadv.2021.107724

Kanematsu, H., & Barry, D. M. (2020). Fundamentals for Biofilms. In *Formation and control of biofilm in various environments* (pp. 13–37). Springer Singapore. https://doi.org/10.1007/978-981-15-2240-6_2

Kaplan, J. B. (2014). Biofilm matrix-degrading enzymes. *Methods in Molecular Biology*, *1147*, 203–213. https://doi.org/10.1007/978-1-4939-0467-9_14

Karande, R., Halan, B., Schmid, A., & Buehler, K. (2014). Segmented flow is controlling growth of catalytic biofilms in continuous multiphase microreactors. *Biotechnology and Bioengineering*, *111*(9), 1831–1840. https://doi.org/10.1002/bit.25256

Kavanaugh, J. S., Flack, C. E., Lister, J., Ricker, E. B., Ibberson, C. B., Jenul, C., Moormeier, D. E., Delmain, E. A., Bayles, K. W., & Horswill, A. R. (2019). Identification of Extracellular DNA-Binding Proteins in the Biofilm Matrix. *MBio*, *10*(3). https://doi.org/10.1128/mBio.01137-19

Kim, S.-K., & Lee, J.-H. (2016). Biofilm dispersion in Pseudomonas aeruginosa. *Journal of Microbiology*, *54*(2), 71–85. https://doi.org/10.1007/s12275-016-5528-7

Krasowska, A., & Sigler, K. (2014). How microorganisms use hydrophobicity and what does this mean for human needs? *Frontiers in Cellular and Infection Microbiology*, *4*, 112. https://doi.org/10.3389/fcimb.2014.00112

Leonov, V. V., Kurlovich, N. A., & Sokolova, T. N. (2014). [Relation of hydrophobicity index of microbial cells to their biofilm-forming ability]. *Biofizika*, *59*(6), 1131–1134.

Li, X., Song, J. L., Culotti, A., Zhang, W., Chopp, D. L., Lu, N., & Packman, A. I. (2015). Methods for characterizing the co-development of biofilm and habitat heterogeneity. *Journal of Visualized Experiments*, *2015*(97). https://doi.org/10.3791/52602

Li, Y., Qu, Y., Yang, H., Zhou, X., Xiao, P., & Shao, T. (2023). Combatting biofilms in potable water systems: A comprehensive overview to ensuring industrial water safety. *Environmental Microbiology Reports*, *15*(6), 445–454. https://doi.org/10.1111/1758-2229.13207

Li, Z., Wang, X., Wang, J., Yuan, X., Jiang, X., Wang, Y., Zhong, C., Xu, D., Gu, T., & Wang, F. (2022). Bacterial biofilms as platforms engineered for diverse applications. *Biotechnology Advances*, *57*, 107932. https://doi.org/10.1016/j.biotechadv.2022.107932

Limoli, D. H., Jones, C. J., & Wozniak, D. J. (2015). Bacterial extracellular polysaccharides in biofilm formation and function. *Microbiology Spectrum*, *3*(3). https://doi.org/10.1128/microbiolspec.MB-0011-2014

Lin, H., Zhu, X., Wang, Y., & Yu, X. (2017). Effect of sodium hypochlorite on typical biofilms formed in drinking water distribution systems. *Journal of Water and Health*, *15*(2), 218–227. https://doi.org/10.2166/wh.2017.141

López, D., Vlamakis, H., & Kolter, R. (2010). Biofilms. *Cold Spring Harbor Perspectives in Biology*, *2*(7), a000398. https://doi.org/10.1101/cshperspect.a000398

Lu, H., Wan, J., Li, J., Shao, H., & Wu, Y. (2016). Periphytic biofilm: A buffer for phosphorus precipitation and release between sediments and water. *Chemosphere*, *144*, 2058–2064. https://doi.org/10.1016/j.chemosphere.2015.10.129

MacIntyre, H.L., Geider, R.J. & Miller, D.C. (1996). Microphytobenthos: The ecological role of the "secret garden" of unvegetated, shallow-water marine habitats. I. Distribution, abundance and primary production. *Estuaries* **19**, 186–201.

Malvankar, N. S., & Lovley, D. R. (2012). Microbial nanowires: A new paradigm for biological electron transfer and bioelectronics. *ChemSusChem*, *5*(6), 1039–1046. https://doi.org/10.1002/cssc.201100733

Mangwani, N., Kumari, S., & Das, S. (2016). Bacterial biofilms and quorum sensing: Fidelity in bioremediation technology. *Biotechnology & Genetic Engineering Reviews*, *32*(1–2), 43–73. https://doi.org/10.1080/02648725.2016.1196554

Mi, Z., Dai, Y., Xie, S., Chen, C., & Zhang, X. (2015). Impact of disinfection on drinking water biofilm bacterial community. *Journal of Environmental Sciences (China)*, *37*, 200–205. https://doi.org/10.1016/j.jes.2015.04.008

Miller, H. C., Wylie, J., Dejean, G., Kaksonen, A. H., Sutton, D., Braun, K., & Puzon, G. J. (2015). Reduced Efficiency of Chlorine Disinfection of Naegleria fowleri in a Drinking Water Distribution Biofilm. *Environmental Science & Technology*, *49*(18), 11125–11131. https://doi.org/10.1021/acs.est.5b02947

Mishra, S., Huang, Y., Li, J., Wu, X., Zhou, Z., Lei, Q., Bhatt, P., & Chen, S. (2022). Biofilm-mediated bioremediation is a powerful tool for the removal of environmental pollutants. *Chemosphere*, *294*, 133609. https://doi.org/10.1016/j.chemosphere.2022.133609

Moller, S., Korber, D. R., Wolfaardt, G. M., Molin, S., & Caldwell, D. E. (1997). Impact of nutrient composition on a degradative biofilm community. *Applied and Environmental Microbiology*, *63*(6), 2432–2438. https://doi.org/10.1128/aem.63.6.2432-2438.1997

Nozaka, S., Furukawa, S., Sasaki, M., Hirayama, S., Ogihara, H., & Morinaga, Y. (2014). Manganese Ion Increases LAB-yeast Mixed-species Biofilm Formation. *Bioscience of Microbiota, Food and Health*, *33*(2), 79–84. https://doi.org/10.12938/bmfh.33.79

Palmer, J., Flint, S., & Brooks, J. (2007). Bacterial cell attachment, the beginning of a biofilm. *Journal of Industrial Microbiology & Biotechnology*, *34*(9), 577–588. https://doi.org/10.1007/s10295-007-0234-4

Pandit, S., Fazilati, M., Gaska, K., Derouiche, A., Nypelö, T., Mijakovic, I., & Kádár, R. (2020). The exo-polysaccharide component of extracellular matrix is essential for the viscoelastic properties of bacillus subtilis biofilms. *International Journal of Molecular Sciences*, *21*(18). https://doi.org/10.3390/ijms21186755

Paulsson, M., Kragh, K. N., Su, Y.-C., Sandblad, L., Singh, B., Bjarnsholt, T., & Riesbeck, K. (2021). Peptidoglycan-binding anchor is a pseudomonas aeruginosa OmpA family lipoprotein with importance for outer membrane vesicles, biofilms, and the periplasmic shape. *Frontiers in Microbiology*, *12*, 639582. https://doi.org/10.3389/fmicb.2021.639582

Rather, M. A., Gupta, K., & Mandal, M. (2021). Microbial biofilm: Formation, architecture, antibiotic resistance, and control strategies. *Brazilian Journal of Microbiology*, *52*(4), 1701–1718. https://doi.org/10.1007/s42770-021-00624-x

Reguera, G. (2018). Microbial nanowires and electroactive biofilms. *FEMS Microbiology Ecology*, *94*(7). https://doi.org/10.1093/femsec/fiy086

Rittmann, B. E. (2018). Biofilms, active substrata, and me. *Water Research*, *132*, 135–145. https://doi.org/10.1016/j.watres.2017.12.043

Rossi, F. (2020). Beneficial biofilms for land rehabilitation and fertilization. *FEMS Microbiology Letters*, *367*(21). https://doi.org/10.1093/femsle/fnaa184

Rumbaugh, K. P., & Sauer, K. (2020). Biofilm dispersion. *Nature Reviews. Microbiology*, *18*(10), 571–586. https://doi.org/10.1038/s41579-020-0385-0

Saba, B., Christy, A. D., Yu, Z., Co, A. C., Islam, R., & Tuovinen, O. H. (2017). Characterization and performance of anodic mixed culture biofilms in submersed microbial fuel cells. *Bioelectrochemistry*, *113*, 79–84. https://doi.org/10.1016/j.bioelechem.2016.10.003

Salisbury, A.-M., & Percival, S. L. (2019). The efficacy of an electrolysed water formulation on biofilms. *Advances in Experimental Medicine and Biology*, *1214*, 1–8. https://doi.org/10.1007/5584_2018_207

Saratale, G. D., Saratale, R. G., Shahid, M. K., Zhen, G., Kumar, G., Shin, H.-S., Choi, Y.-G., & Kim, S.-H. (2017). A comprehensive overview on electro-active biofilms, role of exo-electrogens and their microbial niches in microbial fuel cells (MFCs). *Chemosphere*, *178*, 534–547. https://doi.org/10.1016/j.chemosphere.2017.03.066

Saxena, P., Joshi, Y., Rawat, K., & Bisht, R. (2019). Biofilms: Architecture, resistance, quorum sensing and control mechanisms. *Indian Journal of Microbiology*, *59*(1), 3–12. https://doi.org/10.1007/s12088-018-0757-6

Schluter, J., Schoech, A. P., Foster, K. R., & Mitri, S. (2016). The evolution of quorum sensing as a mechanism to infer kinship. *PLoS Computational Biology*, *12*(4), e1004848. https://doi.org/10.1371/journal.pcbi.1004848

Schmidt, B., Sánchez, L. A., Fretschner, T., Kreps, G., Ferrero, M. A., Siñeriz, F., & Szewzyk, U. (2014). Isolation of Sphaerotilus-Leptothrix strains from iron bacteria communities in Tierra del Fuego wetlands. *FEMS Microbiology Ecology*, *90*(2), 454–466. https://doi.org/10.1111/1574-6941.12406

Schuster, J. J., & Markx, G. H. (2014). Biofilm architecture. *Advances in Biochemical Engineering/Biotechnology*, *146*, 77–96. https://doi.org/10.1007/10_2013_248

Scribani Rossi, C., Barrientos-Moreno, L., Paone, A., Cutruzzolà, F., Paiardini, A., Espinosa-Urgel, M., & Rinaldo, S. (2022). Nutrient sensing and biofilm modulation: The Example of L-arginine in pseudomonas. *International Journal of Molecular Sciences*, *23*(8). https://doi.org/10.3390/ijms23084386

Seike, S., Kobayashi, H., Ueda, M., Takahashi, E., Okamoto, K., & Yamanaka, H. (2020). Outer membrane vesicles released from aeromonas strains are involved in the biofilm formation. *Frontiers in Microbiology*, *11*, 613650. https://doi.org/10.3389/fmicb.2020.613650

Singh, M. P., Singh, P., Li, H.-B., Song, Q.-Q., & Singh, R. K. (2020). Microbial biofilms: Development, structure, and their social assemblage for beneficial applications. In *New and future developments in microbial biotechnology and bioengineering: microbial biofilms* (pp. 125–138). Elsevier. https://doi.org/10.1016/B978-0-444-64279-0.00010-4

Siqueira, V., & Lima, N. (2012). Surface hydrophobicity of culture and water biofilm of Penicillium spp. *Current Microbiology*, *64*(2), 93–99. https://doi.org/10.1007/s00284-011-0037-8

Sivadon, P., Barnier, C., Urios, L., & Grimaud, R. (2019). Biofilm formation as a microbial strategy to assimilate particulate substrates. *Environmental Microbiology Reports*, *11*(6), 749–764. https://doi.org/10.1111/1758-2229.12785

Sonawane, J. M., Rai, A. K., Sharma, M., Tripathi, M., & Prasad, R. (2022). Microbial biofilms: Recent advances and progress in environmental bioremediation. *The Science of the Total Environment*, *824*, 153843. https://doi.org/10.1016/j.scitotenv.2022.153843

Stoodley, P., Debeer, D., & Lewandowski, Z. (1994). Liquid flow in biofilm systems. *Applied and Environmental Microbiology*, *60*(8), 2711–2716. https://doi.org/10.1128/aem.60.8.2711-2716.1994

Tasaki, S., Nakayama, M., & Shoji, W. (2017). Self-organization of bacterial communities against environmental pH variation: Controlled chemotactic motility arranges cell population structures in biofilms. *Plos One*, *12*(3), e0173195. https://doi.org/10.1371/journal.pone.0173195

Tolker-Nielsen, T. (2015). Biofilm development. *Microbiology Spectrum*, *3*(2), MB-0001. https://doi.org/10.1128/microbiolspec.MB-0001-2014

Toyofuku, M., Inaba, T., Kiyokawa, T., Obana, N., Yawata, Y., & Nomura, N. (2015). Environmental factors that shape biofilm formation. *Bioscience, Biotechnology, and Biochemistry*, *80*(1), 7–12. https://doi.org/10.1080/09168451.2015.1058701

Velmourougane, K., Prasanna, R., & Saxena, A. K. (2017). Agriculturally important microbial biofilms: Present status and future prospects. *Journal of Basic Microbiology*, *57*(7), 548–573. https://doi.org/10.1002/jobm.201700046

Vishwanathan, A. S. (2021). Microbial fuel cells: A comprehensive review for beginners. *3 Biotech*, *11*(5), 248. https://doi.org/10.1007/s13205-021-02802-y

Voglauer, E. M., Zwirzitz, B., Thalguter, S., Selberherr, E., Wagner, M., & Rychli, K. (2022). Biofilms in water hoses of a meat processing environment harbor complex microbial communities. *Frontiers in Microbiology*, *13*, 832213. https://doi.org/10.3389/fmicb.2022.832213

Wan, K., Zhang, M., Ye, C., Lin, W., Guo, L., Chen, S., & Yu, X. (2019). Organic carbon: An overlooked factor that determines the antibiotic resistome in drinking water sand filter biofilm. *Environment International*, *125*, 117–124. https://doi.org/10.1016/j.envint.2019.01.054

Wang, G., Zhao, G., Chao, X., Xie, L., & Wang, H. (2020). The Characteristic of Virulence, Biofilm and Antibiotic Resistance of Klebsiella pneumoniae. *International Journal of Environmental Research and Public Health*, *17*(17). https://doi.org/10.3390/ijerph17176278

Wang, Z.-W., & Chen, S. (2009). Potential of biofilm-based biofuel production. *Applied Microbiology and Biotechnology*, *83*(1), 1–18. https://doi.org/10.1007/s00253-009-1940-9

Webster, S. S., Lee, C. K., Schmidt, W. C., Wong, G. C. L., & O'Toole, G. A. (2021). Interaction between the type 4 pili machinery and a diguanylate cyclase fine-tune c-di-GMP levels during early biofilm formation. *Proceedings of the National Academy of Sciences of the United States of America*, *118*(26). https://doi.org/10.1073/pnas.2105566118

Xing, Y., Tan, S., Liu, S., Xu, S., Wan, W., Huang, Q., & Chen, W. (2022). Effective immobilization of heavy metals via reactive barrier by rhizosphere bacteria and their biofilms. *Environmental Research*, *207*, 112080. https://doi.org/10.1016/j.envres.2021.112080

Xiong, X., Xing, Y., He, J., Wang, L., Shen, Z., Chen, W., & Huang, Q. (2021). Keystone species determine the "selection mechanism" of multispecies biofilms for bacteria from soil aggregates. *The Science of the Total Environment*, *773*, 145069. https://doi.org/10.1016/j.scitotenv.2021.145069

Xue, Z., & Seo, Y. (2013). Impact of chlorine disinfection on redistribution of cell clusters from biofilms. *Environmental Science & Technology*, *47*(3), 1365–1372. https://doi.org/10.1021/es304113e

Yang, L., Liu, Y., Wu, H., Hóiby, N., Molin, S., & Song, Z. (2011). Current understanding of multi-species biofilms. *International Journal of Oral Science*, *3*(2), 74–81. https://doi.org/10.4248/IJOS11027

Yang, L., Liu, Y., Wu, H., Song, Z., Høiby, N., Molin, S., & Givskov, M. (2012). Combating biofilms. *FEMS Immunology and Medical Microbiology*, *65*(2), 146–157. https://doi.org/10.1111/j.1574-695X.2011.00858.x

Yin, W., Wang, Y., Liu, L., & He, J. (2019). Biofilms: The microbial "protective clothing" in extreme environments. *International Journal of Molecular Sciences*, *20*(14). https://doi.org/10.3390/ijms20143423

York, A. (2017). Biofilms: The architect of the biofilm. *Nature Reviews. Microbiology*, *15*(11), 642–643. https://doi.org/10.1038/nrmicro.2017.127

Yuan, K., Li, S., & Zhong, F. (2020). Treatment of coking wastewater in biofilm-based bioaugmentation process: Biofilm formation and microbial community analysis. *Journal of Hazardous Materials*, *400*, 123117. https://doi.org/10.1016/j.jhazmat.2020.123117

2 Aquatic Ecosystems and Diversity of Microorganisms

Suseela Lanka, Anitha K., Ashok Kota, and Santhilatha Pandrangi

2.1 INTRODUCTION

An ecosystem is a biological system composed of all organisms that rely on their surroundings and one another to survive. The two primary types of ecosystems are aquatic and terrestrial habitats. Without question, water is a major source of life. In addition, aquatic animals are those that live in water. They also need water to fulfil their basic needs, which include eating and reproducing. This chapter will briefly review the many aquatic ecosystem types, the diversity of microorganisms, and their significance (Poff et al., 2002).

2.2 AQUATIC ECOSYSTEM

An aquatic ecosystem is referred to as a type of habitat that is primarily composed of water. It is the largest of various other ecosystems covering around 75% of our planet's surface. This ecosystem includes living things interacting with the physical and chemical components of the environment along with floral and faunal diversity. Organisms present in the aquatic ecosystem are those living things whose food, shelter, reproduction, and other basic needs are dependent on an aquatic environment (Balasubramanian, 2011).

As a medium for material transfer and the catalyst for chemical interactions between the lithosphere, biosphere, and atmosphere, water is essential to regulating the process of the global ecosystem in aquatic systems. The distinct physicochemical characteristics of water indicate the quality of the aquatic body and dictate how much effectively it works and to what extent it supports living beings. Sediment matter of these ecosystems is similar to soil of land-based ecosystem (terrestrial) in that they offer nutrients, substrate, and a habitat for aquatic organisms. The quality of water dynamics and the food cycles of the environment are significantly accelerated by sediments (Sigee, 2005).

The quality of the sediment directly or indirectly impacts an aquatic ecosystem's ability to operate. The quality of sediment is determined by its numerous physicochemical characteristics. Fishes have a prominent role in the ecosystem throughout

 DOI: 10.1201/9781003487203-2

the food chain. The most prevalent species in the aquatic ecosystems are plankton, benthos, and nekton. A few well-known examples of aquatic ecosystems are lakes, seas, ponds, rivers, swamps, coral reefs, marshes, etc. (Sigee, 2005).

2.2.1 Types of Aquatic Ecosystems

In general, aquatic ecosystems are distinguished into two main types, marine/saltwater and freshwater, which are divided further as described below.

2.2.1.1 Marine/Saltwater

This is the largest, occupying 70% of the total planet's surface and containing water with saltiness. Nevertheless, the aquatic organisms in this ecosystem are well acclimated to salty water, and have difficulty living in freshwater. These ecosystems are further divided as follows.

(a) Ocean Ecosystem

The Pacific, Atlantic, Arctic, Southern, and Indian are the five major abysses on Earth, with the Pacific occupying the first place in terms of size and depth followed by the Atlantic. The Southern Ocean has the highest krill population. Oceanic aquatic ecosystems serve as a habitat for organisms such as plankton, turtles, crustaceans, shellfish, corals, harpies, tube worms, reptiles, etc. (Beck & Lehner, 1974).

(b) Estuaries

Estuaries are the connecting point between sea and river water, with the water slightly saline compared to freshwater and highly diluted compared to seawater. These unique environments are characterized by dynamic and productive ecosystems (Hughes et al., 2017). Biologically, they are productive with stimulation of primary production and they also trap nutrients of plants. River mouths, tidal marshes, and coastal bays are a few types of estuaries.

(c) Coral Reefs

Coral reefs are diverse and complex marine ecosystems characterized by coral polyps, tiny soft-bodied organisms that secrete calcium carbonate to form intricate structures. These ecosystems are known for their exceptional biodiversity, providing a habitat for many marine species. The stony corals with their polyps are clustered into groups by secreting calcium carbonate to make up the mature coral reefs (Hughes et al., 2017). In the animal kingdom, the phylum Cnidarian and the class Anthozoa include sea anemones jellyfish and the corals. Corals contain exoskeletons made of hard carbonate for support and to cover them. Coral reefs are invaluable ecosystems that demand urgent attention and conservation action. Through collaborative efforts at local, national, and international levels, it is possible to safeguard these biodiversity hotspots for future generations (Pandolfi et al., 2003).

(d) Coastal Ecosystem

Coastal ecosystems, the transitional zones where land meets the sea, are vital components of the Earth's biosphere. These ecosystems are unique in terms of structure, inflow of energy, and variety. These ecosystems are rich in many plant species and algae. The fauna consists of fish, snails, crabs, shrimp, lobsters, and other species (Alongi et al., 2008). In this ecosystem, the role of human activities, climate change, and conservation efforts are highlighting the need for sustainable management strategies (Duarte et al., 2013). Coastal ecosystems are critical for global biodiversity, livelihoods, and various ecosystem services.

2.2.1.2 Freshwater Ecosystem

Freshwater ecosystems comprise less than 1% of the planet's surface and are classified as follows.

(a) Swamps & Wetlands

The swamps and wetlands are enchanting and lush ecosystems, characterized by their marshy terrain and their tendency to be submerged in water. These areas are teeming with a plethora of plant and animal life, making them a haven for nature enthusiasts. The water lilies, morasses, and Northern Pikes add bursts of colour to the serene landscape, while the dragonflies zip and dart through the air. The Green Herons, with their striking emerald plumage, are a particularly breathtaking sight, perched elegantly atop the surrounding foliage (Middleton, 1999).

(b) Lentic Ecosystems

The lentic ecosystems are characterized by standing or steady-moving aquatic bodies such as lakes and ponds, and are habitats for both rooting and floating plants, algae, and invertebrates. These ecosystems contrast with lotic ecosystems, which are characterized by flowing water such as rivers and streams (Dodds et al., 2009). These environments support algal members, floating-leaved and rooted plants, and crustaceans such as shrimp and crabs. One can find salamanders and frogs, reptiles such as water snakes, and alligators.

(c) Lotic Ecosystems

The lotic systems are considered to be freshwater ecosystems characterized by a flow of water that will move in one direction such as rivers and streams. These ecosystems are dynamic and support a diverse range of flora and fauna (Allan, 2004). These ecosystems are habitats for an abundant array of insects including mayflies, beetles, stoneflies, etc. These ecosystems also bear species such as beavers, trout, river dolphins, otters, minnows, eel, etc.

2.3 MICROORGANISMS IN AQUATIC ENVIRONMENTS

Microbial consortia are requisite to the sustainment of wholesome aquatic bodies owing to their involvement in several biological and chemical processes, including the nutrients recycle, organic matter breakdown, energy and chemicals transport, etc. (Sang et al., 2018; Chen et al., 2019; Huang et al., 2022, Zhang et al., 2023a,b). The most essential categories of the taxonomic variety seen in aquatic environments are broadly classified into three categories: Bacteria, Eukarya and Archaea, which can be differentiated based on their physiological, anatomical, and molecular characteristics.. There are bacteria in every freshwater habitat and belong to a single kingdom called Eubacteria, which is made up of actinomycetes, bacteria, and blue-green algae. Extreme freshwater settings are often habitats for members of the domain Archaea. There are five distinct size categories for planktonic organisms, which go from femtoplankton to macroplankton. These groups can have a maximum linear dimension of up to 200 μm. The classification includes a range of organisms such as viruses, bacteria, archaea, microalgae, fungi, and protozoa (Sigee, 2005).

Microorganisms in freshwater ecosystems can be divided into two major classes based on their nutritional habits: autotrophs and heterotrophs. Autotrophs produce polymeric carbon compounds by utilizing CO_2. They derive required nutrient supplies of phosphorus and nitrogen from inorganic substances. These include microalgal members and photosynthetic bacteria that inhabit either freshwater or marine ecosystem's and constitute primary producers. The other group, the heterotrophs, depend on organic compounds for their carbon source. In freshwater environments, the majority of microorganisms such as bacteria, protozoa, and fungi are heterotrophic. Heterotrophic nutrition is achieved through saprotrophic, predation mode, and in concert with other organisms. Saprotrophs get their food from dead remains things by the way of directly absorbing soluble compounds or indirectly absorbing them through enzyme secretion (fungal members and bacteria). Protozoa, on the other hand, obtain their nutrients by ingesting particulate matter through phagocytosis (Alexopoulos et al., 2009). Associations with living things, such as parasitism and symbiosis, constitute the third facet of heterotrophy. The parasite in parasitism is strictly dependent upon the host member. The basic structure pertaining to the food web in freshwater systems is similar to that of oceanic environments, with planktonic bacteria consuming detritus, benthic organisms like bacteria, fungi, and protozoa breaking down detritus, and viruses and fungi parasitizing phytoplankton populations (Bezirtzoglou et al. 1996, Maipa et al. 2001). **Figure 2.1** shows the diversity of microorganisms in aquatic ecosystems.

2.3.1 Cyanobacteria (Photosynthetic)

These can exist in varied aquatic systems and vary in size between 1μm to 100μm. These photosynthetic bacteria produce oxygen and use the available CO_2 in water while also being responsible for N_2 fixation. Scientists have been studied the taxonomy of these organisms more extensively than other bacteria due to their size. The

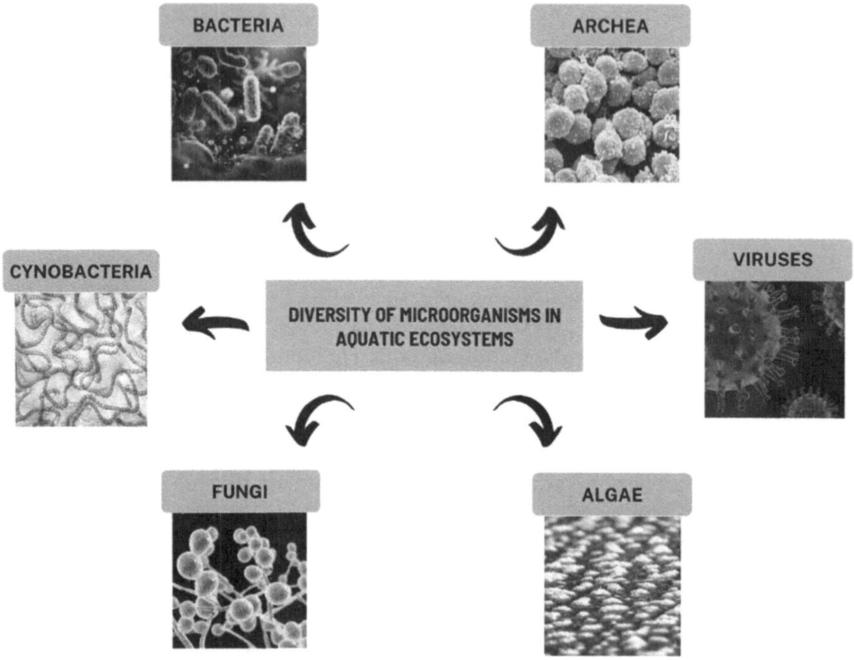

FIGURE 2.1 Diversity of aquatic microorganisms.

ecological impact of these bacteria is concerning, because they result in population rise producing blooms under eutrophic conditions, drastically affecting the quality of seawater and life of aquatic residents. The predominant similarity between photosynthetic land dwellers and phytoplankton includes carbon fixation and in oceanic systems the process is exclusively carried out by the latter, which are free-floating microorganisms.

Autotrophs, phytoplankton, and cyanobacteria primarily use CO_2 as their source of carbon and fix it via Calvin cycle. However, unlike the terrestrial counterparts, most of the abovementioned organisms use carotenoid pigments. Diatoms, which can grow up to 100 μm in size and have distinctive shapes, have been extensively studied due to their ease of quick identification. Their skeletons have piled up on the oceanic floor and have revolutionized our planet by giving birth to other forms. Although not as extensively studied, the role of pico-sized phytoplankton is also important in marine ecosystems.

Varied bacteria, including cyanobacteria, proteorhodopsin-bearing bacteria, and aerobic bacteria, have been found in the oceans. However, with the exception of cyanobacteria, allother groups are not autotrophs and do not involve in primary production to prepare their food (Kirhman, 2008). In lieu, they bestow towards assimilation of dissolved oceanic organic components. This is due to their vast surface-to-volume ratio. The abundance of all microorganisms in marine environments, including heterotrophs, is closely related to this ratio, with viruses being the most abundant of all.

2.3.2 Bacteria

Studies on agile bacterial consortia of aqueous environments reinforce the linkages between environmental conditions and the associations among bacterial assemblages that are functional (Adyasari et al., 2019). Bacteria of heterotrophic nature are essential decomposing agents in underwater systems, while the autotrophs, which are similar to cyanobacteria, are vital managers in nutrient-rich bodies of water. These bacteria are crucial in the cycling of nutrients and the transformation of organic matter (Seymour et al., 2017; Cotner et al., 2002). Due to climate change and violent hurricanes, natural submarine ecosystems have suffered from increased salinization and eutrophication in recent decades. (Monchamp et al., 2018; Dugan et al., 2017; Herbert et al., 2015). It is crucial to comprehend how bacterial communities react to salinization of lakes and nutrient overload (Steffen et al., 2007; Taranu et al., 2015). Studies exploring diversity and consortia composition of bacteria in lakes with varying nutrient status have been conducted frequently over the past two decades (Kiersztyn et al., 2019; Horner- Devine et al., 2003; Zwirglmaier et al., 2015).

The connection between trophic status and bacterial diversity is intricate, but some have proposed that the highest diversity is found in situations with moderate levels of nutrients or productivity (Kassen et al., 2000). The presence of salt is a significant environmental factor that plays a crucial role in shaping bacterial communities. The saltiness factor is responsible for creating a fresh environment that helps to govern the composition of bacterial colonies (Lozupone et al., 2007; Ji et al., 2019; Herlemann et al., 2011). The impact of salt concentration on the variety of bacteria present varies, with studies suggesting a negative correlation (Benlloch et al., 2002; Yang et al., 2016), no impact (Hewson et al., 2004), or a maximum variety at low levels of salinity (Wang et al., 2011).

There has been no exploration to date on how bacterial heterogeneity and bacterial colony composition (BCC) react to simultaneous increases in nutrient levels and salinity. When environmental pressures such as eutrophication and salinization are introduced, bacterial assemblage patterns can be impacted both at indigenous and at original scales. Disentwining the bottom-line mechanisms that drive bacterial consortia assemblage remains a big task in environmental microbiology. It is widely agreed that community assemblage is influenced by a fine balance between deterministic and random exercises (Romina Schiaffino et al., 2011).

The major factors influencing the composition of randomly formed communities are ecological drift, group effects/homogenizing dispersal (in the case of strongly coupled systems), and dispersal limits (large spatial scales). On the other hand, deterministically formed communities emerge as a result of biotic interactions or selection through local environmental filtration (Zhou et al., 2017; Nemergut et al., 2013; Van der Gucht et al., 2007).

The decline in the rush of water and the increase in air temperature, evaporation, over-grazing, and fish husbandry are considered the primary causes of the degradation, saltiness, and nutrient load of most closed/sink lakes, according to Liang et al. (2017) and Zhou et al. (2009). Microscopic aquatic organisms such as bacteria and viruses play a vital role in ecological processes, including carbon cycling and energy flow in underwater environments (Fenchel et al., 2008). Bacteria are crucial agents for the microbiological paths in underwater systems, responsible for transferring dissolved organic matter and nutrients to metazoan trophic levels, thus affecting the inflow of carbon and energy within an ecosystem (Azam et al., 1998; Pomeroy et al., 1974). **Table 2.1** summarizes the bacterial populations or bacterial species of various aquatic ecosystems reported in the literature.

TABLE 2.1
Bacterial Species Reported in Various Aquatic Ecosystems

Bacterial biodiversity	Aquatic ecosystem	References
Rhodobacteraceae family	Malta River and Thakuran River	Dhal et al., 2020
Vibrio sp	Mangrove sediments	Haldar and Nazareth, 2018
Prevotella copri *Bacteroides fragilis*	Mangrove sediments	Chithira et al., 2021
Desulfovibrio *Desulfotomaculum* *Azospirillum* *Azotobacter* *Bacillus sp* *Pseudomonas sp* *Chromatium* *Beggiatoa* *Methanoccoides methylutens sp*	Mangrove ecosystem	Palit and Das, 2021; Mukherji et al., 2020
Macellibacteroides fermentans, *Arcobacter bivolvorium,* *Arcobacter butzleri,* *Methanothrix soenhngeii,* *Bacteroides graminisolvens*	Cooum river	Aiswarya et al., 2021
Acinetobacter baumannii *Prevotella copri*	Ganga river	Reddy and Dubet, 2021
Coxiella burnetii, *Fibrobacter intestinalis,* *Candidatus Cloacamonas acidaminovorans*	Soda lakes	Dudhagara et al., 2015
Methylophaga	Pangong lake	Rathur et al., 2017
Anaerolinea	littoral and benthic regions from the Chilika lake	Raina et al., 2019
Synechococcus sp.	Mooriganga estuary	Bhadury and Singh, 2020
Pseudoalteromonadaceae *Synechococcaceae* *Pelagibacteraceae and* *Caulobacteraceae families*	Marine ecosystems	Rajpathak et al., 2018

2.3.3 Viruses

The smallest natural creatures are known as viruses and invade both prokaryotic and eukaryotic cells. These parasites exist throughout underwater ecosystems, and they have received significant attention due to their abundance and role in submarine food webs. Since these parasites release DOM (dissolved organic matter) and various nutrients by way of lysing their hosts, they have a crucial biogeochemical function that can impact various environmental factors, such as primary productivity, ecosystem respiration, species distribution, and gene exchanges between microbes (Fuhrman et al., 1999). Bacteriophages are viral clusters that infect bacteria and have been extensively studied (Fuhrman et al., 1995; Sawstrom et al., 2008; Barros et al., 2010).

Genetic. Standard techniques are available to evaluate both structural (non-taxonomic) and functional characteristics of algal members. Sampling, interpretation, and monitoring methods for determining algal members in various freshwater systems have been proposed by Stevenson & Lowe (1986); Round (1991); Rott (1991); McCormick & Cairns (1994); Van Dam et al. (1994); Stevenson & Pan (1999).

Anthropogenic stressors affect algal assemblages, which impact relations pertaining to the food web and other factors related to the ecosystem (McCormick & Cairns, 1994). Algal members impact taste of the water and as well the odour. Predicting their growth is crucial for managing drinking water distribution (Whitehead & Hornberger, 1984). Periphyton algae are crucial for substrate association in underwater environments and have been widely used to improve water quality (Leland & Carter, 1985; Newman et al., 1985; Cosgrovea et al., 2004). These organisms are diverse and play a crucial role in energy inflow and nutrient cycling in submarine ecosystems. Their community structure and function can serve as aquatic body ecological matrices (Hill et al., 1999). The communities of Periphyton act as sensors to ecological changes and hence can be used to monitor water quality. These organisms have fixed territories and can quickly recolonize after disturbances. Sample collection and analysis are easy, and their pollution tolerance is well-known due to their wide, common taxa (Biggs, 1985). According to Round (1991), diatoms play a considerable role in water quality determination. They inhabit diverse environments and can indicate environmental shifts (Stevenson & Bahls, 1999). Diatoms provide precise water quality data compared to chemical and zoological assessments. Algae can serve as an index of water quality and have been used to assess environmental status (Kolkwitz and Marsson, 1908). The categorization of algae by settlers was based on their tolerance to pollution. Certain species of algae were found to indicate areas of decline (Palmer, 1969) of water quality. Patrick (1949) created community indices for presenting data that show that a diverse range of species from several kinds of creatures, including algae, comprise a healthy community. Patrick (1971) suggested utilizing artificial substrates made of diatom flora affixed to glass slides as a numeric-based method for studying the quality of water. Diatom foliage was also used as an important index of environmental change, as suggested by (Dixit et al., 1992), and emerged as a favored index in related studies. In laboratory bioassays, algae are utilized to assess water quality using various media into which varying quantity of the

TABLE 2.2
Algal Species Reported in Various Aquatic Ecosystems

Algal biodiversity	Aquatic ecosystems	References
Chlorella vulgaris (green algae)	Marine ecosystems	Kalhor et al., 2017
Westiellopsis prolific, Anabaena variabilia, Oscillatoria pranceps, Phormidium mucicola, Lyngbya digueti, A. quadruplicatum,	Marine ecosystems	Al-Hussieny et al., 2020
Microcoleus Chthonoplastes, Phormidium corium	Coastal ecosystems	Ghasemi et al., 2011
Nannochloropsis oculata (eustitgmatophyte) Isochrysis galbana	Sea water	Ammar et al., 2018
Agardhiella subulata (red algae)	Sea water	Hung et al., 2020
Pavlova lutheri Nannochloropsis oculata	Marine ecosystems	Chernikova et al., 2020
Tribonema sp Chlamydomonas sp Frustilia sp Pinnularia Sp Hydrodicton sp Cyanobacteria Chlorella sp	Fresh water ecosystems	Maneesh and Siva, 2022

contaminants are incorporated (Guckert et al., 1992; Knauer et al., 1997). **Table 2.2** provides a summary of algal populations or species in various aquatic ecosystems reported by various authors.

2.3.4 Fungi

Fungal diversity in aquatic ecosystems includes around 2000 different species from submerged decaying materials, spores in water, animals, and decaying macrophytes in washes. Only around 5% of the world's eukaryotes are known to date, making fungi one of the largest groups (after insects). There are 138 non-fungal oomycetes and about 3,000 fungus species that have been found to exist in watery environments (Ittner et al., 2018). It is believed that the true variety of fungi in freshwater environments is far higher than the 2000 species known. From brackish regions, fungi belonging to all major taxonomic groups have been observed, isolated, and characterized, including Zygomycota, Ascomycota, Basidiomycota, and Chytridiomycota. The only class of fungi that can survive in submerged environments and produce both zoospores and genuine mycelium are chytrids. Despite their ability to break down resistant composites like chitin or cellulose, little is known about how they work in the degradation of organic materials. We do know certain chytrid taxa like Batrachochytrium dendrobatidis are well-known saprophytes that cause skin diseases to amphibians like frogs, toads, and other amphibians and causes global losses in amphibian populations. Moreover, saprotrophic Zygomycota in freshwater are poorly understood, despite the fact that molecular techniques can identify them on submerged substrates. Within this phylum are the Trichomycetes, a class of rather frequent freshwater symbionts that are essentially technical obligatory symbionts in arthropod digestive tracts. In watery environments, basidiomycetes or their mitospores are rarely found. In freshwater environments, ascomycetes and their anamorphs (i.e., fungal species whose sexual or meiosporic stages are unknown) are the most prevalent and ecologically significant group of fungi. This category of fungi is mostly covered in the sections that follow about fungal traits and processes and their importance in corruption. Oomycetes are extremely prevalent in freshwater and are typically investigated by mycologists, despite their phylogenetically distant relationship to genuine fungus.

According to Woese and Fox (1977) and Woese et al. (1990), fungi are a kingdom under the domain Eukaryota. Voigt & Kirk (2011) have emphasized that over the past few decades, genetic investigations have considerably altered the taxonomy of fungi, and Lutzoni et al. (2004), James et al.. (2006), and Hibbett et al. (2007) have developed a fungal tree of life. Hawksworth (1991, 2001) estimated that there are roughly 1.5 million fungal species; however, only about 7 percent of them have been traced (Mueller & Schmit, 2007). There are 138 non-fungal oomycetes and about 3,000 fungus species that have been found to exist in watery environments. Of these groups, temperate climates have the most biodiversity (Shearer et al., 2007).

Different classification schemes have been used for freshwater fungi (Shearer et al., 2007). Four phyla are known to contain the bulk of species found in freshwater environments: basidiomycetes, chytridiomycetes, ascomycetes, and glomeromycetes. More than 600 freshwater species were documented by Goh and Hyde (1996), comprising roughly ascomycetes and mitosporic fungus of 300 each, a few chytridiomycetes, and oomycetes (non-fungal). It is likely that the number of newly discovered species will increase rapidly, as only a small percentage of the aquatic fungus population has been characterized to date.

A classification approach used by Krauss et al. (2011) focuses more on the functional traits of freshwater fungus in aquatic habitats than it does on phylogeny. The suggested taxonomy consists of four main categories: aquatic hyphomycetes (also known as freshwater hyphomycetes or Ingoldian fungus), yeasts, glomeromycetes, chytridiomycetes, or chytrids.

The aqueous hyphomycetes primarily associate with the ascomycetes group, with a minor percentage in the basidiomycetes group, even though the majority of these groups are thought to be monophyletic. Conversely, Shearer et al. (2007) report that yeasts are a polyphyletic category with both the above groups. It is possible to distinguish between hyphomycetes and yeasts mostly by looking at their unique morphologies. Furthermore, because oomycetes are not classified as mushrooms, they are handled differently (Shearer et al., 2007). Nonetheless, it makes sense to include them in this review because they function in freshwater environments similarly to fungi and occupy niches that are comparable to those of aquatic fungus. **Table 2.3** summarizes fungal populations or fungal species in various aquatic ecosystems.

TABLE 2.3
Fungal Species Reported in Various Aquatic Ecosystems

Fungal biodiversity	Aquatic ecosystem	References
Aspergillus sydowii *Aspergillusflavus* *Aspergillus niger* *Aspergillus terreus*	Marine water ecosystems	Barnes et al., 2018
Acremonium sclerotigenum *Penicillium citrinum*	Mangrove sediment	
Penicillium mononematosum Penicillium chrysogenum *Penicillium polonicum* *Penicillium cyclopium*	Marine sediments	Maamar et al., 2020
Aspergillus sclerotiorum *Mucor racemosus*	Coastal Atlantic Ocean	Passarini et al., 2011
Gracilibacillus sp	Salt lake	Yu and Li, 2015
Massarina ingoldiana Aniptodera chesapeakensis Aniptodera lignatilis Annulatascus velatisporus Halosarpheia retorquens Nais inornata	Freshwater ecosystems	Shearer, 2001

2.4 AQUATIC MICROBES IN THE WASTEWATER MANAGEMENT

In the age of climate change, wastewater treatment and reuse are crucial to meet the world's water needs (Yadav et al., 2020). According to Ferrera and Sanchez (2016), microorganisms are in charges of biotic wastewater treatment activities such as the elimination of organic components, hazardous metallic compounds, and odor, including phosphorus and nitrogen. For this reason, understanding the composition and roles of microbial populations and communities in wastewater is essential to the design and development of wastewater treatment systems. Criado Monleon et al. (2022) examined characteristics of the microbial consortia in wastewater treatment plants that set them apart and determined the impact on the microbial communities of pre-treatment raw household wastewater. The use of pretreatment wastewater in this investigation revealed notable alterations in the structure of the microbial consortium, and possible consequences in microbial function and possible threats to the environment and community health. Huang et al. (2021) examined the composition and potential function of bacterial consortia in wastewater stabilization ponds (WSP), frequently employed methods for treating Canadian Arctic municipal wastewater. Variations in WSP composition over time and space were also studied. Temperature was found to be a significant factor affecting the composition of the bacterial community and the possibility of nutrient removal in the WSPs. Cruz et al. (2021) conducted a preliminary investigation to compare bacterial consortia in wastewater coming from

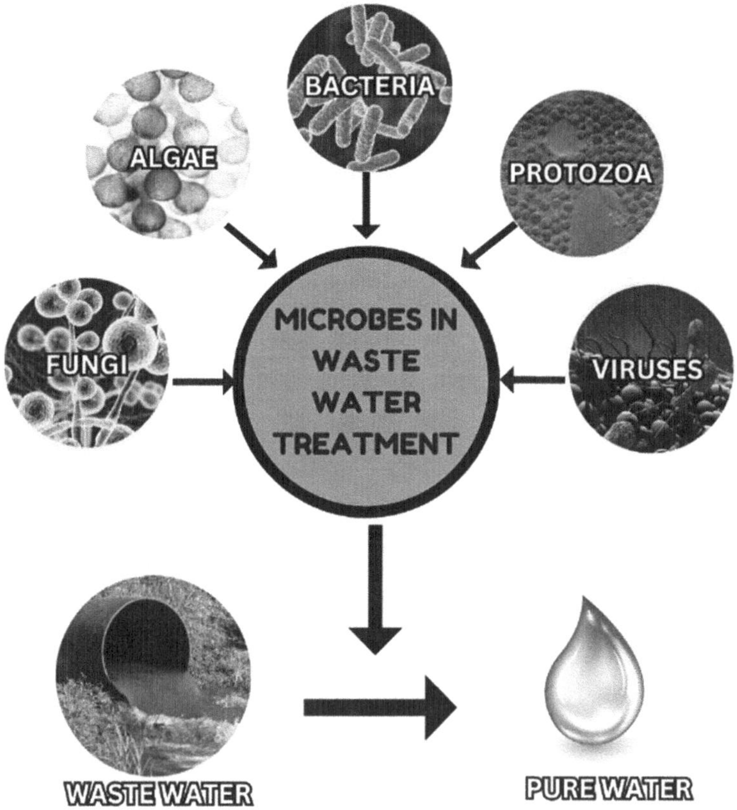

FIGURE 2.2 Role of aquatic microbes in wastewater treatment.

various sources, before and after treatment (e.g., surface waters, hospitals, sewage effluents, and fish farms) in the Sri Lanka and Philippines.

Wei et al. (2022) used Alcanivorax's ability to fix carbon and oxidize iron (Fe(II) in an autotrophic manner. The authors postulated that rich chemoautotrophic or chemoheterotrophic iron-oxidizing bacteria may have been promoted by the features and makeup of microbes residing in deep-ocean hydrothermal sulfides and saltwater. The strain MM125-6 was shown to be able to chemoautotrophically oxidize Fe(II) and fix carbon in an autotrophic manner, according to the results. Hou et al. (2022) assessed the stages of organic matter breakdown and calculated the compositions of bacterial communities in sediments. According to their research, *P. clarkia* significantly increased the sediments' oxidation-reduction potential values and created an environment that was conducive to the breakdown of organic materials. Furthermore, *P. clarkia* altered the makeup and function of bacterial communities to boost their ability to break down organic materials. **Figure 2.2** shows the role of aquatic microbes in wastewater treatment.

2.5 INDUSTRIAL IMPORTANCE OF AQUATIC MICROORGANISMS

The microorganisms in submarine ecosystems not only function as scavengers but are also being studied for their potential to produce useful chemicals. Hydrolytic enzymes, including lipases, proteases, and amylases, are especially abundant in these bacteria. These microbes are increasingly recognized for their industrial significance in biotechnology. Mudryk and Podgorskar (2006) brought to light the ability of bacteria in lakes to produce extracellular enzymes of industrial importance. Moreover, the protease enzyme produced by microorganisms has diverse biochemical functions in the food, medical, cleaning, and waste management industries, as highlighted by Saurabh (2007).

The alkaline proteases are currently the most widely used enzymes due to their versatility, and these enzymes are mainly obtained from the Bacillus species. Additionally, the enzyme phospholipase is produced by the lipolytic bacteria, which plays a vital role in the bakery industry, as well as in egg processing, animal feed production, and vegetable oil refining.

Syed et al. (2009) discovered the importance of an enzyme called α-amylase in various industrial procedures. This particular enzyme accounts for 25% of global enzyme demand. Amylase enzymes are utilized for hydrolyzing complex starch molecules into simpler glucose units (Sindhu, 1997). The industrial enzymes are given names based on the capability of the microbes implicit and how their genes are expressed in the hosts of microorganisms or reproduced. These qualities make them commercially viable in significant quantities (De Maria et al., 2007).

Understanding of the diversity of aquatic microorganisms, both genetically and biochemically, is still in its early stages, and there is a significant potential for enzymes with various applications that have yet to be discovered. Many researchers have used the API ZYM kit and other commercial kits to extract these commercially important enzymes. The study conducted by Mudryk and Podgorskar (2006) examined the enzymatic activity of bacterial strains and found that the degradation of organic macromolecules in an Estuarine Lake was carried out by both neustonic and planktonic bacteria. Other research groups around the world have conducted similar studies. Numerous aquatic microbes are able to eliminate heavy toxic metals as per Dixit et al. (2015). They can clear petroleum layers according to Zaki et al. (2015), and acts as accelerators of effluent treatment processes (Daims et al., 2006). Wu et al. (2022) elucidated degradation of *Bellamya* spp. shells in the aqueous bodies and determined the cause of "biological shell erosion" in aqueous bodies that proved to have potential in the aquatic industry. Clarifying the reason for the biological deterioration of the aquaculture environment was made easier by the investigation.

While there have been many novel technologies used to investigate the microbial diversity in aquatic ecosystems, still a large number of microorganisms have yet to be discovered and their ecological functions are still unknown.

Enzymes have gained global recognition for their potential use in breaking down agricultural waste and are increasingly being employed in the textile, pharmaceutical, and food industries. Advanced molecular and biochemical methods have revolutionized microbial metabolism (Overbeck et al., 1990; Kemp et al., 1994;

Cooksey, 1998). Through the study of microbial processes, we have gained new knowledge and understanding regarding their functions and significance in the operation of aquatic ecosystems.

Our current focus is on exploring the aquatic prokaryotic metagenomic sequencing that will yield a substantial amount of data and information on the microorganisms. This will allow us to identify their metabolic capabilities and study microbial biodiversity in new ways as we gain more evidence of the multiple interactions between microorganisms and other aquatic ecosystems, as well as their role in ecosystem functions. Based on the results of several studies, it is evident that these tiny organisms are a major contributor to nutrient cycles and ecosystem structures. Therefore, they can be studied further to determine their varied uses, including the production of extracellular enzymes as a byproduct. Industrial enzymes produced by the microorganisms are proving their potential in the market and as a result, the cost of enzymes is also on the rise owing to their increasing demand. Therefore, it is crucial to optimize the growth conditions and production of these microorganisms to explore their prospects thoroughly (Joshi et al., 2016).

2.6 CONCLUSION

Aquatic ecosystems encompass various bodies of water, such as oceans, lakes, rivers, and estuaries, and are known for their dynamic nature. These ecosystems are habitats for different types of microorganisms, including bacteria, archaea, protozoa, algae, fungi, and viruses, which play crucial roles in maintaining ecological balance and sustaining life within aquatic environments. The presence of a variety of microorganisms in aquatic ecosystems is essential for the overall health and functionality of these systems. Understanding the complex connections between microorganisms and aquatic ecosystems is crucial for managing water resources sustainably and conserving biodiversity. As research continues to uncover the complexities of microbial communities in aquatic environments, it becomes increasingly evident that the health of these ecosystems is intricately linked to the well-being of the entire planet. Multi-omics, sophisticated sequencing methods, bioinformatics, and a combination of all these techniques have contributed to the recent rapid development of molecular microbial ecological research.

REFERENCES

Adyasari, D., Hassenrück, C., Oehler, T., Sabdaningsih, A., Moosdorf, N. 2019. Microbial community structure associated with submarine groundwater discharge in northern Java (Indonesia). *Sci. Total Environ.* 689, 590–601. doi:10.1016/j.scitotenv.2019.06.193

Aishwarya, S., Gunasekaran, K., Kumar, P.S., Begum, A., Shantha, E., Jeevitha, V., Gayathri, K.V. 2021. Structural, functional, resistome and pathogenicity profling of the Cooum river. *Microb. Pathog.* 158, 105048.

Alexopoulos, A., Plessas, S., Bezirtzoglou, E. 2009. Water microbial ecology—An overview. Encyclopedia of Life Sciences; Wiley: Chichester, UK, pp. 1–24.

Al-Hussieny, A.A., Imran, S.G., Jabur, Z.A. 2020. The use of local blue-green algae in the bioremediation of hydrocarbon pollutants in wastewater from oil refineries. *Plant Arch.* 20, 797–802.

Allan, J.D. 2004. Landscapes and riverscapes: The influence of land use on stream ecosystems. *Annu. Rev. Ecol. Evol. Syst.* 35, 257–284.

Alongi, D.M. 2008. Mangrove forests: Resilience, protection from tsunamis, and responses to global climate change. *Estuar. Coast. Shelf Sci.* 76(1), 1–13.

Ammar, S.H., Khadim, H.J., Mohamed, A.I. 2018. Cultivation of Nannochloropsis oculata and Isochrysis galbana microalgae in produced water for bioremediation and biomass production. *Environ. Technol. Innova.* 10, 132–142.

Azam, F. 1998. Microbial control of oceanic carbon flux: The plot thickens. *Sci.* 280, 694–6.

Balasubramanian, A. 2011. Aquaticecosystems–Marine types. University of Mysore.

Barnes, N.M., Khodse, V.B., Lotlikar, N.P., Meena, R.M., Damare, S.R. 2018. Bioremediation potential of hydrocarbon-utilizing fungi from select marine niches of India. *3 Biotech,* 8, 21.

Barros, N.O., Farjalla, V.F., Soares, M.C., Melo, R.C.N., Roland, F. 2010. Virus-bacterium coupling driven by both Turbidity and hydrodynamics in an Amazonian floodplain lake. *Appl. Environ. Microbiol.* 76(21), 7194–201.

Beck, R.H., Lehner, P. 1974. Oceans, new frontier in exploration. *AAPG Bulletin*, 58(3), 376–395.

Benlloch, S., Lopez-Lopez, A., Casamayor, E.O., Ovreas, L., Goddard, V., Daae, F.L., Smerdon, G., Massana, R., Joint, I., Thingstad, F., et al. 2002. Prokaryotic genetic diversity throughout the salinity gradient of a coastal solar saltern. *Environ. Microbiol.* 4, 349–360.

Bezirtzoglou, E., Dimitriou, D., Panagiou, A. 1996. Occurrence of Clostridium perfringens in river water by using a new procedure. *Anaerobe*, 2, 169–173.

Bhadury, P., Singh, T. 2020. Analysis of marine planktonic cyanobacterial assemblages from Mooriganga Estuary, Indian Sundarbans using molecular approaches. *Front. Mar. Sci.* 7, 22.

Biggs, B.J.F. 1985. The use of periphyton in the monitoring of water quality. In R.D Pridmore, A.B Cooper (eds.). *Biological monitoring in freshwaters: Proceedings of a seminar. Water and soil miscellaneous publication 82* (pp. 117–142). Wellington, New Zealand: Ministry of Works and Development.

Bradley, D.E. 1967. Ultrastructure of bacteriophage and bacteriocins. *Bacteriol. Rev.* 31(4), 230.

Chen, L., Hu, B.X., Dai, H., Zhang, X., Xia, C.-A., Zhang, J. 2019. Characterizing microbial diversity and community composition of groundwater in a salt-freshwater transition zone. *Sci. Total Environ.* 678, 574–584. doi:10.1016/j.scitotenv.2019.05.017

Chernikova, T.N., Bargiela, R., Toshchakov, S.V., Shivaraman, V., Lunev, E.A., Yakimov, M.M., Thomas, D.N., Golyshin, P.N. 2020. Hydrocarbon-degrading bacteria alcanivorax and marinobacter associated with microalgae pavlova lutheri and nannochloropsis oculata. *Front. Microbiol.* 11, 572931.

Chithira, M.S., Aishwarya, P.V., Mohan, A.S., Antony, S.P. 2021. Profling bacteriome associated with the sediments of a mangrove ecosystem in Kerala India. *Ecol. Genet. Genom.* 21, 100103.

Cooksey, K.E. 1998. Molecular approaches to the study of the Ocean. USA. Springer Netherlands, pp. 549.

Cosgrovea, J.D.W, Morrison, P., Hillman, K. 2004. Periphyton indicate effects of wastewater discharge in the near-coastal zone, Perth, Western Australia. *Estuar. Coast. Shelf Sci.* 61(2), 331–338.

Cotner, J.B., Biddanda, B.A. 2002. Small players, large role: Microbial influence on biogeochemical processes in pelagic aquatic ecosystems. *Ecosyst.* 5, 105–121.

Criado Monleon, A.J., Knappe, J., Somlai, C., Betancourth, C.O., Ali, M., Curtis, T.P., Gill, L.W. 2022. Spatial variation of the microbial community structure of on-site soil treatment units in a temperate climate, and the role of pre-treatment of domestic effluent in the development of the biomat community. *Front. Microbiol.* 13, 915856. doi:10.3389/fmicb.2022.915856

Cruz, A.F., Wijesekara, R.G.S., Jinadasa, K.B.S.N., Gonzales, B.J., Ohura, T., Guruge, K.S. 2021. Preliminary investigation of microbial community in wastewater and surface waters in Sri Lanka and the Philippines. *Front. Water*, 3. doi:10.3389/frwa.2021.730124

Daims, H., Taylor, M.W., Wagner, M. 2006. Wastewater treatment: A model system for microbial ecology. *Trends in Biotechnology*, 24(11), 483–489. doi:10.1016/j.tibtech.2006.09.002

De Maria, L., Vind, J., Oxenbøll, K.M., et al. 2007. Phospholipases and their industrial applications. *Appl Microbiol Biotechnol.* 74(2), 290–300.

Dhal, P.K., Kopprio, G.A., Gärdes, A. 2020. Insights on aquatic microbiome of the Indian Sundarbans mangrove areas. *PLoS One.* 15(2), e0221543.

Dixit, R., Wasiullah, Malaviya, D., Pandiyan, K., Singh, U., Sahu, A., … Paul, D. 2015. Bioremediation of heavy metals from soil and aquatic environment: An overview of principles and criteria of fundamental processes. *Sustainability*, 7(2), 2189–2212. doi:10.3390/su7022189

Dixit, S. S, Smol, J. P, Kingston, J.C., Charles, D.F. 1992. Diatoms: Powerful indicators of environmental change. *Environ. Sci. Technol.* 26, 23–33.

Dodds, W.K., Bouska, W.W., Eitzmann, J.L., Pilger, T.J., Pitts, K.L., Riley, A.J., … Thornbrugh, D.J. 2009. Eutrophication of U.S. freshwaters: Analysis of potential economic damages. *Environ. Sci. Technol.* 43(1), 12–19.

Duarte, C.M., Losada, I.J., Hendriks, I.E., Mazarrasa, I., Marbà, N. 2013. The role of coastal plant communities for climate change mitigation and adaptation. *Nat. Clim. Change*, 3(11), 961–968.

Dudhagara, P., Ghelani, A., Patel, R., Chaudhari, R., Bhatt, S. 2015. Bacterial tag encoded FLX titanium amplicon pyrosequencing (bTEFAP) based assessment of prokaryotic diversity in metagenome of Lonar soda lake, India. *Genom. Data*, 4, 8–11.

Dugan, H.A., Bartlett, S.L., Burke, S.M., Doubek, J.P., Krivak-Tetley, F.E., Skaff, N.K., Summers, J.C., Farrell, K.J., McCullough, I.M., Morales-Williams, A.M., et al. 2017. Salting our freshwater lakes. *Proc. Natl. Acad. Sci. USA*, 114, 4453–4458.

Fenchel, T. 2008. The microbial loop–25 years later. *J. Exp. Mar. Bio. Ecol.* 366(1), 99–103.

Ferrera, I., Sánchez, O. 2016. Insights into microbial diversity in wastewater treatment systems: How far have we come? *Biotechnol. Adv.* 34(5), 790–802. doi:10.1016/j.biotechadv.2016.04.003

Fuhrman, J.A. 1999. Marine viruses and their biogeochemical and ecological effects. *Nature*, 399, 541–8.

Fuhrman, J.A., Noble, R.T. 1995. Viruses and protists cause similar bacterial mortality in coastal seawater. *Limnol. Oceanogr.* 40(7), 1236–42.

Ghasemi, Y., Rasoul-Amini, S., Fotooh-Abadi, E. 2011. The biotransformation, biodegradation, and bioremediation of organic compounds by microalgae. *J. Phycol.* 47, 969–980.

Goh, T.K., Hyde, K.D. 1996. Biodiversity of freshwater fungi. *J. Ind. Microbiol. Biot.* 17, 328–345. doi: 10.1007/BF01574764

Guckert, J.B., Nold, S.C, Boston, H.L., White, D.C. 1992. Periphyton response in an industrial receiving stream: Lipid-based physiological stress analysis and pattern recognition of microbial community structure. *Can J. Fish. Aquat. Sci.* 49(12), 2579–2587.

Haldar, S., Nazareth, S.W. 2018. Taxonomic diversity of bacteria from mangrove sediments of Goa: metagenomic and functional analysis. *3 Biotech.* 8(10), 1–10.

Hawksworth, D.L. 1991. The fungal dimension of biodiversity: Magnitude, significance, and conservation. *Mycol. Res.* 95, 641–655. doi: 10.1016/S0953-7562(09)80810-1

Hawksworth, D.L. 2001. The magnitude of fungal diversity: The 1.5 million species estimate revisited. *Mycol. Res.* 105, 1422–1432. doi: 10.1017/S0953756201004725

Herbert, E.R., Boon, P., Burgin, A.J., Neubauer, S.C., Franklin, R.B., Ardon, M., Hopfensperger, K.N., Lamers, L.P.M., Gell, P. 2015. A global perspective on wetland salinization: Ecological consequences of a growing threat to freshwater wetlands. *Ecosphere.* 6, 206.

Herlemann, D.P.R., Labrenz, M., Jurgens, K., Bertilsson, S., Waniek, J.J., Andersson, A.F. 2011. Transitions in bacterial communities along the 2000 km salinity gradient of the Baltic Sea. *ISME J.* 5, 1571–1579.

Hewson, I., Fuhrman, J.A. 2004. Richness and diversity of bacterioplankton species along an estuarine gradient in Moreton Bay, Australia. *Appl. Environ. Microbiol.* 70, 3425–3433.

Hibbett, D. S., Binder, M., Bischoff, J. F., Blackwell, M., Cannon, P. F., Eriksson, O. E.,... & Zhang, N. (2007). A higher-level phylogenetic classification of the Fungi. *Mycological Research*, 111(5), 509–547.

Hill, M. O, Mountford, J.O., Roy, D.B., Bunce, R.G.H. 1999. Ellenberg's indicator values for British plants, *ECOFACT*, vol 2, Technical Annex. Huntingdon, UK: Institute for Terrestrial Ecology.

Horner-Devine, M.C., Leibold, M.A., Smith, V.H., Bohannan, B.J.M. 2003. Bacterial diversity patterns along a gradient of primary productivity. *Ecol. Lett.* 6, 613–622.

Hou, Y., Jia, R., Ji, P., Li, B., Zhu, J. 2022. Organic matter degradation and bacterial communities in surface sediment influenced by Procambarus clarkia. *Front.Microbiol.* 13, 985555. doi:10.3389/fmicb.2022.985555

Huang, Y., Ragush, C.M., Johnston, L.H., Hall, M.W., Beiko, R.G., Jamieson, R.C., Truelstrup Hansen, L. 2021. Changes in bacterial communities during treatment of municipal wastewater in arctic wastewater stabilization ponds. *Front. Water*, 3. doi:10.3389/frwa.2021.710853

Huang, Y., Zhang, H., Liu, X., Ma, B., Huang, T. 2022. Iron-activated carbon systems to enhance aboriginal aerobic denitrifying bacterial consortium for improved treatment of micro-polluted reservoir water: Performances, mechanisms, and implications. *Environ. Sci. Technol.* 56(6), 3407–3418. doi:10.1021/acs.est.1c05254

Hughes, T.P., Barnes, M.L., Bellwood, D.R., Cinner, J.E., Cumming, G.S., Jackson, J.B., … Scheffer, M. 2017. Coral reefs in the Anthropocene. *Nature*, 546(7656), 82–90.

Hung, C.M., Huang, C.P., Hsieh, S.L., Tsai, M.L., Chen, C.W., Dong, C. 2020. Di Biochar derived from red algae for efficient remediation of 4-nonylphenol from marine sediments. *Chemosphere*, 254, 126916.

Ittner, L.D., Junghans, M., Werner, I. 2018. Aquatic fungi: A disregarded trophic level in ecological risk assessment of organic fungicides. *Front. Environ. Sci.* 6, 105.

James, T.Y., Kauff, F., Schoch, C.L., Matheny, P.B., Hofstetter, V., Cox, C.J., et al. 2006. Reconstructing the early evolution of fungi using a six-gene phylogeny. *Nature,* 443, 818–822. doi: 10.1038/nature05110

Ji, M., Kong, W., Yue, L., Wang, J., Deng, Y., Zhu, L. 2019. Salinity reduces bacterial diversity, but increases network complexity in Tibetan Plateau lakes. *FEMS Microbiol. Ecol.* 95, fiz190.

Joshi, P., Pande, V., Joshi, P. 2016. Microbial diversity of aquatic ecosystem and its industrial potential. *J Bacteriol Mycol Open Access*. 3(1), 177–179. DOI: 10.15406/jbmoa.2016.03.00048

Kalhor, A.X., Movafeghi, A., Mohammadi-Nassab, A.D., Abedi, E., Bahrami, A. 2017. Potential of the green alga Chlorella vulgaris for biodegradation of crude oil hydrocarbons. *Marine Pollution Bulletin.* 123, 286–290.

Kassen, R., Buckling, A., Bell, G., Rainey, P.B. 2000. Diversity peaks at intermediate productivity in a laboratory microcosm. *Nature*, 406, 508–512.

Kemp, P., Sherr, B.F., Sherr, E.B., et al. 1994. Handbook of methods in aquatic microbial ecology. USA: Lewis Publishing, pp. 1496–1497.

Kiersztyn, B., Chróst, R., Kalínski, T., Siuda, W., Bukowska, A., Kowalczyk, G., Grabowska, K. Structural and functional microbial diversity along a eutrophication gradient of interconnected lakes undergoing anthropopressure. *Sci. Rep.* 2019, 9, 11144.

Kirchman, D.L. 2008. Microbial ecology of the Oceans 2nd Ed., USA: John Wiley & Sons, Inc., [A comprehend book about marine microbial ecology].

Knauer, K., Behra, R. , Sigg, L. 1997. Effects of free Cu2+ and Zn2+ ions on growth and metal accumulation in freshwater algae. *Environ. Toxicol. Chem.* 16(2), 220–229.

Kolkwitz, R., dan Marsson, M. 1908. Oekologie der pflanzlichen Saprobien. *Berichte der Deutschen Botanischen Gesellschaft*, 26, 505–519.

Krauss, G.-J., Solé, M., Krauss, G., Schlosser, D., Wesenberg, D., Bärlocher, F. 2011. Fungi in freshwaters: Ecology, physiology and biochemical potential. *FEMS Microbiol. Rev.* 35, 620–651. doi: 10.1111/j.1574-6976.2011. 00266

Leland, H.V., Carter, J.L. 1985. Effects of copper on production of periphyton, nitrogen fixation and processing of leaf litter in a Sierra Nevada. *California stream. Freshw. Biol.* 15(2), 155–173.

Liang, W., Chun, X., Liu, J., Bai, X., Siqin, B., Wu, Z., Song, J. 2017. Research on the area change processes in the past 40a of Daihai Lake. *J. Arid Land Resour. Environ.* 31, 93–98.

Lozupone, C.A., Knight, R. 2007. Global patterns in bacterial diver

Lutzoni, F., Kauff, F., Cox, C.J., McLaughlin, D., Celio, G., Dentinger, B., et al. 2004. Assembling the fungal tree of life: Progress, classification, and evolution of subcellular traits. *Am. J. Bot.* 91, 1446–1480. doi: 10.3732/ajb.91.10.1446

Ma, Z., Gao, L., Sun, M., Liao, Y., Bai, S., Wu, Z., & Li, J. (2022). Microbial diversity in groundwater and its response to seawater intrusion in Beihai City, southern China. *Frontiers in Microbiology*, *13*, 876665.

Maamar, A., Lucchesi, M.E., Debaets, S., van Long, N.N., Quemener, M., Coton, E., Bouderbala, M., Burgaud, G., MatallahBoutiba, A. 2020. Highlighting the crude oil bioremediation potential of marine fungi isolated from the Port of Oran (Algeria). *Diversity,* 12, 196.

Maipa, V., Alamanos, Y., Bezirtzoglou, E. 2001. Seasonal fluctuation of bacterial indicators in coastal waters. *Microb.Ecol. Health Dis.* 13, 143–146.

Maneesh, G., Siva, G.V. 2022. Biodiversity of freshwater algae from temple tank. *IJCSRR,* 5, 760–765.

McCormick, P.V., Cairns, J. Jr. 1994. Algae as indicators of environmental change. *J. Appl. Phycol.* 6, 509–526.

Middleton, B.A. 1999. *Wetland restoration, flood pulsing, and disturbance dynamics*. John Wiley & Sons.

Monchamp, M.E., Spaak, P., Domaizon, I., Dubois, N., Bouffard, D., Pomati, F. Homogenization of lake cyanobacterial communities over a century of climate change and eutrophication. *Nat. Ecol. Evol.* 2018, 2, 317–324.

Mudryk, Z.J., Podgorska, B. 2006. Enzymatic activity of bacterial strains isolated from Marine beach sediments. *Polish J Environ Stud.* 15(3), 441–448.

Mueller, G.M., Schmit, J.P. 2007. Fungal biodiversity: What do we know? What can we predict? *Biodivers. Conserv.* 16, 1–5. doi: 10.1007/s10531-006-9117-7

Mukherji, S., Ghosh, A., Bhattacharyya, C., Mallick, I., Bhattacharyya, A., Mitra, S., Ghosh, A. 2020. Molecular and culture-based surveys of metabolically active hydrocarbon-degrading archaeal communities in Sundarban mangrove sediments. *Ecotox. Environ. Safe*. 195, 110481.

Nemergut, D.R., Schmidt, S.K., Fukami, T., O'Neill, S.P., Bilinski, T.M., Stanish, L.F., Knelman, J.E., Darcy, J.L., Lynch, R.C., Wickey, P., et al. 2013. Patterns and processes of microbial community assembly. *Microbiol. Mol. Biol. Rev*. 77, 342–356.

Newman, M. C, Alberts, J.J., Greenhut, V.A. 1985. Geochemical factors complicating the use of aufwuchs to monitor b10-accumulation of arsenic, cadmium, chromium, copper, zinc. *Water Res*. 19(9), 1157–1165.

Overbeck, J., Chrost, R.J. 1990. *Aquatic microbial ecology: Biochemical and molecular approaches*. USA: Springer–Verlag, pp. 190.

Palit, K., Das, S. 2021. Community structure, taxonomic diversity and spatio-temporal variation of sediment and water bacteria in Bhitarkanika mangrove ecosystem India. *Int. J. Environ. Sci. Technol*. 18(5), 1147–1166.

Palmer, C.M. 1969. A composite rating of algae tolerating organic pollution. *J. Phycol*. 5(1), 78–82.

Pandolfi, J.M., Bradbury, R.H., Sala, E., Hughes, T.P., Bjorndal, K.A., Cooke, R.G., … Jackson, J.B. 2003. Global trajectories of the long-term decline of coral reef ecosystems. *Science,* 301(5635), 955–958.

Passarini, M.R.Z., Rodrigues, M.V.N., da Silva, M., Sette, L.D. 2011. Marine-derived filamentous fungi and their potential application for polycyclic aromatic hydrocarbon bioremediation. *Marine Pollution Bulletin*. 62, 364–370.

Patrick, R. 1949. A proposed biological measure of stream conditions, based on a survey of the Conestoga Basin, Lancaster County, Pennsylvania. *Proceedings of the Academy of Natural Sciences of Philadelphia* 101, 277–341.

Patrick, R. 1971. Diatom communities. In J Cairns (ed.). *The structure and function of freshwater microbial communities*. Blacksburg, Virginia, USA: Virginia Polytechnic Institute and State University.

Poff, N.L., Brinson, M.M., Day, J.W. 2002. Aquatic ecosystems and global climate change. Pew Center on Global Climate Change, Arlington, VA, 44, 1-36.Pomeroy LR. 1974. The ocean's food web, a changing paradigm. *Bioscience*. 24(9), 499–504.

Pomeroy L.R. 1974. The ocean's food web, a changing paradigm. *Bioscience*. 24(9), 499–504

Raina, V., Panda, A.N., Mishra, S.R., Nayak, T., Suar, M. 2019. Microbial biodiversity study of a brackish water ecosystem in eastern India: The Chilika Lake. In *Microbial diversity in the Genomic Era* (pp. 47–63). Academic Press.

Rajpathak, S.N., Banerjee, R., Mishra, P.G., Khedkar, A.M., Patil, Y.M., Joshi, S.R., Deobagkar, D.D. 2018. An exploration of microbial and associated functional diversity in the OMZ and non-OMZ areas in the Bay of Bengal. *J. Biosci*. 43(4), 635–648.

Rathour, R., Gupta, J., Kumar, M., Hiloidhari, M., Mehrotra, A.K., Thakur, I.S. 2017. Metagenomic sequencing of microbial communities from brackish water of Pangong Lake of the northwest Indian Himalayas. *Genome Announce*. 5(40), e01029–17.

Reddy, B., Dubey, S.K. 2021. Exploring the allochthonous pollution infuence on bacterial community and co-occurrence dynamics of River Ganga water through 16S rRNA-tagged amplicon metagenome. *Environ. Sci. Pollut. Res*. 28(21), 26990–27005.

Romina Schiaffino, M., Unrein, F., Gasol, J.M., Massana, R., Balague, V., Izaguirre, I. Bacterial community structure in a latitudinal gradient of lakes: The roles of spatial versus environmental factors. *Freshw. Biol*. 2011, 56, 1973–1991.

Rott, E. 1991. Methodological aspects and perspectives in the use of periphyton for monitoring and protecting rivers. In B.A. Whitton, E. Rott, G. Friedrich (eds.). Use of algae for monitoring rivers. Innsbruck, Austria: Institut für Botanik, Universitat Innsbruck.

Round, F.E. 1991. Diatoms in river water-monitoring studies. *J. Appl. Phycol*. 3, 129–145.

Sang, S., Zhang, X., Dai, H., Hu, B.X., Ou, H., Sun, L. 2018. Diversity and predictive metabolic pathways of the prokaryotic microbial community along a groundwater salinity gradient of the Pearl River Delta, China. *Sci. Rep.* 8(1), 17317. doi:10.1038/s41598-018-35350-2

Saurabh, S. 2007. Enhanced productivity of serine alkaline protease by Bacillus sp. using soybean as substrate. *Malaysian J Microbiol.* 3(1), 1–6.

Sawstrom, C., Pearce, I., Davidson, A.T., Rosen, P., Laybourn-Parry, J. 2008. Influence of environmental conditions, bacterial activity and viability on the viral component in 10 Antarctic lakes. *FEMS Microbiol. Ecol.* 63(1), 12–22.

Seymour, J.R., Amin, S.A., Raina, J.-B., Stocker, R. 2017. Zooming in on the phycosphere: The ecological interface for phytoplankton bacteria relationships. *Nat. Microbiol.* 2, 17065.

Shearer, C.A. 2001. The distribution of freshwater filamentous ascomycetes. In J.K. Misra, B.W. Horn (eds) Trichomycetes and other fungal groups: Robert W Lichtwardt Commemoration (pp. 225–292). Science Publishers, Inc., Enfield, New Hampshire, USA.

Shearer, C.A., Descals, E., Kohlmeyer, B., Kohlmeyer, J., Marvanov,á, L., Padgett, D., et al. 2007. Fungal biodiversity in aquatic habitats. *Biodivers. Conserv.* 16, 49–67. doi: 10.1007/s10531-006-9120-z

Sigee, D.C. 2005. *Freshwater microbiology: Biodiversity and dynamic interactions of microorganisms in the aquatic environment.* John Wiley & Sons.

Sindhu, G.S., Sharma, P., Chakrabarti, T., et al. 1997. Strain improvement for the production of a thermostable α–amylase. *Enz Microbiol Technol.* 24(7), 584–589.

Steffen, W., Crutzen, P.J., McNeill, J.R. 2007. The Anthropocene: Are humans now overwhelming the great forces of nature. *Ambio.* 36, 614–621.

Stevenson, R.J., Bahls, L.L. 1999. Periphyton protocols. In M. T Barbour, J. Gerritsen, B.D. Snyder, J.B. Stribling (eds.). Rapid bioassessment protocols for use in streams and Wadeable Rivers: Periphyton, Benthic Macroinvertebrates, and Fish, 2nd ed. (pp. 6.1–6.22). EPA 841-B99–002. Washington D.C.: US Environmental Protection Agency, Office of Water.

Stevenson, R.J., Lowe, R.L. 1986. Sampling and interpretation of algal patterns for water quality assessment. In R.G Isom (ed.). Rationale for sampling and interpretation of ecological data in the assessment of freshwater ecosystems (pp. 118–149). Philadelphia, USA: American Society for Testing and Materials.

Stevenson, R.J., Pan, Y. 1999. Assessing ecological conditions in rivers and streams with diatoms. In E.F Stoemer, J.P Smol (eds.). The diatom: Applications to the environmental and earth science (pp. 11–40). Cambridge, UK: Cambridge University Press.

Syed, D.G., Agasar, D., Pandey, A. 2009. Production and partial purification of alpha–amylase from a novel isolate Streptomyces gulbargensis. *J. Ind. Microbiol. Biotechnol.* 36(2), 189–194.

Taranu, Z.E., Gregory-Eaves, I., Leavitt, P.R., Bunting, L., Buchaca, T., Catalan, J., Domaizon, I., Guilizzoni, P., Lami, A., McGowan, S., et al. 2015. Acceleration of cyanobacterial dominance in north temperate-subarctic lakes during the Anthropocene. *Ecol. Lett.* 18, 375–384.

Van Dam, H., Mertens, A., Sinkeldam, J. 1994. A coded checklist and ecological indicator values of freshwater diatoms from the Netherlands. *Aquat Ecol.* 28(1): 117–133.

Van der Gucht, K., Cottenie, K., Muylaert, K., Vloemans, N., Cousin, S., Declerck, S., Jeppesen, E., Conde-Porcuna, J.-M., Schwenk, K., Zwart, G., et al. 2007. The power of species sorting: Local factors drive bacterial community composition over a wide range of spatial scales. *Proc. Natl. Acad. Sci. USA*, 104, 20404–20409.

Voigt, K., Kirk, P.M. 2011. Recent developments in the taxonomic affiliation and phylogenetic positioning of fungi: Impact in applied microbiology and environmental biotechnology. *Appl. Microbiol. Biot.* 90, 41–57. doi: 10.1007/s00253-011-3143-4

Wang, J.J., Yang, D.M., Zhang, Y., Shen, J., Van der Gast, C., Hahn, M.W., Wu, Q.L. 2011. Do patterns of bacterial diversity along salinity gradients differ from those observed for macroorganisms? *PLoS ONE.* 6, e27597.

Wei, M., Zeng, X., Han, X., Shao, Z., Xie, Q., Dong, C., … Qiu, Z. 2022. Potential autotrophic carbon-fixer and Fe(II)-oxidizer Alcanivorax sp. MM125-6 isolated from Wocan hydrothermal field. *Front. Microbiol.* 13, 930601. doi:10.3389/fmicb.2022.930601

Weinbauer, M.G. 2004. Ecology of prokaryotic viruses. *FEMS Microbiol. Rev.* 28(2), 127–81.

Whitehead, P.G., Hornberger, G.M. 1984. Modelling algal behaviour in the River Thames. *Water Res.* 18(8), 945–953.

Woese, C.R., Fox, G.E. 1977. Phylogenetic structure of the prokaryotic domain: The primary kingdoms. *Proc. Natl. Acad. Sci. U.S.A.* 74, 5088–5090. doi: 10.1073/pnas.74.11.5088

Woese, C.R., Kandler, O., Wheelis, M.L. 1990. Towards a natural system of organisms: Proposal for the domains Archaea, Bacteria and Eucarya. *Proc. Natl. Acad. Sci. U.S.A.* 87, 4576–4579. doi: 10.1073/pnas.87.12. 4576

Wu, G., Huang, A., Wen, Y., Wang, H., Wang, J., Luo, F., Wu, M. 2022. Euendolithic Cyanobacteria and Proteobacteria together contribute to trigger bioerosion in aquatic environments. *Front. Microbiol.* 13, 938359. doi:10.3389/fmicb.2022.938359

Yadav, B., Pandey, A.K., Kumar, L.R., Kaur, R., Yellapu, S.K., Sellamuthu, B., et al. 2020. 1 – Introduction to wastewater microbiology: Special emphasis on hospital wastewater. In R.D. Tyagi, B. Sellamuthu, B. Tiwari, S. Yan, P. Drogui, X. Zhang, et al (eds.). Current developments in biotechnology and bioengineering (pp. 1–41). Amsterdam; Oxford; Cambridge: Elsevier.

Yang, J., Ma, L., Jiang, H.C., Wu, G., Dong, H.L. 2016. Salinity shapes microbial diversity and community structure in surface sediments of the Qinghai-Tibetan Lakes. *Sci. Rep.* 6, 25078.

Yu, H.Y., Li, X. 2015. Alkali-stable cellulase from a halophilic isolate, Gracilibacillus sp. SK1 and its application in lignocellulosic saccharification for ethanol production. *Biomass and Bioenergy,* 81, 19–25.

Zaki, M.S., Authman, M.M., Abbas, H.H. 2015. Bioremediation of petroleum contaminants in aquatic environments. *Life Sci. J.* 12, 127–39.

Zhang, H., Shi, Y., Huang, T., Zong, R., Zhao, Z., Ma, B., … Liu, M. 2023a. NirS-type denitrifying bacteria in aerobic water layers of two drinking water reservoirs: Insights into the abundance, community diversity and co-existence model. *J. Environ. Sci. (China),* 124, 215–226. doi:10.1016/j.jes.2021.10.013

Zhang, H., Yang, Y., Liu, X., Huang, T., Ma, B., Li, N., … Zhao, K. 2023b. Novel insights in seasonal dynamics and co-existence patterns of phytoplankton and micro-eukaryotes in drinking water reservoir, Northwest China: DNA data and ecological model. *Sci. Total Environ.* 857(Pt 1), 159160. doi:10.1016/j.scitotenv.2022.159160

Zhou, J., Ning, D. 2017. Stochastic community assembly: Does it matter in microbial ecology? *Microbiol. Mol. Biol. Rev.* 81, e00002–17.

Zhou, Y., Jiang, J. 2009. Changes in the ecological environment in the Daihai Lake basin over the last 50 years. *Arid Zone Res.* 26, 162–168.

Zwirglmaier, K., Keiz, K., Engel, M., Geist, J., Raeder, U. 2015. Seasonal and spatial patterns of microbial diversity along a trophic gradient in the interconnected lakes of the Osterseen Lake District, Bavaria. *Front. Microbiol.* 6, 1168.

3 Microbial Biofilms in Drinking Water and Water Quality

Hiral Desai, Megha Barot, Sachidanand Singh, and Arpan Modi

3.1 INTRODUCTION

In the early 1970s, microbiologist J. William Costerton and associates coined the term "biofilm" in their investigation of the bacterial communities that grow on surfaces in aquatic environments. Looking back, it is believed that the first biofilm observation was made by the Dutch scientist Antonie van Leeuwenhoek in the 17th century. He observed communities of microbes in dental plaque and other habitats using a basic homemade microscope. Using the term "animalcules," he pointed out that this material was made up of a variety of microorganisms. The ability of microorganisms to embed themselves inside a matrix of extracellular polymeric substances created by the constituent cells, which also happen to accumulate materials from the surrounding environment such as salts and metals, is now recognized as the distinguishing feature of biofilms. The extracellular DNA and RNA, polysaccharides, lipids, and proteins make up the biofilm matrix along with other biomolecules, which work together to create a sticky and protective covering around the community (Bamford et al., 2023).

Because biofilms have improved mineralization, bioaccumulation, and biosorption capabilities, they are becoming increasingly important in wastewater and drinking water treatment processes. Water professionals face a constant challenge due to the occurrence of biofilms in drinking water networks and reservoirs, even with the positive effects of biological filters like schmutzdecke (known as dirt cover) or bio-sand filters in slow sand filtration (Farkas, A et al., 2012). The biofilms linked to drinking water can lead to the degradation of residual disinfectants and cause visual problems such as taste, odor, and color deterioration. These biofilms can also be harmful to human and animal health because they harbor protozoa, fungi, viruses, invertebrates, algae, and toxin-producing bacteria. Pathogenic microorganisms in these microbial networks are shielded from environmental stressors like temperature, shear stress, and chlorine, allowing them to survive (Simoes and Simões, 2013). When contaminated drinking water is consumed, it can lead to a variety of illnesses and health issues in healthy individuals as well as those who are more vulnerable, such as ill individuals, young children, or those with weakened immune systems. It is well-established that one of the main microbial issues in drinking water distribution systems (DWDSs),

DOI: 10.1201/9781003487203-3

which leads to the decline in water quality, is biofilms (Momba et al., 2000). Still, it is nearly impossible to remove them from these systems. To stop and manage their growth, however, a number of factors can be taken into account. These include the residual disinfectant, water's nutrient content, the pipe materials, the network's hydro-dynamic conditions, and environmental variables like temperature and pH of water (Simões et al., 2012).

3.2 MICROBIAL BIOFILM IN WATER INDUSTRY

In the water industry, biofilms in the DWDS, also known as biofouling, are a well-known issue. The undesired buildup of biotic matter on a surface is commonly referred to as "biofouling." Numerous issues with DWDS are caused by microbes, such as the development of biofilms, nitrification, corrosion caused by microorganisms, and the presence and persistence of pathogens (Teng et al., 2008). When it comes to treated water-fed DWDSs without pipeline breaches, biofilms are thought to be the main source of microorganisms. These systems are especially concerning when they are older (Wingender et al., 2011). Because studies of microorganisms from a small number of lineages for which cultivation conditions have been established and for some process conditions that do not replicate real environments, studies of microbial physiology and genetics have largely limited our current understanding of the structure and activities in biofilm communities. The dynamics of microbial reproduction in drinking water systems are highly complicated due to the involvement of numerous interacting processes. Whether or not there is a residual disinfectant present, biofilms always colonize the insides of drinking water pipes. Biofilms regulate the microbiological levels of the dispersed water and have the potential to harbor pathogens in addition to posing a risk for erosion, taste, and odor issues (Szewzyk et al., 2000). *L. pneumophila* and *coliforms* with both intestinal and extraintestinal origins are among the bacterial pathogens found in biofilms that grow in potable water systems (September et al., 2007). Although Payment et al. (2011) did not find a correlation between the presence of biofilm in DWDSs and the development of disease,it is known that contaminated water is a means of transmission for pathogens like *Burkholderia spp., Mycobacterium spp., L. pneumophila, Klebsiella spp., P. aeruginosa, Cryptosporidium,* and *Giardia,* among others. As multiple authors have demonstrated, biofilms are a good fit because they can serve as a shield for the pathogen's persistence in drinking water. Epidemics of waterborne illnesses do not only occur in developing nations; wealthy nations are also impacted (Blasi et al., 2008).

3.2.1 Biofilm Formation

Numerous microbial species are able to approach a surface more closely, adhere to it securely, encourage cell-to-cell contacts, and develop into complex structures by a variety of methods. There are several processes involved in biofilm development.

1) Prior conditioning of the adhesion surface by macromolecules that are either purposefully covered on the surface or exist in the bulk liquid.
2) Planktonic bacterial migration from bulk liquid to the surface
3) Cells adhering to the surface

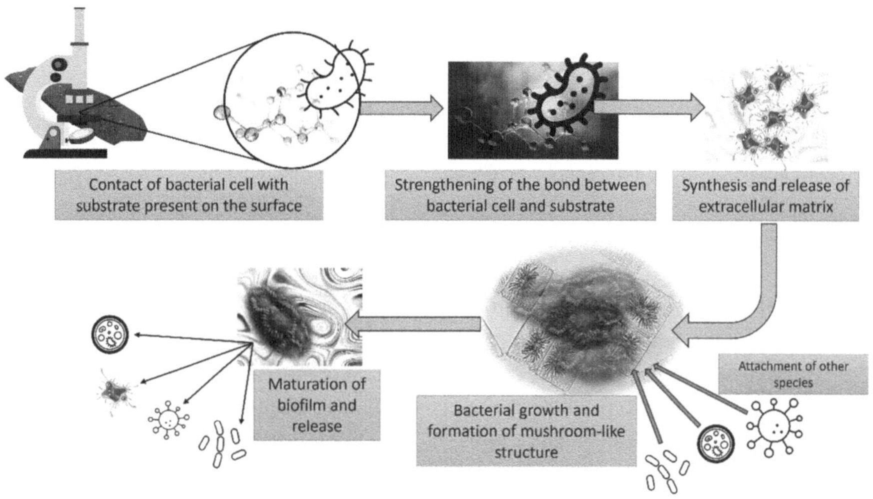

FIGURE 3.1 Biofilm formation.

4) Removal of cells that have been reversibly attached
5) Bacterial cells adhering irrevocably to a surface
6) Manufacturing of signaling molecules between cells
7) Substance movement into and through the biofilm
8) The biofilm-bound cells undergoing substance metabolism, and product transferring out of the biofilm
9) Shedding or detaching biofilms to remove them (Simoes and Simões, 2013)

The process of biofilm formation is illustrated in Figure 3.1.

3.2.1.1 Conditioning Film

The conditioning film forms on the adhesion surface, a light covering of organic molecules and ions, before any microbes adhere to the surface. These particles may bind chemically or physically to the surfaces. Physical adsorption typically involves nonspecific bonds and is a reversible process that forms a monolayer. Chemical adsorption results in the formation of multiple adsorbed molecular layers that involve hydrophobic, dipole, and electrostatic chemical bonds, among others (Marshall, 1996).

3.2.1.2 Adhesion

The process begins with the shifting of microbial cells to the adhesion surface, which can occur via active cell motility (such as flagella) or using fluid dynamics, gravitational forces, and Brownian motion. The environmental factors that affect bacterial adhesion include things like bacterial orientation, pressure, temperature, available energy, and surface functionality (Garrett et al., 2008). If repulsive forces surpass attractive forces, the bacteria will separate from the surface. Long and short/intermediate distance forces can be established as a result of interactions between the cells as they get closer to the surface. The Van der Waals attractive force and the electrostatic double-layer repulsive force make up the long-distance forces that are explained by the Derjaguin-Landau-Verwey-Overbeek (DLVO) theory (Simoes and Simões, 2013). Among the short- and intermediate-distance forces are polymer bridges, steric forces, hydrophobic interactions born repulsion forces, and hydrophobic pressure (Simoes and Simões, 2013).

3.2.1.3 Irreversible Adhesion

Many of the cells that are adsorbed reversibly stay immobile and become irreversibly adsorbed. According to certain theories, the physical repulsion of the electrical double layer is subdued by the bacterial appendages (pili, fimbriae, and flagella). The bacterial surface bond is then strengthened and chemical reactions like oxidation and hydration are sparked when the appendages come into contact with the conditioning layer's bulk lattice (Kumar and Anand, 1998). According to certain data, the hydrophobic and hydrophilic characteristics of interacting surfaces have a significant impact on microbial adhesion (Liu et al., 2004).

3.2.1.4 Growth and Maturation Phase

The microorganisms that are attached begin to grow and form microcolonies. They release organic polymers and initiate the development of the biofilm network. It has been demonstrated that exopolysaccharide production is crucial for the development of microcolonies (Watnick and Kolter, 1999). Transporting nutrients from the external liquid medium to the inner layers of the biofilm and excreting metabolites in the opposite direction are crucial for biofilm stability as the thickness of the biofilm increases. During the maturation phase, a complex and well-organized consortium of microorganisms develops within an organic matrix that shields them from external stresses (Simoes and Simões, 2013). At this stage, a significant amount of extracellular polymeric substances (EPS) is produced by microbes. An established biofilm's structure is influenced by its temperature, hydrodynamic conditions, microbial composition, EPS production, and availability of nutrients. Multiple processes may take place concurrently in an established biofilm:

- Bacteria that separate and enter water
- Planktonic bacterial attachment
- Growth and death

Though there are sporadic variations, these processes are in equilibrium at this point, and the quantity of cells attached per unit surface area remains constant over time (Bryers, 2008). The biofilm should now have reached its maximum thickness, which is mostly determined by mass transport, cohesion within the biofilm, and hydrodynamic conditions.

3.2.1.5 Detachment

The biofilm uses different cell signaling pathways at high cell concentrations; these are known as quorum-sensing mechanisms (Bassler, 1999). Quorum sensing is the process of using a variety of auto-inducers (highly concentrated chemical and peptide signals, such as homoserine lactones) to promote the alginates' genetic expression and mechanical and enzymatic processors. Alginates are an essential component of the extracellular matrix (Garrett et al., 2008). During the death phase, the biofilm disintegrates.

The community itself produces the enzymes that degrade the polysaccharides that keep the biofilm together, actively emitting surface bacteria to colonize new substrates. Enzymes involved in the degradation of the biofilm matrix include hyaluronidase from *Streptococci equi*, N-acetyl-heparosan lyase from *Escherichia coli*, and alginate lyase from *Pseudomonas fluorescens* and *Pseudomonas aeruginosa* (Sutherland, 1999). Different mechanisms lead to detachment: sloughing, abrasion, erosion, and predator grazing. Nutritional concentration variations and a chemical shift in EPS caused by chelating substances (Ca2+) that reduces the cohesive force of the connected cells (Simoes and Simões, 2013).

3.3 MICROBIOLOGICAL FACTORS AND BIOFILMS IN AN ANALYSIS OF DRINKING WATER QUALITY

It is assumed that tap water is not sterile, as the bulk water's microorganisms primarily consist of non-aggressive heterotrophs that are likely released during dispersion from related biofilms. European regulations state that the following microbial indices are measured using standardized conventional culturing methods: total coliforms, *Clostridium perfringens,* intestinalEnterococciand *Escherichia coli,* and colony counts at 37°C and 22°C (Farkas et al., 2012). The most significant microbiological risk is linked to drinking water contaminated by human or animal excrement, which raises the possibility of pathogenic bacteria, viruses, and protozoan parasite cysts. The presence of fecal indicators, such as *C. perfringens*, intestinal Enterococci, *E. coli* is regularly tested.

The following factors are taken into consideration when discussing the flaws of surveillance of water quality that undervalues the numbers and composition of drinking water microbial populations. These indicators rely on fecal indicators and heterotrophic plate count.

- The vast majority of bacteria in drinking water systems (roughly 95%) are thought to be attached to the surfaces, with only 5% being found in the water phase and found through sampling, which is frequently employed for quality control (Flemming et al., 2002).
- According to Servais et al. (2004), additional research indicates that the biomass of bacteria in pipe biofilms is approximately 25 times more plentiful than that of suspended cells.
- However, Srinivasan's (2008) research shows that bulk bacteria may be more prevalent in network sections with residence times longer than 12 hours and chlorine residuals less than 0.1 mg/L, proving that the commonly held notion that biofilms dominate distribution systems is not always true.
- Routine monitoring only looks at a tiny volume, roughly equivalent to 2×10^{-7} to 5×10^{-7}% of delivered drinking water (Allen, 2011).

- According to Moritz et al. (2010), a sizable portion of water and biofilm biomass may be in a viable but non-cultivable state. This means that while they cannot develop on artificial growth media, they are still alive, capable of renewed activity, and thus relevant to hygiene. It is estimated that culturable heterotrophic bacteria make up 0.01% of waterborne microorganisms (Watkins and Jian, 1997; Exner et al., 2003).
- The most accurate predictors of possible hazards are fecal indices; however, their concentrations do not always match exactly with those of pathogens. While there are notable associations between fecal indices and pathogenic species in freshwater environments, the corresponding relationships with several pathogenic or opportunistic pathogenic bacteria, protozoa, fungi, and viruses in drinking water are either negligible or nonexistent (Payment and Locas, 2011). There is a list of disease-causing pathogens found in water belonging to protozoa, helminths, bacteria, viruses, fungi, and algae in **Table 3.1**.

3.4 INTERACTIONS BETWEEN MICROBES

There are many intricate relationships involving interactions between and within species that arise from the diversity of microbial communities. The early phases of biofilm creation and growth may be significantly impacted by interactions between different bacterial species. Cell-to-cell communication through quorum sensing helps the biofilm community by reducing unwanted overgrowth and nutrient competition (Davey et al., 2000). It has been shown that quorum sensing is involved in the adherence and dissociation of cells from biofilms (Daniels et al., 2004; Donlan, 2002). Bacteria must be able to detect, react, and modify gene expression in response to their surrounding environment for them to successfully adapt to shifting natural conditions (Daniels, 2004). Surfaces offer a niche that encourages the development of intricate relationships between bacterial cells. Following cell adhesion, the metabolism and growth of each species within the community are reliant on the surface conditions of their respective locations. The behavior of cells both within and between populations defines the biological complexity of a system. When microorganisms attach themselves to a surface, their metabolic processes lead to the evolution of interfacial gradients of chemicals in space and time, resulting in conditions that are not typically found in the bulk aqueous phase (Geesey, 2001).

The presence of interspecies relationships and microbial heterogeneity in drinking water can lead to improved methods for controlling microbial growth (Simões et al., 2010a; Rasmussen et al., 2005). One of the main evolving driving factors in the microbial world is thought to be competition for substrate, and experimental data collected in a lab setting demonstrated how various microorganisms may effectively outcompete others due to more efficient use of a particular energy source (Christensen et al., 2002). A complex web of interactions lies at the heart of every community's composition, structure, and operation. The establishment of a structured community for interspecies interactions between *P. putida* KT2440 and *Acinetobacter sp.* strain C6 was facilitated by spatial structure as the primary environmental factor (Hansen et al., 2007). The majority of studies on interactions between different species in biofilms have concentrated on the advantages of these connections. Even so, not every

TABLE 3.1
Waterborne Disease-Causing Pathogens

Protozoa	Helminths	Bacteria	Viruses	Fungi	Algae
Sarcocytis spp.	*Schistosoma spp*	*Aeromonas spp.*	*Enterovirus*	*Cladosporium spp.*	*Microcystis spp.*
Giardia intestinalis	*Dracunculus medinensis*	*Escherichia coli pathogenic*	*Coxsackie vírus B*	*Acremonium spp.*	*Chroococcusdispersus*
Microsporidia	*Ascaris lumbricoides*	*Burkholderiapseudomallei*	*Hepatite A virus*	*Fusarium spp.*	*Oocystis spp.*
Giardia duodenalis	*Fasciola spp*	*Francisella tularensis*	*Poliovirus*	*Aspergillus spp.*	*Oscillatoria spp.*
Naegleria fowleri		*E. coli enterohaemorrhagic*	*Sapovirus*	*Phialophora spp.*	*Anabaena spp.*
Entamoeba histolytica		*Salmonella paratyphi*	*Norovirus*	*Epicoccum spp.*	*Microcystis aeruginosa*
Toxoplasma gondii		*Staphylococcus aureus*	*Astrovirus*	*Mucor spp.*	*Xenococcus spp.*
Cyclospora cayetanensis		*Pseudomonas aeruginosa*	*Coxsackie vírus A*	*Exophiala spp.*	
Entamoeba histolytica		*Legionella pneumophila*	*Hepatite E virus*	*Candida spp.*	
Blastocystis hominis		*Salmonella spp.*	*Adenovirus*	*Trichoderma spp.*	
Acanthamoeba castellani		*Mycobacterium spp. (non- tubercolous)*	*Echovirus*	*Paenicillium spp.*	
Giardia lamblia		*Klebsiella spp.*		*Alternaria spp.*	
		Shigella spp.			
		Helicobacter pylori			

interaction will be advantageous because antagonistic interactions can be crucial to the growth of microbial communities. The majority of bacteria appear to be universally capable of producing antimicrobial compounds, such as antibiotics, toxins, bacteriocins, and bacteriolytic enzymes (Tait andSutherland, 2002). Interaction between species is explained in Table 3.2.

3.5 FACTORS AFFECTING BIOFILM

Numerous factors influence the intricate processes of microorganisms adhering to surfaces and forming highly complex biofilms. The DWDS possesses various factors such as nutrient nature and concentration, sediment accumulation, microbial interactions and variety of microorganisms, the concentration of free residual disinfectants, pH and temperature fluctuations, water residence time, hydrodynamic conditions, network design (including dead ends and pipe diameter), features of the material covering the distribution pipes, and age.

3.5.1 Environmental Factors (Temperature and pH)

Temperature and pH are thought to be two significant elements that influence life by altering a variety of characteristics, including metabolism in microbes, enzyme activity, kinetics and equilibrium of reactions, and the electrostatic interactions across surfaces and microorganisms (Simoes and Simões, 2013). The efficiency of disinfection is also impacted by temperature and pH. When the temperature rises and pH falls, the amount of chlorine residuals in drinking water is significantly decreased. Pinel et al. (2021) conducted an experiment ontheeffect of different temperatures (from 20 to 27 °C) on biofilm formation and observed that the community composition and EPS formation have strong positive correlation with the tested temperature regime.

3.5.2 Hydrodynamic

There are fluctuating hydrodynamic conditions in DWDSs. Although there are times when the flow is laminar and other times when it is turbulent, stagnant waters can also happen in areas with low water consumption, such as buildings and reservoirs. Various impacts on biofilm formation and dissociation may result from variations in flow velocity (Melo, 2003). After reaching a maximum value, nutrient transport rates within the biofilm rise with flow velocity and subsequently fall when additional velocity increases. In the biofilm, this transport rate encourages growth of bacteria. Conversely, the flow velocity increases the biofilm dissociation and thickness. It has been noted by many authors that the development of biofilm rises with flow velocity (Lehtola et al., 2006), whereas others had the opposite outcome (Tsai, 2005). Increased flow rates make cells more hydrophobic, which encourages cell accumulation and, ultimately, the buildup of biofilms. In a controlled environment exhibiting two different hydrodynamic conditions (2300 and 4600 Reynolds numbers), *E. coli* strains with the expression of green fluorescent protein (to monitor the growth) were

TABLE 3.2
Biofilm Communities: Pertinent Interspecies Interactions

Strains	Relationships between species	References
Lactococcus lactis ssp. cremoris/Pseudomonas fluorescens	Commensalism	Yu, Vinarov, Robysheva, Smirnov and Sokolov (2002).
Acinetobacter sp./Pseudomonas putida		Christensen et al., (2002).
Pseudomonas sp. strain GJ1/*Pseudomonas putida* DMP1		Cowan, Gilbert, Liepmann and Keasling (2000).
Schizosaccharomyces sp.	Neutralism	Yu, Vinarov, Robysheva, Smirnov and Sokolov (2002).
Saccharomyces sp.		
Pseudomonas sp.		
Corynebacterium sp.		
Candida sp.		
Bacillus cereus/Pseudomonas fluorescens	Antagonism	Simões et al., 2008.
Marine pelagic bacteria		Bhattarai et al., 2006.
DW-isolated bacteria		Simões et al., 2007.
Enteric bacteria		Tait and Sutherland (2002).
Marine epiphytic bacteria	Mutualism	Burmølle, Webb, Rao, Hansen, Sørensen andKjelleberg (2006).
Soil bacteria		Wolfaardt, Lawrence, Robarts, Caldwell and Caldwell (1994).
DW-isolated bacteria		Simões, Simões andVieira (2007b).
E. coli PHL565/ *P. putida* MT2; *E. coli* PHL565/ environmental *E. coli* from DWDS		Castonguay, van der Schaaf, Koester, Krooneman, van der Meer, Harmsen and Landini (2006)
Denitrifying bacteria	Competition	Andersson et al. (2009).
Klebsiella oxytoca/Burkolderia cepacia		Komlos et al. (2005).
Acinetobacter sp./Pseudomonas putida		Christensen,et al.,(2002).
Marine epiphytic bacteria		Rao, Webb, and Kjelleberg (2005).

examined. It was observed that expression of proteins was two-and three-fold higher in 2300 and 4600 Reynolds numbers, respectively (Soares et al., 2022).

3.5.3 Sediment Accumulation

Organic material, including microorganisms, and insoluble material, primarily iron and manganese, can both be found in sediment. Sediments that have accumulated may have an enormous microbial population. In low-flow regions or dead ends of the DWDS, both organic and inorganic particles can build up and promote the activity of bacteria by offering nutrients and protection (Swaileh et al., 2008). Sloughing biofilms have the potential to build up on the edges of distribution networks, causing sedimentation and the growth of certain microbes (Simoes and Simões, 2013). Decreases in the amount of disinfectant residual in water can also result from sediment buildup. Sand is one type of inorganic particle that will encourage biofilm erosion, whereas clay may cause biofilms to become thicker and more resilient.

3.5.4 Nutrients

Two major factors that affect biofilm growth are nutrients and hydrodynamic conditions, specifically the density, thickness, and structure of the film (Wasche et al., 2002). Thick and dense biofilms with decreased internal nutrient diffusion and enhanced cohesion and resistance to removal resulted from high shear stress and nutrient limitations (Paris et al., 2007). Biofilms with a low-density structure and numerous pores, channels, and protuberances grow rapidly in conditions of low flow velocities and high nutrient content (Van Loosdrecht et al., 1995).

In general, environments with low concentration of nutrients such as phosphorus, nitrogen, and carbon known as oligotrophic environments are thought to exist in DWDS. Development of biofilm is encouraged by an increase in nutrients in water. Polluted water from industries, households waste, agriculture, and rivers contribute to contaminating aquatic ecosystems (Barot and Kumar, 2021). Research has indicated a favorable correlation between bacterial regrowth in DWDS and the percentage of nutrients in drinking water (Simoes and Simões, 2013). The organic carbon level was the limiting nutrient in numerous DWDS studies conducted worldwide. Batté et al., 2003, noted that even though phosphorus was being incorporated into biofilm, it did not affect the development of biofilm. Finding the limiting nutrient in DWDS is crucial because it has been suggested that adding compounds based on phosphate will stop erosion of pipes and bacterial regrowth (Appenzeller et al., 2002). The amount of nutrients, for example, the carbon/nitrogen ratio, influences the adherence of microorganisms to surfaces by influencing the synthesis of extracellular polymers (Veiga et al., 1997).

3.5.5 Microorganisms

Each type of microorganism in the bulk water will have a different extracellular appendage, surface characteristics, and capacity to produce EPS, which will impact the attachment process. The hydrophobic nature of the cell surface and the existence of extracellular filamentous appendages can affect how quickly and how deeply bacteria attach themselves. The microbial cell surface and the adhesion surface are the two surfaces involved in adhesion, and their hydrophobicity is crucial because hydrophobic interactions tend to increase as these surfaces become more non-polar (Donlan, 2002). Numerous microorganisms generate filamentous appendages that are extracellular. This suggests that they might be involved in the mechanism of attachment. Their interaction radius with the surface is much smaller than the cell's own. There are several known examples of these structures, including holdfast, pili or fimbriae, flagella, stalks, and prothecae (Harbron and Kent, 1988). These kinds of structures play a key role in adhesion, cell-surface interactions, and motility (Simões et al., 2010b). The ability of microorganisms to produce EPS to bind cells and other particulate matter together is called cohesion and to the surface is known as adhesion (Allison, 2003; Sutherland, 2001a). Polymers, lipids, proteins, humic substances, and nucleic acids make up the bulk of bacterial EPS (Sutherland, 2001). Proteins and polysaccharides make up 75–89% of the structure of the biofilm EPS, suggesting that they are the main constituents (Tsuneda et al., 2003). It has been demonstrated that bacterial grazing, which limits the accumulation of biofilms, is primarily caused by the protozoa in the DWDS (Snelling et al., 2006; Berry et al., 2006).

3.5.6 Materials

Previously, cement or iron-based materials were used to construct almost all of the pipelines in DWDS. Polymeric resources, particularly polyvinyl chloride (PVC) and polyethylene (PE), have gained popularity recently due to their ease of handling and application. The literature has extensive documentation on the impact of support materials on the development of biofilm (Yu et al., 2010; Momba et al., 2000). Comparing polymeric and metallic surface materials, there is still debate regarding how surface materials affect the formation of biofilms. Barot et al. (2021) has reported that the microbes also play a major role in bioremediation of such metal and metalloids in water. According to certain reports, drinking water biofilms grew more slowly on iron matrices than they did on polymeric materials (Hem et al., 2002; Chang et al., 2003). This fact was linked to iron corrosion products that support biofilm defense against chemical and mechanical stressors. Some research found that PVC and PE surfaces produced more biofilm than materials made of galvanized steel (Bachmann et al., 2006; Cloete et al., 2003). Biofilms develop more quickly on PE than on copper pipes; however, such variations were not present in earlier piping systems (Lehtolaet al., 2004). Materials' texture and surface physicochemical characteristics such as hydrophobicity, solid surface tension, chemical nature, and surface charge are crucial for the development of biofilm. When polymeric materials come into contact with drinking water, biodegradable compounds may be released, which would promote the growth of biofilms (Simoes and Simões, 2013). Additionally, the materials' ability to resist erosion may be a key consideration when selecting a material for the DWDS. In addition to releasing unwanted materials into the water, corroded iron pipes may provide a multitude of places for bacteria to bind themselves and shield them from the effects of flow rate and disinfectants (Niquette et al., 2000).

Functionally varied microbial species within the biofilms may be the cause of the corrosion on metallic surfaces (Beech, 2004).

3.5.7 Disinfecting Agents

The amount of disinfecting residual in DWDSs is another significant factor in the development of biofilm. The most popular disinfectants are UV light, ozone, chlorine dioxide, chloramines, and chlorine (Lee and Nam, 2005). Chlorine is the disinfectant that is used in the DWDS the most frequently. A widely used tactic to stop bacterial regrowth during water transportation is disinfection with chemicals and maintaining chlorine residual through distribution systems (Codony et al., 2005). According to certain studies, chlorine can regulate the development of biofilms by slowing down the rate of biofilm growth, encouraging biofilm separation, and lowering microbial activity (Codony et al., 2005; Codony et al., 2002). The distribution of high-quality drinking water despite the addition of residual chlorine is the strategy adopted by some European nations, including Germany, Switzerland, Netherlands, and Austria. In these nations, the regulation of microbial growth is achieved by producing biologically stable drinking water through more suitable drinking water treatments (UV disinfection, peroxide, sedimentation, ozone, filtration), which minimize the nutrients necessary for survival (Simoes and Simões, 2013). The potential disinfectants being used and their mechanisms are summarizedin Table 3.3. The pathogens responsible for waterborne diseases and their response to chlorine disinfectant areelaborated on in Table 3.4.

3.6 WATER QUALITY PARAMETERS

Biofilms developed in the drinking water system are affected by several factors like organic matter, pH, temperature, supply system, and infrastructure. These three important factors are described in Table 3.5. These factors reportedly affect growth of microbes and the effect of disinfectant in DWDSs. The effect of microbial growth in DWDS was studied by Calero et al. (2021); they found that *Pseudomonas* and *Fusarium*, in particular, were favored by increase in temperature and have enhanced capacity to develop biofilms and compromised water quality. The variation in pH directly indicates the chemical composition of water and has a significant role in development of biofilm (Characklis and Cooksey, 1983). The input of organic and inorganic matter is responsible for turbidity in water, which provides a conditioned surface for microbial biofilm.

3.7 STRATEGIES TO CONTROL BIOFILMS IN DRINKING WATER SYSTEMS

Microbes can enter the drinking water system or reservoir through infrastructure and untreated and contaminated water. Mainly, pipes are the main source where nutrients accumulate and provide suitable adhesion for the microbes. Apart from reservoirs and water quality many factors influence the formation of biofilms.

TABLE 3.3
The Primary Disinfectants Employed in Drinking Water Distribution System and Their Characteristics

Disinfectant	Disinfecting residuals in DWDS	Efficiency in opposition to	Disinfection by-products	Mechanisms of deactivation
Ozone	No	Giardia Cryptosporidium Viruses Bacteria	Haloacetic acids Aldehydes Aldo- and Ketoacids Brominated DBPs Hydrogen peroxide	It targets amino acids, glycolipids, and glycoproteins found in bacterial cell membranes. It also acts on the sulfhydryl groups of some enzymes, causing disruptions to enzymatic activity. It may impact both of purines and pyrimidines in nucleic acids and act on the nuclear material in addition to the cell membrane and wall. It harms the nucleic acids and viral capsid proteins in viruses.
Chloramines	Yes	Bacteria	Haloacetic acids Trihalomethanes	Amino acids and it react easily together. Thus, it is believed that the process of deactivation entails a blocking of proteins or protein-mediated functions, including breathing.
UV radiation	No	Protozoa Viruses Bacteria	No	breaks down the genetic material within the microbial cell wall, preventing replication.
Chlorine dioxide	Yes	Viruses Giardia Cryptosporidium Bacteria	Haloacetic acids Chlorate Chlorite	In contrast to viral ribonucleic acids, it reacts easily with nucleic acids, free fatty acids, amino acids. Through changes to viral capsid proteins, viruses become inactive. Besides, it prevents the production of proteins and messes with the permeability of bacteria's outer membrane.
Chlorine	Yes	Viruses Bacteria	Haloacetic acids Trihalomethanes	Able to damage DNA and cause fatal events at or close to the cell membrane. Chlorine has been shown to negatively impact potentially DNA activity, transport, and bacterial cell respiration. The effects of it include an instantaneous reduction in oxygen consumption, damage to the cell wall membrane, increased permeability of the cell membrane, and decreased the synthesis of DNA.

TABLE 3.4
Pathogens Responsible for Waterborne Illnesses and Their Resistance to Chlorine Disinfection

Microorganisms	Chlorine resistance
Bacteria	
Escherichia coli pathogenic	Low
Francisella tularensis	Moderate
Acinetobacter spp.	Low
Campylobacter coli	Low
Helicobacter pylori	Low
Pseudomonas aeruginosa	Moderate
Klebsiella spp.	Low
Shigella spp.	Low
Salmonella typhi	Low
Yersinia enterocolitica	Low
Mycobacterium spp.	High
Legionella pneumophila	Low
Staphylococcus aureus	Moderate
Vibrio cholera	Low
Burkholderia	Low
Viruses	
Enterovirus	Moderate
Hepatitis E virus	Moderate
Rotavirus	Moderate
Norovirus	Moderate
Sapovirus	Moderate
Coxsackie virus A	Moderate
Adenovirus	Moderate
Hepatitis A virus	Moderate
Echovirus	Moderate
Astrovirus	Moderate
Coxsackie virus B	Moderate
Protozoa	
Toxoplasma gondii	High
Cryptosporidium parvum	High
Acanthamoeba castellani	High
Entamoeba histolytica	High
Giardia lamblia	High
Sarcocytis spp.	High
Microsporidia	Moderate
Balantidium coli	High
Naegleria fowleri	Low
Blastocystis hominis	High
Giardia intestinalis	High
Giardia duodenalis	High
Helminths	
Schistosoma spp.	Moderate
Dracunculus medinensis	Moderate
Free-living nematodes other than *Dracunculus medinensis*	High
Fasciola spp.	High

TABLE 3.5
Effect of pH, Temperature, and Turbidity on Biofilms

Parameters	Effects on Biofilm
pH	• A pH higher than 9 may encourage scaling, which may provide microbes with a place to live and develop biofilms.
Temperature	• The diversity and flourishing of microorganisms in pipes are enhanced by increased water temperatures. • In general, the development of bacteria is favoured by water temperatures of 15°C or higher (USEPA, 2008). • Reduction in chlorine residue due to higher temperature promotes the growth of biofilms in DWDS (Li et al., 2018).
Turbidity	• Higher turbidity can reduce chlorine residual and shelter microbes from disinfection.

To control the biofilm formation in DWDS, approaches are employed at three different levels. First is the prevention in which safety standards are applied to the water sources to protect the water from contaminants. As biofilm formation is promoted with the rough surfaces of the pipes, smooth materials can be used in the manufacturing of the pipes. Alternatively, the inner coating of the wall can also be done. To prevent contamination based on human activities, training and awareness campaigns should be done. The second strategy comprises monitoring and maintenance. Water quality is frequently checked under surveillance programs. The quality check parameters include the presence and load of microbes, EPS, and enzyme activities in DWDS. Modern biotechnological tools like 16s rRNA gene-based identification of the microbes can also be done to accurately predict the type of bacteria and their further mitigation strategies. Frequent replacement of pipes, coating the pipes with phosphates, and pH adjustment of the water entering the distribution system are several factors contributing to the corrosion control in DWDSs (Liu et al., 2016). Along with the control strategies, maintenance of the water reservoir and pipes equally contributeto the prevention of biofilms. Such maintenance includes the replacement of faulty modules within the distribution system,the prevention of anthropogenic activities near the reservoir, and the design and implementation of drinking water safety plans. The last strategy includes a reduction in nutrient load, and appropriate treatments applied to water and distributing apparatus. Nitrogen is a backbone of genetic materials and proteins and thus has a direct correlation with the growth of biofilms. Reducing the amount of nitrogen that is accessible as nutrients can help stop the production of biofilms (Kowalchuk and Stephen, 2001) as well as copper (Dwidjosiswojo et al., 2011) and iron (Musk et al., 2005). Finally, appropriate treatments such as chlorination, ozonization, and UV radiation to the water can be given to retard bacterial growth during biofilm formation. However, UV radiation is ineffective against certain bacteria as they are resistant to them (Hijnen et al., 2006). Physical treatment such as flushing (forcing high-speed water to remove particulates and contaminants) and air-water

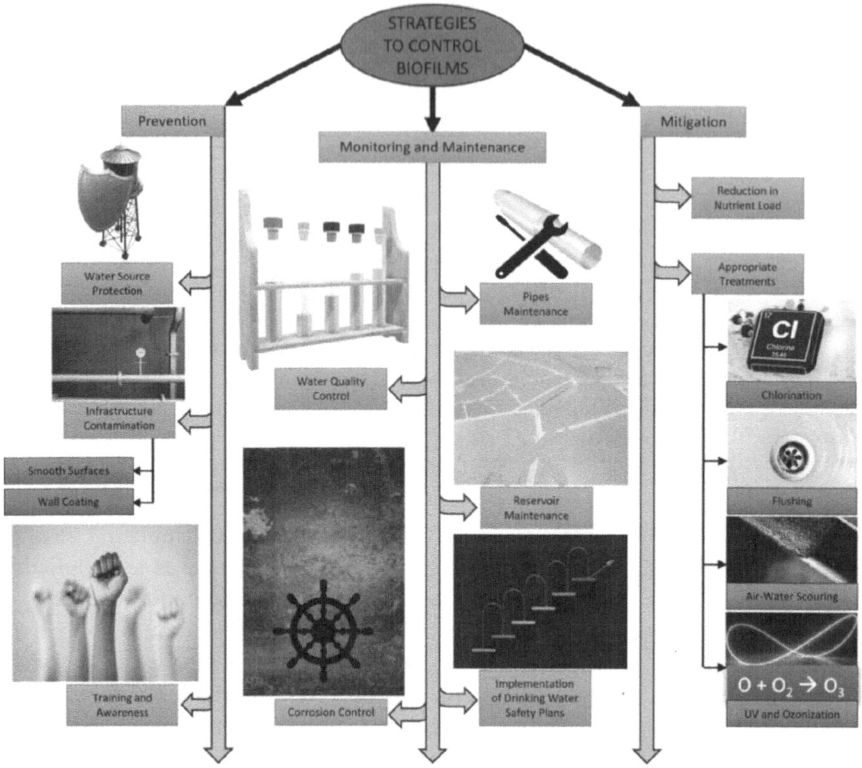

FIGURE 3.2 Strategies to control biofilms in drinking water system.

scouring can also be given to the pipes. They are found to be highly reliable and effective in certain cases. A graphical representation of the strategies employed to prevent biofilm proliferation is givenin Figure 3.2.

3.8 CONCLUSION

The presence of microbial biofilms in drinking water presents serious problems for both public health and water quality maintenance. To manage and control the presence of biofilms in water distribution systems, it is essential to comprehend the composition, formation, and effects of these organisms. The intricate relationships between microbial biofilms and drinking water quality require continuous research and the application of innovative methods in water treatment.

REFERENCES

Allen, M.J. (2011). Escherichia coli – the most relevant microbial public health indicator for drinking water. Faecal indicators: Problem or solution? Conference presentation, Edinburgh, UK.

Allison, D. G. (2003). The biofilm matrix. *Biofouling*,19, 139–150.

Andersson, S., Dalhammar, G., Land, C. J., & Kuttuva Rajarao, G. (2009). Characterization of extracellular polymeric substances from denitrifying organism Comamonas denitrificans. Applied Microbiology and Biotechnology, 82, 535–543.

Appenzeller, B. M., Duval, Y. B., Thomas, F., & Block, J. C. (2002). Influence of phosphate on bacterial adhesion onto iron oxyhydroxide in drinking water. *Environmental Science & Technology*, 36(4), 646–652.

Bachmann, R. T.,&Edyvean, R. G. J. (2006). AFM study of the colonisation of stainless steel by Aquabacterium commune. *International Biodeterioration & Biodegradation*, 58, 112–118.

Bamford, N. C., MacPhee, C. E., & Stanley-Wall, N. R. (2023). Microbial Primer: An introduction to biofilms–what they are, why they form and their impact on built and natural environments. *Microbiology*, 169(8), 001338.

Barot, M., &Kumar, N. (2021). Remediation action on persistent organic pollutants by wonder weeds and associated microbiomes. In *Microbe Mediated Remediation of Environmental Contaminants*(pp.355–368). Woodhead Publishing.

Barot, M., Modi, A., &Kumar, A. (2021). Application of plant–microbe systems in bioremediation of metalloid-contaminated soils. In *Microbe Mediated Remediation of Environmental Contaminants*(pp. 227–240). Woodhead Publishing.

Bassler, B.L.(1999). How bacteria talk to each other: Regulation of gene expression by quorum sensing. *Current Opinion in Microbiology*, 2(6), 582–587.

Batté, M., Koudjonou, B., Laurent, P., Mathieu, L., Coallier, J., & Prévost, M. (2003). Biofilm responses to ageing and to a high phosphate load in a bench-scale drinking water system. *Water Research*, 37(6), 1351–1361.

Beech, I. B. (2004). Corrosion of technical materials in the presence of biofilms-current understanding and state-of-the art methods of study. *International Biodeterioration & Biodegradation*,53, 177–183.

Berry, D., Xi, C.,&Raskin, L. (2006). Microbial ecology of drinking water distribution systems. *Current Opinion in Biotechnology*,17, 297–302.

Bhattarai, H. D., Lee, Y. K., Cho, K. H., Lee, H. K.,&Shin, H. W. (2006). The study of antagonistic interactions among pelagic bacteria: A promising way to coin environmental friendly antifouling compounds. *Hydrobiology*, 568, 417–423.

Blasi, M.F., Carere, M., Pompa, M.G., Rizzuto, E.,&Funari, E.(2008). Water-related diseases outbreaks reported in Italy. *Journal of Water and Health*, 6(3), 423–432.

Bryers, J.D.(2008). Medical biofilms. *Biotechnology and Bioengineering*, 100(1), 1–18. https://doi.org/10.1002/bit.21838

Burmølle, M., Webb, J. S., Rao, D., Hansen, L. H., Sørensen, S. J.,&Kjelleberg, S. (2006). Enhanced biofilm formation and increased resistance to antimicrobial agents and bacterial invasion are caused by synergistic interactions in multispecies biofilms. *Applied and Environmental Microbiology*,72, 3916–3923.

Calero Preciado, C., Boxall, J., Soria-Carrasco, V., Martínez, S., &Douterelo, I. (2021). Implications of climate change: How does increased water temperature influence biofilm and water quality of chlorinated drinking water distribution systems?. *Frontiers in Microbiology*,12, 658927.

Castonguay, M.-H., van der Schaaf, S., Koester, W., Krooneman, J., van der Meer, W., Harmsen, H.,&Landini, P. (2006). Biofilm formation by Escherichia coli is stimulated by synergistic interactions and co-adhesion mechanisms with adherence-proficient bacteria. *Research in Microbiology*,157, 471–478.

Chang, Y. C., Puil, M. L., Biggerstaff, J., Randall, A. A., Schulte, A.,&Taylor, J. S. (2003). Direct estimation of biofilm density on different pipe material coupons using a specific DNA-probe. *Molecular and Cellular Probes*, 17, 237–243.

Characklis, W. G., &Cooksey, K. E. (1983). Biofilms and microbial fouling. In *Advances in Applied Microbiology*(Vol. 29, pp. 93–138). Academic Press.
Christensen, B.B., Haagensen, J. A. J., Heydorn, A.,&Molin, S. (2002). Metabolic commensalism and competition in a two-species microbial consortium. *Applied and Environmental Microbiology*, 68, 2495–2502.
Cloete, T. E., Westaard, D.,&van Vuuren, S. J. (2003). Dynamic response of biofilm to pipe surface and fluid velocity. *Water Science and Technology*, 47, 57–59.
Codony, F., Morato, J., Ribas, F.,&Mas, J. D. (2002). Effect of chlorine, biodegradable dissolved organic carbon and suspended bacteria on biofilm development in drinking water systems. *Journal of Basic Microbiology*, 42, 311–319.
Codony, F., Morato, J.,&Mas, J. (2005). Role of discontinuous chlorination on microbial production by drinking water biofilms. *Water Research*,39, 1986–1906.
Cowan, S. E., Gilbert, E., Liepmann, D.,&Keasling, J. D. (2000). Commensal interactions in a dual-species biofilm exposed to mixed organic compounds. *Applied and Environmental Microbiology*, 66, 4481–4485.
Daniels, R., Vanderleyden, J.,&Michiels, J. (2004). Quorum sensing and swarming migration in bacteria. *FEMS Microbiology Reviews*, 28, 261–289.
Davey, M. E.,&O`Toole, G. A. (2000). Microbial biofilms: From ecology to molecular genetics. *Microbiology and Molecular Biology Reviews*, 64, 847–867.
Donlan, R. M. (2002). Biofilms: Microbial life on surfaces. *Emerging Infectious Diseases*, 8, 881–890.
Dsouza, F.P., Dinesh, S. and Sharma, S., 2024. Understanding the intricacies of microbial biofilm formation and its endurance in chronic infections: a key to advancing biofilm-targeted therapeutic strategies. *Archives of Microbiology*, 206(2), p.85.
Dwidjosiswojo, Z., Richard, J., Moritz, M.M., Dopp, E., Flemming, H.C.,&Wingender, J. (2011). Influence of copper ions on the viability and cytotoxicity of *Pseudomonas aeruginosa* under conditions relevant to drinking water environments. *International Journal of Hygiene and Environmental Health*,214(6), 485–492.
Exner, M., Vacata, V.,&Gebel, J. (2003). Public health aspects of the role of HPC – an introduction, In: *Heterotrophic Plate Counts and Drinking Water Safety. The Significance of HPCs for Water Quality and Human Health.*Bartram, J.; Cotruvo, J.; Exner, M.; Fricker, C. &Glasmacher, A. (Ed.), IWA Publishing, London, UK, pp. 12–19.ISBN 1 84339 025 6
Farkas, A., Ciataras, D.,&Brandus, B.(2012). Biofilms impact on drinking water quality. *Ecological Water Quality—Water Treatment and Reuse. Rijeka: InTech*, 141–154.
Flemming, H.C.,Percival, S.L.,&Walker, J.T. (2002). Contamination potential of biofilms in water distribution systems. *Water Supply*, 2(1), 271–280.
Garrett, T.R., Bhakoo, M.,&Zhang, Z.(2008). Bacterial adhesion and biofilms on surfaces. *Progress in Natural Science*,18(9), 1049–1056.
Geesey, G. G. (2001). Bacterial behaviour at surfaces. *Current Opinion in Microbiology*,4, 296–300.
Hansen, S. K., Rainey, P. B., Haagensen, J. A. J.,&Molin, S. (2007). Evolution of species interactions in a biofilm community. *Nature*,445, 533–536.
Harbron, R. S., & Kent, C. A. (1988). Aspects of cell adhesion. In *Fouling Science and Technology* (pp. 125–140). Dordrecht: Springer Netherlands.
Hem, L. J.,&Skjevrak, I. (2002). Potential water quality deterioration of drinking water caused by leakage of organic compounds from materials in contact with the water. In *Copenhagen: 20th NoDig Conference*. https://doi.org/10.1007/s00253-017-8223-7
Hijnen, W.A.M., Beerendonk, E.F.,&Medema, G.J.(2006). Inactivation credit of UV radiation for viruses bacteria and protozoan (oo) cysts in water: A review. *Water Research*,40(1), 3–22.

Komlos, J., Cunningham, A. B., Camper, A. K., & Sharp, R. R. (2005). Interaction of Klebsiella oxytoca and Burkholderia cepacia in dual-species batch cultures and biofilms as a function of growth rate and substrate concentration. *Microbial Ecology*, 49, 114–125.

Kowalchuk, G.A.,&Stephen, J.R.(2001). Ammonia-oxidizing bacteria: A model for molecular microbial ecology. *Annual Reviews in Microbiology*,55(1), 485–529.

Kumar, C.G.,&Anand, S.K.(1998). Significance of microbial biofilms in food industry: A review. *International Journal of Food Microbiology*, 42(1–2), 9–27.

Lee, Y., & Nam, S. (2005, January). Comparative Study of Chemical Disinfection in Drinking Water Supply. In *Disinfection and Reuse Symposium 2005* (pp. 556–581). Water Environment Federation.

Lehtola, M. J., Miettinen, I. T., Keinanen, M. M., Kekki, T. K., Laine, O., Hirvonen, A., Vartiainen, T.,&Martikainen, P. J. (2004). Microbiology, chemistry and biofilm development in a pilot drinking water distribution system with copper and plastic pipes. *Water Research*,38, 3769–3779.

Lehtola, M.J., Laxander, M., Miettinen, I.T., Hirvonen, A., Vartiainen, T.,&Martikainen, P.J. (2006). The effects of changing water flow velocity on the formation of biofilms and water quality in pilot distribution system consisting of copper or polyethylene pipes. *Water Research*, 40(11), 2151–2160.

Li, W., Zhang, J., Wang, F., Qian, L., Zhou, Y., Qi, W., &Chen, J. (2018). Effect of disinfectant residual on the interaction between bacterial growth and assimilable organic carbon in a drinking water distribution system. *Chemosphere*, 202, 586–597.

Liu, S., Gunawan, C., Barraud, N., Rice, S.A., Harry, E.J.,&Amal, R.(2016). Understanding, monitoring, and controlling biofilm growth in drinking water distribution systems. *Environmental Science &Technology*,50(17), 8954–8976.

Liu, Y., Yang, S.F., Li, Y., Xu, H., Qin, L.,&Tay, J.H. (2004). The influence of cell and substratum surface hydrophobicities on microbial attachment. *Journal of Biotechnology*, 110(3), 251–256.

Marshall, K.C.(1996). *Adhesion as a Strategy for Access to Nutrients*(pp. 59–87). Wiley: New York.

Melo, L.F.(2003). Biofilm formation and its role in fixed film processes. *book: Water Wastewater Microbiology*, 337–349.

Momba, M., Kfir, R., Venter, S. N.,&Cloete, T. E. (2000). An overview of biofilm formation in distribution systems and its impact on the deterioration of water quality. *Water SA*, 26, 59–66.

Moritz, M.M.,Flemming, H.C.,&Wingender, J. (2010). Integration of *Pseudomonas aeruginosa* and *Legionella pneumophila* in drinking water biofilms grown on domestic plumbing materials. *International Journal of Hygine Environment and Health*,213(3), 190–197.

Musk, D.J., Banko, D.A.,&Hergenrother, P.J.(2005). Iron salts perturb biofilm formation and disrupt existing biofilms of *Pseudomonas aeruginosa*. *Chemistry & Biology*,12(7), 789–796.

Niquette, P., Servais, P.,&Savoir, R.(2000). Impacts of pipe materials on densities of fixed bacterial biomass in a drinking water distribution system. *Water Research*, 34(6), 1952–1956.

Paris, T., Skali-Lami, S.,&Block, J. C. (2007). Effect of wall shear rate on biofilm deposition and grazing in drinking water flow chambers. *Biotechnology and Bioengineering*, 97, 1550–1561.

Payment, P,&Locas, A. (2011). Pathogens in water: Value and limits of correlation with microbial indicators. *Ground Water*,49(1), 4–11.

Pinel, I., Biškauskaitė, R., Pal'ová, E., Vrouwenvelder, H.,&vanLoosdrecht, M.(2021). Assessment of the impact of temperature on biofilm composition with a laboratory heat exchanger module. *Microorganisms*, 9(6), 1185.

Pugazhendhi, A.S., Wei, F., Hughes, M. and Coathup, M., 2022. Bacterial adhesion, virulence, and biofilm formation. In *Musculoskeletal Infection* (pp. 19–64). Cham: Springer International Publishing.

Rao, D., Webb, J. S.,&Kjelleberg, S. (2005). Competitive interactions in mixed species biofilms containing the marine bacterium Pseudoalteromonas tunicata. *Applied and Environmental Microbiology*, 71, 1729–1736.

Rasmussen, T. B., Skindersoe, M. E., Bjarnsholt, T., Phipps, R. K., Christensen, K. B., Jensen, P. O., Anderson, J. B., Koch, B., Larsen, T. O., Hentzer, M., Eberl, L., Høiby, N.,&Givskov, M. (2005). Identity and effects of quorum-sensing inhibitors produced by Penicilium species. *Microbiology*,151, 1325–1340.

September, S.M., Els, F.A., Venter, S.N.,&Brözel, V.S.2007. Prevalence of bacterial pathogens in biofilms of drinking water distribution systems. *Journal of Water and Health*, 5(2), 219–227.

Servais, P.,Anzil, A.,Gatel, D.,&Cavard, J. (2004). Biofilm in the Parisian suburbs drinking water distribution system. *Journal of Water Supply Research and Technology*,53, 313–324.

Simoes, L.C.,& Simões, M. (2013). Biofilms in drinking water: Problems and solutions. *Rsc Advances*, *3*(8), 2520–2533.

Simões, L. C., Simões, M., & Vieira, M. J. (2012). A comparative study of drinking water biofilm monitoring with flow cell and Propella™ bioreactors. *Water Science and Technology: Water Supply*, 12(3), 334–342.

Simões, L. C., Simões, M., Oliveira, R., & Vieira, M. J. (2007). Potential of the adhesion of bacteria isolated from drinking water to materials. *Journal of Basic Microbiology*, 47(2), 174–183.

Simões, L. C., Simões, M.,&Vieira, M. J. (2007b). Biofilm interactions between distinct bacterial genera isolated from drinking water. *Applied and Environmental Microbiology*,73, 6192–6200.

Simões, L. C., Simões, M.,&Vieira, M. J. (2010a). Influence of the diversity of bacterial isolates from drinking water on resistance of biofilms to disinfection. *Applied and Environmental Microbiology*,76, 6673–6679.

Simões, M., Simões, L. C., Pereira, M. O.,&Vieira, M. J. (2008). Antagonism between Bacillus cereus and Pseudomonas fluorescens in planktonic systems and in biofilms. *Biofouling*, 24, 339–349.

Simões, M., Simões, L. C.,&Vieira, M. J. (2010b). A review of current and emergent biofilme control statagies. *LWT – Food Science and Technology*, 43, 573–583.

Simões, L. C., Simões, M., & Vieira, M. J. (2012). A comparative study of drinking water biofilm monitoring with flow cell and Propella™ bioreactors. *Water Science and Technology: Water Supply*, 12(3), 334–342.

Snelling, W. J., Moore, J. E., McKenna, J. P., Lecky, D. M.,&Dooley, J. S. G. (2006). Bacterial-protozoa interactions: An update on the role these phenomena play towards human illness. *Microbes and Infection*, 8, 578–587.

Soares, A., Gomes, L.C., Monteiro, G.A.,&Mergulhão, F.J.(2022). Hydrodynamic effects on biofilm development and recombinant protein expression. *Microorganisms*, 10(5), 931.

Srinivasan, S. (2008). *Managing Bacterial Regrowth and Presence in Drinking Water Distribution Systems*, Ph.D. thesis, University of Wisconsin-Madison, USA.

Sutherland, I. W. (2001). Biofilm exopolysaccharides: a strong and sticky framework. *Microbiology*, 147(1), 3–9.

Sutherland, I. W. (2001a). The biofilm matrix – an immobilized but dynamic microbial environment. *Trends in Microbiology*, 9, 222–227.
Sutherland, I.W.(1999). Polysaccharases for microbial exopolysaccharides. *Carbohydrate Polymers*, 38(4), 319–328.
Swaileh, K., Hussein, R.M., Al-Sa'ed, R.,&Abu Heleu, R.(2008). *Health Risks from Microbial Growth and Biofilms in Drinking Water Distribution Systems in Palestine*.489–496. URI: http://hdl.handle.net/20.500.11889/3982
Szewzyk, U., Szewzyk, R., Manz, W.,&Schleifer, K.H.(2000). Microbiological safety of drinking water. *Annual Reviews in Microbiology*, 54(1), 81–127.
Tait, K.,&Sutherland, I. W. (2002). Antagonistic interactions amongst bacteriocinproducing enteric bacteria in dual species biofilms. *Journal of Applied Microbiology*,93, 345–352.
Teng, F., Guan, Y.T.,&Zhu, W.P.(2008). Effect of biofilm on cast iron pipe corrosion in drinking water distribution system: Corrosion scales characterization and microbial community structure investigation. *Corrosion Science*, 50(10), 2816–2823.
Tsai, Y.P.(2005). Impact of flow velocity on the dynamic behaviour of biofilm bacteria. *Biofouling*, 21(5–6), 267–277.
Tsuneda, S., Aikawa, H., Hayashi, H., Yuasa, A.,&Hirata, A.(2003). Extracellular polymeric substances responsible for bacterial adhesion onto solid surface. *FEMS Microbiology Letters*, 223(2), 287–292.
USEPA (2008) Water quality in small community distribution systems. *A Reference Guide for Operators*. EPA/600/R-08/039 March 2008.
Van Loosdrecht, M.C.M., Eikelboom, D., Gjaltema, A., Mulder, A., Tijhuis, L.,&Heijnen, J.J. (1995). Biofilm structures. *Water Science and Technology*, 32(8), 35–43.
Veiga, M. C., Jain, M. K., Wu, W. M., Hollingsworth, R. I.,&Zeikus, J. G. (1997). Composition and role of extracellular polymers in methanogenic granules. *Applied and Environmental Microbiology*, 63, 403–407.
Wasche, S., Horn, H.,&Hempel, D. C. (2002). Influence of growth conditions on biofilm development and mass transfer at the bulk/biofilm interface. *Water Research*,36, 4775–4784.
Watkins, J.,&Jian, X. (1997). Cultural methods of detection for microorganisms: recent advances and successes. In: *The Microbiological Quality of Water*, Sutcliffe, D.W (Ed.), *Freshwater Biological Association*, Ambleside, UK, pp. 19–27.ISBN 0 900386 57 6
Watnick, P.I.,&Kolter, R.(1999). Steps in the development of a Vibrio cholerae El Tor biofilm. *Molecular Microbiology*, 34(3), 586–595.
Wingender, J.,&Flemming, H.C.(2011). Biofilms in drinking water and their role as reservoir for pathogens. *International Journal of Hygiene and Environmental Health*, 214(6), 417–423.
Wolfaardt, G. M., Lawrence, J. R., Robarts, R. D., Caldwell, S. J.,&Caldwell, D. E. (1994). Multicellular organization in a degradative biofilm community. *Applied and Environmental Microbiology*, 60, 434–446.
Yu, A., Vinarov, Z., Robysheva, N., Smirnov, V. N.,&Sokolov, D. P. (2002). Studies of the stability of microbial association use in industrial biofiltering of gaseous discharges. *Applied Biochemistry and Microbiology*,38, 445–449.
Yu, J., Kim, D.,&Lee, T. (2010). Microbial diversity in biofilms on water distribution pipes of different materials. *Water Science Technology*, 61,163–171.
Zhao, A., Sun, J. and Liu, Y., 2023. Understanding bacterial biofilms: From definition to treatment strategies. *Frontiers in Cellular and Infection Microbiology*, 13, p.1137947.

4 Marine Ecosystems and Biofilms

Santhilatha Pandrangi, Sudha Charita Chillarige, Gutta Rishitha Chowdhary, and Suseela Lanka

4.1 INTRODUCTION TO MARINE ECOSYSTEMS AND BIOFILMS

Marine ecosystems have a vast environment that covers 71% of the Earth's surface dynamically. In marine ecosystems, there are a variety of habitats such as oceans, seas, estuaries, salt marshes, mangroves, and coral reefs. Each of these habitats supports a diverse array of marine life, ranging from microscopic organisms to large marine mammals. The marine environment plays a crucial role in regulating the Earth's climate. Also, it supports various ecosystems that are essential for humans. Organisms are interconnected through food webs, nutrient cycles, temperature, salinity, and currents. Biofilms are communities of microorganisms that adhere to surfaces and produce a slimy, protective matrix of extracellular polymeric substances (EPSs). In marine ecosystems, biofilms play a fundamental role in the functioning of various environments, from the surfaces of rocks and sand grains to the hulls of ships and the skin of marine organisms. They impact the fouling of marine structures, influence the recruitment of marine organisms, and participate in the degradation of pollutants. Additionally, biofilms contribute to the development and maintenance of healthy coral reefs, where microbial interactions are critical for the functioning of these diverse and delicate ecosystems. Biofilms consist of bacteria, algae, fungi, and other microorganisms that collaborate in a symbiotic manner. They contribute to the nutrient cycling, decomposition of organic matter, and the overall health of marine ecosystems. Biofilms are also involved in processes like nutrient retention, sediment stabilization, and the formation of microbial mats in intertidal zones (Allard et al., 2020). Figure 4.1 depicts different types of biofilms.

4.1.1 ESTUARIES

Estuaries are unique and productive ecosystems formed at the interface between freshwater and saltwater environments. They occur in areas where rivers meet the sea, creating a transition zone characterized by fluctuating salinity levels (Cloern, 2001). The term "biofilm" refers to a community of microorganisms that adhere to a surface and produce a slimy, extracellular matrix composed of polysaccharides, proteins, and DNA. Biofilms primarily consist of bacteria, where bacteria are the

 DOI: 10.1201/9781003487203-4

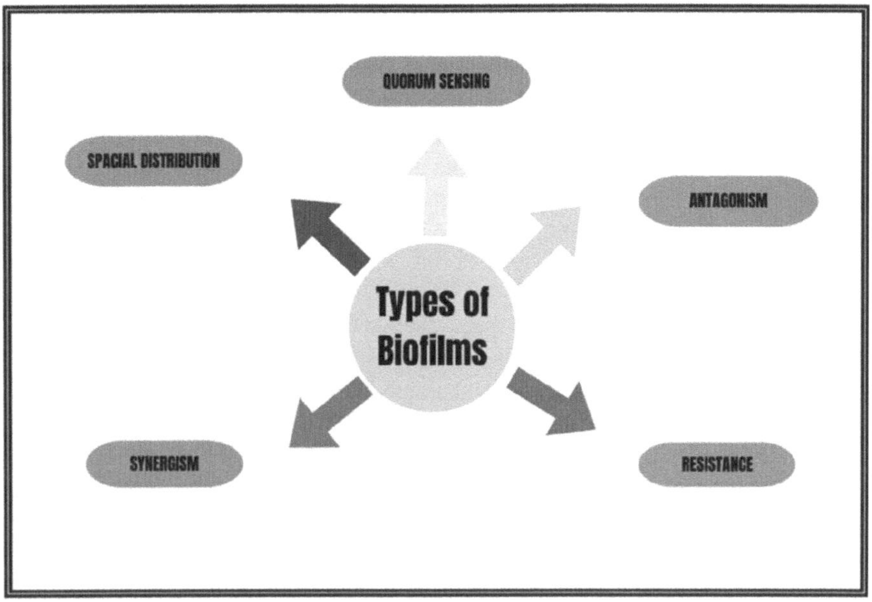

FIGURE 4.1 Different types of marine biofilms.

predominant members of biofilm communities. Although they can incorporate other microorganisms, where certain environments may allow for the coexistence of algae or fungi within the matrix. (Britannica, 2023) They contribute to the nutrient cycling, decomposition of organic matter, and the overall health of marine ecosystems. Biofilms play a significant role in estuarine ecosystems, contributing to various ecological processes and functions. Biofilms in estuaries contribute to the breakdown of organic matter, releasing nutrients back into the water, and sustaining the estuarine food web (Underwood & Paterson, 2003).

4.1.2 Mangrove Ecosystems

Mangroves are coastal ecosystems characterized by the presence of salt-tolerant trees, shrubs, and other plant species. These resilient ecosystems thrive in the challenging interface between land and sea, providing crucial habitats for diverse marine life and offering valuable ecosystem services, such as coastal protection and nutrient cycling. They grow in intertidal zones and are found in tropical and subtropical regions. Mangrove plants have adaptations to tolerate high salinity levels in their habitat. These adaptations include specialized root structures, such as prop roots and pneumatophores, that help them survive in waterlogged, saline conditions. Microorganisms, including bacteria and algae, colonize the root surfaces, forming biofilms that play a role in nutrient cycling (Zedler et al., 2008). Biofilms contribute to nutrient cycling by breaking down organic matter and releasing nutrients into the surrounding environment. Mangrove forests are highly effective at sequestering carbon dioxide from the atmosphere. The organic material in mangrove soils stores large amounts of carbon, making mangrove ecosystems significant in the global carbon cycle. Biofilms are integral components of mangrove ecosystems, contributing to various ecological functions within this unique coastal habitat. There are some aspects of the relationship between biofilms and mangrove ecosystems such as the complex structures of mangrove roots and the surrounding sediment that create microhabitats that support a diverse microbial community. Biofilms on mangrove roots contribute to this microbial diversity, influencing the overall ecological balance of the ecosystem (Smithsonian Ocean Portal, 2018).

4.1.3 Salt Marshes

Salt marshes are coastal ecosystems characterized by the presence of halophytic (salt-tolerant) plants. They are found in intertidal areas, where saltwater from the ocean mixes with freshwater from rivers and streams (Silliman, 2014). Salt marshes are vital transitional zones between terrestrial and marine environments, providing numerous ecological benefits. Salt-tolerant plants, such as grasses and sedges, dominate salt marshes (Topke K., 2023). These plants have adaptations to thrive in saline conditions and play a crucial role in stabilizing sediments (Adam, 1990). Salt marshes support fishes, crustaceans, mollusks, and numerous bird species. The dense root systems of salt marsh plants help stabilize sediments, preventing coastal erosion. Salt marshes provide a natural buffer against storm surges and contribute to the overall coastal areas, because the dense root systems of salt marsh plants help stabilize sediments (Holguin et al., 2001). Microorganisms, including bacteria and algae, form biofilms on sediment surfaces, creating a matrix that helps bind particles together. This stabilization is crucial for maintaining the integrity of salt marsh habitats. Biofilms contribute to the detoxification of pollutants in salt marsh sediments by facilitating the microbial degradation of contaminants (Wolińska, & Błaszczyk, 2020).

4.1.4 Coral Reefs and Biofilms

Coral reefs are vibrant and diverse marine ecosystems formed by the accumulation of calcium carbonate skeletons. They thrive in warm, clear, shallow waters, and are found in tropical and subtropical regions. Coral reefs play a critical role in supporting marine life. Coral reefs host a remarkable diversity of marine organisms, including corals, fish, invertebrates, and algae. Corals form symbiotic relationships with photosynthetic algae known as zooxanthellae. These algae live within coral tissues and provide the corals with nutrients through photosynthesis. Coral reefs face numerous threats, including climate change, overfishing, pollution, and habitat destruction (Rohwer et al., 2002). Healthy biofilms can act as a barrier, preventing the colonization of harmful pathogens and helping to maintain the overall health of coral colonies. The matrix produced by biofilms provides a suitable substrate for coral larva to attach and develop into new coral colonies, contributing to the regeneration of coral reefs.

4.1.5 Open Sea and Deep Sea – Biofilms

The open sea includes both surface waters and the water extending to great depths. The open sea encompasses a dynamic environment characterized by ocean currents, temperature variations, and the continual movement of marine organisms. This expansive marine realm is shaped by complex interactions among physical, chemical, and biological factors, contributing to its biodiversity and ecological significance. The open sea is an integral part of the global ocean circulation system, influencing climate patterns and heat distribution across the planet. Marine snow is a collection of organic detritus and microorganisms that clump together. Biofilms can form on the surfaces of these particles, contributing to the breakdown of organic matter and nutrient cycling in the open sea. Biofilms play a role in the microbial loop, a process where microorganisms are consumed by larger organisms (Danovaro, & Middelboe, 2010). Sunlight does not penetrate the deep sea, and organisms in these depths have adapted to rely on alternative sources of energy, such as chemosynthesis near hydrothermal vents. Hydrothermal vents also support unique ecosystems by providing chemical energy in the form of minerals released from the Earth's interior. Biofilms can form around hydrothermal vent structures, where high-temperature, mineral-rich fluids are released into the deep sea (Figure 4.2).

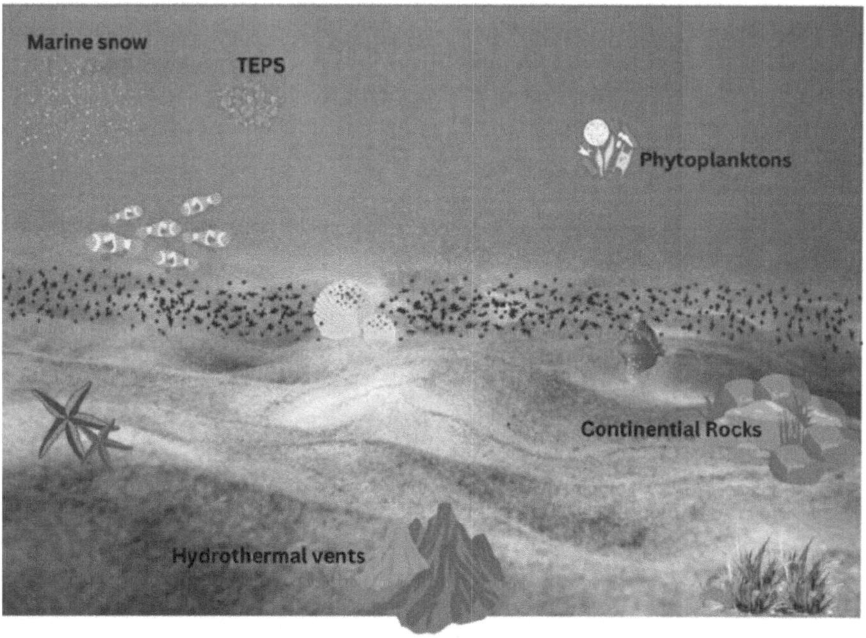

FIGURE 4.2 Pictorial representation of deep sea.

4.2 BIOFILM DEVELOPMENT IN MARINE ENVIRONMENTS

Biofilm development is a natural and ubiquitous process in marine environments. The biofilm development process begins with the initial attachment of microorganisms to a submerged surface. This can include natural substrates such as rocks, sand grains, or plant surfaces. This involves the accumulation of primary colonizers, which adhere to the surface and create a microenvironment that facilitates further colonization. Microbial cells within the biofilm communicate through quorum sensing, a process by which they release signaling molecules to coordinate their activities. The biofilm matrix, composed of (EPSs), provides structural stability to the biofilm and protects microorganisms (Di Martino, 2018).

Biofilm detachment can occur due to various factors, including physical disturbances and changes in environmental conditions. Biofilms provide substrate and habitat for a variety of marine organisms, including larval stages of invertebrates and small fish. In addition to natural substrates, biofilm formation on artificial surfaces can lead to fouling, where the accumulation of organisms on ship hulls and marine infrastructure can have economic and ecological consequences (Dobretsov et al., 2009).

4.3 MARINE BIOFILMS – DIVERSITY, DYNAMICS, ABUNDANCE, AND FUNCTIONS

The interactions among different microbial species within biofilms are complex. Microorganisms in marine biofilms exhibit adaptations to the unique conditions of their environment, such as high salinity, fluctuations in temperature, and exposure to ultraviolet (UV) radiations. Biofilm dispersal can occur through the detachment of microbial cells or biofilm fragments. Detached biofilm particles can contribute to the seeding of new surfaces. Marine biofilms are ubiquitous and colonize natural substrates like rocks, sand, and seagrasses, as well as artificial structures such as ship hulls, piers, and buoys (Lotze et al., 2006). Biofilms contribute to primary productivity, nutrient cycling, and the recruitment of marine organisms. Biofilms offer attachment sites for larval stages of invertebrates, contributing to the recruitment and settlement of benthic species. Biofilms can contribute to the detoxification of pollutants in marine environments. Biofilms facilitate microbial interactions through processes like quorum sensing. Microorganisms within biofilms communicate with each other, coordinating activities and responding to environmental cues.

4.3.1 Marine Biofilms and Biofouling

Marine biofilms are complex communities of microorganisms that form on surfaces submerged in seawater. Microbial biofilms form in a multi-step process of signaling and regulatory events, all of which lead to the development of sessile microbial communities (in a physiological state that is distinct from planktonic cultures) encased in an extracellular polymer matrix (Raeid et al., 2013). These biofilms are composed of bacteria, algae, fungi, and other microorganisms embedded in a slimy matrix of (EPSs). The composition of marine biofilms varies, but they commonly include bacteria, diatoms, algae, protozoa, and other microorganisms. Biofouling, the accumulation of microorganisms, plants, and animals on submerged surfaces, presents significant challenges in marine industries. Initiated by microorganisms attaching to surfaces, the process begins with the formation of a biofilm, a slimy layer of microorganisms and (EPSs). Two categories of organisms contribute to biofouling: microorganisms, including bacteria and algae, in the early stages, and larger macroorganisms like barnacles and mussels that settle on the biofilm, exacerbating fouling (Hideyuki Kanematsu, Dana M. Barry, 2020). Biofouling progresses through stages: attachment, where microorganisms attach to submerged surfaces, and biofilm formation, involving the development of a slimy layer that acts as a foundation for fouling organisms (Stanczak, Marianne, 2004). There are two major categories present in biofouling. Microfouling: Attachment (hours to days): Microorganisms, including bacteria, algae, and protozoa, attach to submerged surfaces within hours to days of exposure to seawater. Initial attachment is a crucial step in biofilm formation (Figure 4.3).

Biofilm formation (days to weeks): Following attachment, microorganisms begin producing (EPS), forming a slimy matrix. This process can take days to weeks, and the biofilm acts as a foundation for subsequent fouling organisms (Ahmed et al., 2018). Macrofouling: Attachment of larger organisms (weeks to months): Once the biofilm has formed, larger macrofouling organisms, such as barnacles, mussels, and other marine invertebrates, can begin to attach to the biofilm. This stage typically occurs over weeks to months, depending on the availability of suitable settlement sites and environmental conditions (Marc W. Mittelman). Formation of complex

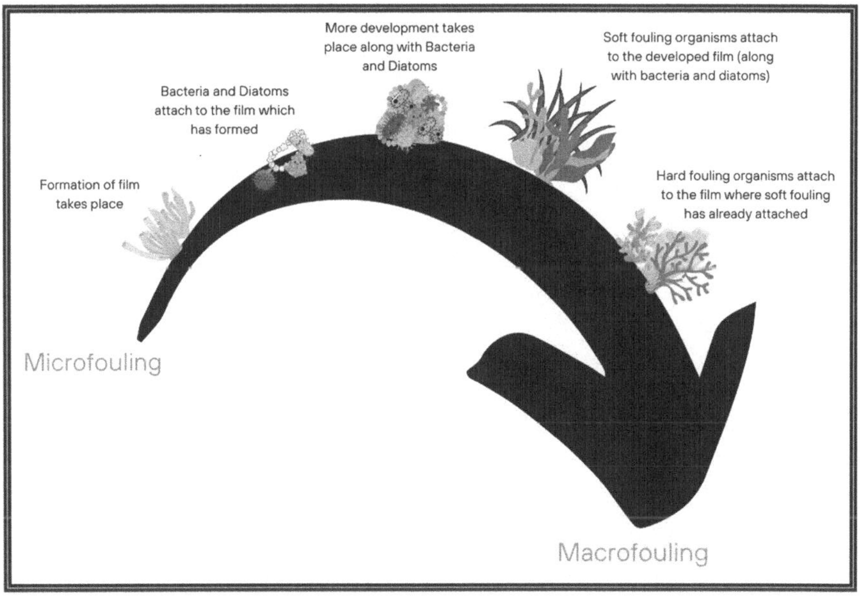

FIGURE 4.3 Micro- and macrofouling.

biofouling communities (months to years): As macrofouling organisms continue to settle and grow on the biofilm, the biofouling community becomes more complex and layered. This process can extend over months to years, resulting in mature biofouling communities (Hayek et al., 2021).

The relationship between biofilms and biofouling is symbiotic. Biofilms initiate biofouling by providing a substrate for microorganism attachment (Callow and Callow, 2002). The biofilm matrix, composed of EPS produced by microorganisms, enhances adhesion and promotes microbial interactions. As the biofilm matures, larger fouling organisms attach, leading to the formation of complex and layered biofouling communities (Marc W. Mittelman, 1999). This complexity poses challenges in various industries, including shipping, aquaculture, and marine infrastructure, impacting vessel efficiency, causing corrosion, and reducing operational efficiency (Callow and Callow, 2002). Understanding and addressing the intertwined dynamics of biofilms and biofouling are essential for developing effective antifouling strategies and mitigating the economic and ecological consequences of biofouling in marine environments.

4.3.2 Marine Biofilms and Biocorrosion

Biocorrosion, also known as microbiologically influenced corrosion (MIC), is a phenomenon where the degradation or corrosion of materials, particularly metals, is influenced or accelerated by the activities of microorganisms. These microorganisms, including bacteria, archaea, algae, and fungi, play a significant role in initiating and promoting the corrosion process. The corrosion caused by microorganisms can manifest in various ways, including pitting corrosion, crevice corrosion, and general corrosion (Butterworth-Heinemann, 2004). Qualitative measurements through various analysis methods, coupled with quantitative assessments of material and composition changes using surface chemical techniques, offer insights into the complex processes of biofilm and (MIC) (Dang et al., 2022). Biocorrosion is a complex process influenced by the interaction between microorganisms, the material being corroded, and the surrounding environment. The biocorrosion has two major mechanisms in it such as the electrochemical corrosion is a chemical process characterized by the transfer of electrons from a zero-valent metal to an external electron acceptor. This transfer leads to the release of metal ions into the surrounding medium, ultimately resulting in the degradation of the metal. The corrosion mechanism involves a sequence of oxidation (anodic) and reduction (cathodic) reactions of chemical species that are either in direct contact with or in close proximity to the metallic surface (Little and Ray, 2002; Iwona B Beech, 2004) and Enzymatic activities: Certain microorganisms produce enzymes that accelerate corrosion. For example, sulfate-reducing bacteria (SRB) are known to produce hydrogen sulfide, which can lead to the formation of metal sulfides and contribute to corrosion. Microbial attachment and colonization: Initial attachment of microorganisms to the material surface is a critical step in biocorrosion. Adhesion is facilitated by factors such as surface roughness and the presence of organic material, allowing the formation of biofilms (Beech Iwona, 2003; Iwona B Beech, Jan Sunner, 2004). Microorganisms involved in biocorrosion produce corrosive substances as by-products of their metabolic activities. For example, certain bacteria can produce organic acids, hydrogen sulfide, and other chemicals that contribute to the degradation of metals. Biocorrosion often leads to localized corrosion, where the corrosion occurs at specific points or in specific areas rather than uniformly across the material surface. This can result in the formation of pits or other structural weaknesses. Biofilms, which are communities of microorganisms encased in a slimy matrix of (EPSs), are commonly associated with biocorrosion. The biofilm provides a protective environment for microorganisms and facilitates their attachment to the material surface. The severity of biocorrosion is influenced by environmental factors such as temperature, pH, oxygen levels, and nutrient availability. Different microorganisms thrive under specific environmental conditions, impacting the corrosion process. Biocorrosion is a significant concern in various industries, particularly those with structures exposed to aquatic environments. This includes marine structures (ships, offshore platforms), water distribution systems, oil and gas pipelines, and other facilities where microbial activity and metal surfaces intersect (Iwona B Beech, Jan Sunner, 2004).

4.3.3 Marine Biofilms and Their Role in Bioremediation

Bioremediation in marine biofilms harnesses the unique capabilities of microorganisms to mitigate pollutants in marine environments. Biofilms, consisting of microorganisms adhering to surfaces, exhibit diverse microbial communities and play a pivotal role in pollutant degradation. Biofilms form on surfaces, creating a matrix of (EPSs). Diverse communities within marine biofilms include bacteria, algae, fungi, and protozoa (Yuvraj, 2022). Microorganisms possess metabolic capabilities to metabolize and degrade pollutants. Biofilms facilitate biodegradation, transforming pollutants into less harmful substances. Biofilms contribute to the natural attenuation of oil spills, with specific bacteria known for hydrocarbon degradation. Bacteria like Alcanivorax and Marinobacter within biofilms are adept at degrading hydrocarbons. Biofilms retain and recycle nutrients, influencing microbial metabolic activities. The availability of nutrients within biofilms enhances overall microbial activity for effective bioremediation. Biofilm-forming microorganisms adapt to changing environmental conditions, ensuring long-term pollutant degradation. Biofilm-based technologies, like biofilters, are used to treat contaminated water. The introduction of specific microbial strains enhances biofilm capabilities, especially for targeted pollutants. Biofilms absorb and sequester contaminants, reducing their mobility and preventing spread in the marine environment. Biofilm-associated organic matter serves as a substrate for microbial processes contributing to contaminant removal. Phytoremediation involving plants and associated microbes integrates into marine biofilms for contaminant removal. The root zone hosts biofilms that aid in pollutant degradation (Canak et al., 2019). Bioremediation in marine biofilms is an interdisciplinary approach that integrates microbiology, environmental science, and engineering to address marine pollution. Understanding biofilm interactions and responses to pollutants is crucial for developing effective strategies to enhance bioremediation in marine ecosystems. Some additional aspects and developments related to marine biofilm bioremediation include: quorum sensing and communications. Microorganisms within marine biofilms often communicate through a process called quorum sensing. This cell-to-cell communication allows them to coordinate their activities and respond collectively to environmental changes. Researchers are exploring ways to manipulate quorum sensing mechanisms to enhance the efficiency of pollutant degradation within biofilms (Kalia, 2018). Advances in genetic engineering have also opened up possibilities for modifying microorganisms within biofilms to improve their pollutant-degrading capabilities. Scientists are working on developing genetically engineered strains with enhanced metabolic pathways for the degradation of specific pollutants, thereby customizing biofilm communities for targeted remediation (Sharma et al., 2021). Nanotechnology is also being incorporated into bioremediation strategies to enhance the performance of marine biofilms. Nano-sized materials, such as nanoparticles and nanocomposites, can be designed to improve the adsorption of pollutants, increase microbial activity, and provide sustained release of nutrients, thereby optimizing biofilm-mediated bioremediation processes (Yadav, 2017).

4.4 DISCUSSION

The exploration of marine ecosystems and biofilms reveals the intricate relationships shaping the vast, dynamic seas covering 71% of Earth's surface. Diverse habitats, from estuaries to coral reefs, host a rich tapestry of life interconnected through food webs and environmental dynamics. Biofilms, symbiotic communities of microorganisms, play pivotal roles in nutrient cycling, sediment stabilization, and pollutant detoxification across habitats. From coastal ecosystems to the open sea and deep sea, biofilms contribute to ecological processes, while their involvement in biofouling and biocorrosion highlights challenges in marine industries. Understanding these complexities is vital for effective environmental management and sustainable marine ecosystems.

4.5 CONCLUSION

The intricate tapestry of marine ecosystems, from estuaries to the deep sea, relies heavily on the symbiotic relationships within biofilms. These microorganisms contribute significantly to nutrient cycling, habitat stability, and pollutant mitigation across diverse habitats. From estuarine processes to coral reef regeneration, and from the open sea to biocorrosion challenges, biofilms play crucial roles. Their involvement in biofouling and bioremediation showcases the complexity of their interactions in marine environments. Understanding these dynamics is imperative for addressing economic and ecological challenges in various industries, ensuring the sustainable management of marine ecosystems, and developing effective strategies for pollution control and habitat preservation.

REFERENCES

Abed, R. M. M. & Teplitski M. (2013). Mini-review: Inhibition of biofouling by marine microorganisms. *Biofouling*, 29 (4): 423–441.

Adam, P. (1990). *Saltmarsh Ecology*. Cambridge University Press, Cambridge.

Ahmed A. B., Adnan J. S., & Sathianeson S. (2018). Microfouling development on artificial substrates deployed in the central Red Sea. *Oceanologia*, 60 (2): 219–231.

Allard S. M., Costa M. T., Bulseco A. N., et al. (2020). Introducing the mangrove microbiome initiative: identifying microbial research priorities and approaches to better understand, protect, and rehabilitate mangrove ecosystems. *mSystems*, 5 (5): e00658–e00720. doi: 10.1128/mSystems.00658-20. PMID: 33082281; PMCID: PMC7577295.

"Bacterial Biofilms and Bifouling: Translational Research in Marine Biotechnology." National Research Council. 2000. Opportunities for Environmental Applications of Marine Biotechnology: Proceedings of the October 5–6, 1999, Workshop. Washington, DC: The National Academies Press. doi: 10.17226/9988.

Beech, I. (2003). Sulfate-reducing bacteria in biofilms on metallic materials and corrosion. *Microbiol Today,* 30: 115–117.

Beech, I. B. (2004). Corrosion of technical materials in the presence of biofilms-current understanding and state-of-the-art methods of study. *International Biodeterioration and Biodegradation*, 53: 177–183.

Beech, I. B., & Sunner, J. (2004). Biocorrosion: Towards understanding interactions between biofilms and metals. *Current Opinion in Biotechnology*, 15 (3): 181–186.

Britannica. The Editors of Encyclopaedia. "estuary". Encyclopedia Britannica, 15 Dec. 2023, www.britannica.com/science/estuary. Accessed 1 February 2024.

Callow, M. E, & Callow, J. E. (2002). Marine biofouling: A sticky problem. *Biologist (London, England)*, 49(1): 10–14.

Canak, S., Berezljev, L., Borojevic, K., Asotic, J., & Ketin, S. (2019). Bioremediation and "green chemistry". *Fresenius Environmental Bulletin*, 28 (4): 3056–3064.

Cloern, J. E. (2001). Our evolving conceptual model of the coastal eutrophication problem. *Marine Ecology Progress Series*, 210: 223–253.

Dang, Y. T. H., Power, A., Cozzolino, D. et al. (2022). Analytical characterisation of material corrosion by biofilms. *Journal of Bio- and Tribo-Corrosion,* 8: 50

Dahiya, D., et al. (2024). AIMS Microbiology, 10 (2): 288–310. https://doi.org10.3934/microbiol.2024015.

Danovaro, R., & Middelboe, M., S. W. Wilhelm, M. G. Weinbauer, and C. A. Suttle (eds.), (2010). Separation of free virus particles from sediments in aquatic systems. In *Manual of Aquatic Viral Ecology* (pp. 74–81). ASLO. DOI 10.4319/mave.2010.978-0-9845591-0-7.74.

Dobretsov, S., Teplitski, M., & Paul, V. J. (2009). Mini-review: Quorum sensing in the marine environment and its relationship to biofouling. *Biofouling*, 25 (5): 413–427.

Gale, W. F., & Totemeier, T. C. (2004). *Smithells Metals Reference Book* (8th ed.). Butterworth-Heinemann, 31–1–31–13.

Hayek, M., Salgues, M., Souche, J.-C., Cunge, E., Giraudel, C., & Paireau, O. (2021). Influence of the intrinsic characteristics of cementitious materials on biofouling in the marine environment. *Sustainability*, 13: 2625. https://doi.org/10.3390/su13052625

Holguin, G., Vazquez, P., & Bashan, Y. (2001). The role of sediment microorganisms in the productivity, conservation, and rehabilitation of mangrove ecosystems: An overview. *Biology and Fertility of Soils*, 33(4): 265–278.

Kaiser, M. J., Attrill, M. J., Jennings, S., & Thomas, D. (Eds.) (2020). Marine Ecology (3rd ed.). London, England: Oxford University Press

Kalia, V. C. (2018). *Quorum Sensing: Molecular Mechanism and Biotechnological Application.* Springer Singapore.

Kanematsu, H., & Barry, D. M. (2020). A sequence between microfouling and macrofouling in marine biofouling. *Marine Ecology: Current and Future Developments* Monitoring Artificial Materials and Microbes in Marine Ecosystems: Interactions and Assessment Methods., 2 (14): 67–80.

Little, B., & Ray, R. (2002). A perspective on corrosion inhibition by biofilms. *Corrosion,* 58: 424–428.

Lotze, H. K., Lenihan, H. S., Bourque, B. J., Bradbury, R. H., Cooke, R. G., Kay, M. C., … & Jackson, J. B. (2006). Depletion, degradation, and recovery potential of estuaries and coastal seas. *Science*, 312 (5781): 1806–1809.

Mangroves. Plants and algae, Mangroves coasts and shallow water, Mangrove Biodiversity – Smithsonian Ocean Portal. 30 April 2018. https://ocean.si.edu/ocean-life/plants-algae/mangroves

Mittelman, M. W. (2000). Bacterial Biofilms and Bifouling: Translational Research in Marine Biotechnology. In: Vaupel, S. (Ed), *Opportunities for Environmental Applications of Marine Biotechnology* (pp. 3–7). NIH.

Rohwer, F., Seguritan, V., Azam, F., & Knowlton, N. (2002). Diversity and distribution of coral-associated bacteria. *Marine Ecology Progress Series*, 243: 1–10.

Sharma, N., Singh Sodhi, A., & Batra, N. (Eds.). (2021). *Basic Concepts in Environmental Biotechnology* (1st ed.). CRC Press.

Silliman, B. R. (2014). Saltmarshes. *Current Biology*, 24(9): R348.

Stanczak, M. (March 2004). Biofouling: It's not just barnacles anymore, Barnacles and Biofouling Eric R Holm. Integr Comp Biol. 2012 Sep. retrieved 21 May 2012.

Töpke, K. (2023). Salt marshes. Available from www.coastalwiki.org/wiki/Salt_marshes.

Underwood, G. J., & Paterson, D. M. (2003). The importance of extracellular carbohydrate production by marine epipelic diatoms. *Advances in Botanical Research*, 40: 183–240.

Wolińska, A., & Błaszczyk, M. (2020). Microbial mats in hypersaline environments: the upper surface and underlying sediments. *Frontiers in Microbiology*, 11: 2071.

Yadav, K. K., Singh, J. K., Gupta, N., & Kumar, V. (2017). A review of nanobioremediation technologies for environmental cleanup: A novel biological approach. *Journal of Materials and Environmental Sciences*, 8(2), 740–757.

Yuvraj. (2022). Microalgal bioremediation: A clean and sustainable approach for controlling environmental pollution. In: *Innovations in Environmental Biotechnology* (Vol. 1, pp. 305–318). Springer Singapore. doi:10.1007/978-981-16-4445-0_13. ISBN 978-981-16-4445-0.

Zedler, J. B., Bonin, C. L., Larkin, D. J., Varty, A. (2008). Salt Marshes. In: Sven Erik Jørgensen, Brian D. Fath (Eds.), *Encyclopedia of Ecology* (pp. 3132–3141). Academic Press.

5 Bacterial Biofilms and Biofouling

Subhaswaraj Pattnaik and Pradeep Kumar Naik

5.1 INTRODUCTION

In the last few decades, the world has not only witnessed a revolution in antibiotic discovery to tackle several chronic infections and other serious health implications, but has also witnessed the emergence of resistance to last-resort antibiotics leading to serious socio-economic and environmental hazards across the globe (Li et al., 2022). The antibiotic resistance phenomena have the elevated risk of therapeutic limitations with emergence of relapsing infections due to high rate of mortality and morbidity (Huemer et al., 2020). No doubt several environmental factors promote antibiotic resistance in the global market; the adaptive feature of several bacterial pathogens is to regulate the differential expression of antibiotic resistance genes (ARGs) through the acquisition of mobile genetic elements (MGEs) (Larsson and Flach, 2022). Among several biological factors associated with microbes' resistance to antibiotics and other antimicrobial agents are biofilms, which facilitate bacterial survival mechanisms to antimicrobial treatment thereby promoting antibiotic resistance (Johnston et al., 2023). Apart from inferring antibiotic resistance, biofilms are also the root cause of biological fouling with devastating effects on several parameters across the globe. Therefore, we need to understand the basic course of biofilms and biofouling and their long-lasting effects and develop strategic approaches to counter them.

5.1.1 Definition of Biofilms

Biofilm is the assemblage of surface-associated microbial cells encased in an extracellular polymeric substance (EPS) matrix secreted by the encased microbial communities as a part of their arsenal against environmental stress (Donlan, 2002; Zhao et al., 2023). Biofilms are the most dynamic and adaptive mode of microbes in terms of their role in pathophysiological behavior, persistence to host immunity, and survival tactics to withstand the environmental stress conditions (Sentenac et al., 2022). The concept of biofilm was first introduced by J. William Costerton and his colleagues as the adaptive functional attributes of the bacterial communities to thrive on different surfaces in complex environments. The protective biofilm matrix is composed of extracellular polysaccharides (e.g., Alginate), proteins (e.g., Fibrin), extracellular DNA (eDNA), lipids, membrane vesicles, and other biomolecules (Bamford et al.,

DOI: 10.1201/9781003487203-5

2023; Flemming et al., 2023; Nesse et al., 2023). The rigidity, viscosity, and stability of the formed biofilms is conferred by the protective extracellular matrix (ECM) by holding the polymeric constituents and the encased bacterial community together (Jana et al., 2020). The ECM not only acts as the shield against extreme environmental conditions but also promotes the embedded microbial community to respond to the environmental cues accordingly (Srinivasan et al., 2021).

Biofilms have the inherent ability to exhibit collective behavior as showcased by differentiated cells in multicellular individuals and are considered to take an 80% share of the chronic microbial infections including biomedical device-associated infections and tissue implant-associated infections (Penesyan et al., 2021). Biofilms also facilitate the embedded microbial communities for conjugation and transformation for horizontal transfer of ARGs by providing a stable physical environment. As a consequence, biofilms provide a suitable environment for the relative expression of ARGs leading to their resistance towards available antibiotics and other antimicrobial drugs (Berlanga and Guerrero, 2016). Since biofilms enable the encased microbial communities to survive the hostile conditions by evading host immunity and environmental stress (i.e., antimicrobial treatment), biofilms remain a pivotal source of severe hospital-acquired infections with high rate of mortality and morbidity. Apart from public health hazards, biofilms also have a catastrophic influence on economies, industry, and the environment (Allewell, 2016; Sharma et al., 2023).

5.1.2 Characteristics of Biofilms

The highly heterogenous, complex, and dynamic stature of biofilm matrices in pathogenic microorganisms strictly adhere to three characteristic domains: social interactions, physiological heterogeneity, and adaptive measures for protection against stress (Parrilli et al., 2022). The dynamic nature of biofilms and their response to environmental cues remain high as compared to planktonic forms, and as a result biofilms have distinct physicochemical and physiological characteristics in terms of their adhesion capabilities, phenotypic variability, differential metabolic activities, spatial physiological adaptation, differential gene expression, and temporal variation in tolerance to environmental stress (Lichtenberg et al., 2023). Biofilms are also characterized by the unique viscoelastic properties of their matrices, which facilitate the intercellular cohesion among the embedded microbial community for better survival to withstand the host immune system (Wells et al., 2023).

Biofilms also enable the encased microbial community to achieve nutritional acquisition, maintenance of highly specific signalling cascade between the embedded bacterial cells, and horizontal transfer of ARGs for expression of resistance profiles (Valen and Scheie, 2018). Within the biofilm matrix, the relative expression of specific virulence genes are also concomitantly regulated by a highly controlled bacterial communication signal system called quorum sensing (QS). The QS-mediated virulence phenotype expression within biofilm provides a distinctive dimension to the pathogenicity of biofilms (Chandki et al., 2011). The highly dynamic and complex biological systems of bacterial biofilms provide a suitable microenvironment to the encased microbes to sustain in severe environmental circumstances such as extreme temperature, nutritional deficiency, exposure to high radiation, stringent acidic and/or alkaline condition, exposure to heavy metal pollution, hypersaline conditions, and other physiochemical parameters (Parrilli et al., 2022; Patwardhan et al., 2023).

5.1.3 Stages of Biofilm Formation

The establishment and development of biofilm matrix in pathogenic microorganisms is a continuous process of sequential events occurring in each cycle of events characterized by initial attachment of free form planktonic bacteria onto biotic/abiotic surfaces, followed by anchoring phase depicting the adhesion and aggregation process, followed by formation of highly differentiated microcolonies with successive multiplication and cell division, followed by maturation phase with the release of specific signalling molecules, and finally dispersion to establish new biofilm matrices (Sharma et al., 2023) (Figure 5.1). The microbes' attachment to different polymeric surfaces depends on several physicochemical parameters such as repulsive electrostatic, Van der Waals and cohesive chemical forces, cell surface hydrophobicity, and relative expression profile of surface appendages like flagella, pili, and fimbriae (Nandakumar et al., 2013; Achinas et al., 2019). For the adherence to biotic surfaces, pathogenic bacterium *Staphylococcus aureus* and other microbes secrete specific adhesion proteins called cell well-anchored proteins (e.g., microbial surface components recognizing adhesive matrix molecules [MSCRAMMs]) (Vani et al., 2023). The lipopolysaccharides (LPS) and cell wall glyco-polymers of Gram negative and Gram positive microbes also substantially promote the initial adhesion to the different surfaces with differential topography (Ruhal and Kataria, 2021). Apart from biological factors, several surface topographical features such as surface topography,

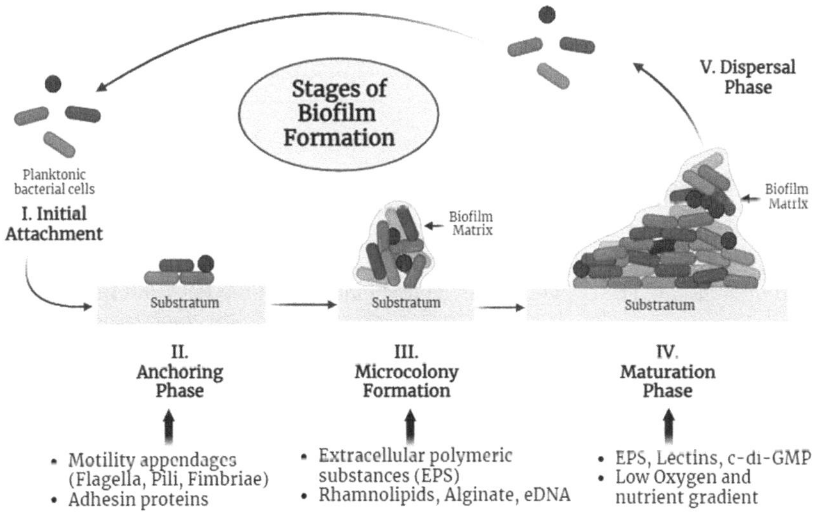

FIGURE 5.1 Schematic representation of the different stages of formation, development, and maturation of biofilms and the factors responsible for the process.

roughness and stiffness of the surface, surface charge density, and wettability also govern the biofilm establishment by promoting initial bacterial adhesion to the surface (Zheng et al., 2021).

5.1.4 Role of Biofilms in Bacterial Pathogenesis

The EPS-rich biofilm matrices in pathogenic microorganisms not only promote biofilm development and maintenance of its stability, but also promote bacterial pathogenicity profile and antimicrobial resistance by triggering persistent infections in clinical healthcare settings (Caldara et al., 2022; Costa et al., 2023). The heterogeneity in biofilm, and the presence of a sub-population of highly persistent persister cells within biofilm matrix, are responsible for the antibiotic resistance leading to more severe chronic infections (Thite et al., 2023). The biomedical device and implant associated chronic infections are generally considered to be the clinical hotspots for the establishment and development of polymicrobial biofilms composed of Gram negative *Pseudomonas aeruginosa*, *Klebsiella pneumoniae*, *Escherichia coli*, etc., and Gram positive *S. aureus*, *Enterococcus faecalis*, etc. (Nandakumar et al., 2013). Several virulence factors associated with pathogenic microorganisms such as stress response regulators, adhesion proteins, QS cascade-derived pathogenic factors, surface- associated proteins, and other virulence factors facilitate the different stages of biofilm establishment and development (Bai et al., 2021). The biofilms are generally associated with the onslaught of several pathophysiological conditions affecting different parts of body systems such as sensory (e.g., otitis media, dental plaque, chronic tonsillitis, contact lenses); respiratory (e.g., cystic fibrosis, respiratory tract infections, pertussis etc.); urinogenital (e.g., urinary tract infections, chronic prostatitis, vaginosis, chronic endometritis etc.); cardiovascular (e.g., infective endocarditis); musculoskeletal (e.g., prosthetic joints, osteomyelitis); and integumentary (e.g., chronic wound infections) (Figure 5.2). The widespread occurrence of bacterial biofilms affecting several body parts with pathophysiological inference is a public health issue and requires urgent attention (Vestby et al., 2021; Abdelhamid and Yousef, 2023).

5.1.5 Biofilms as Potentiators for Antibiotic Resistance

Biofilms provide a sophistically controlled microenvironment for the encased microbial community to survive under different environmental stress conditions including antibiotic treatment. The ecological niche associated with the biofilm microenvironment provides an impetus to the embedded microbes to confer tolerance and resistance towards antibiotic therapies. The stature of tolerance to antibiotics depends on several factors such as biofilm matrix composition, stage of biofilm development, growth conditions, and biofilm-mediated acquisition of foreign genetic elements (Uruén et al., 2021; Roy et al., 2022). Since more than 80% of chronic and recurrent bacterial infections and resistance to antibiotics are mediated by microbes with biofilm mode, biofilms are the key drivers for the microbial community to confer resistance against antibiotics and antimicrobial agents. Among the several factors responsible for biofilm-mediated antibiotic resistance are the capsules or glycocalyx

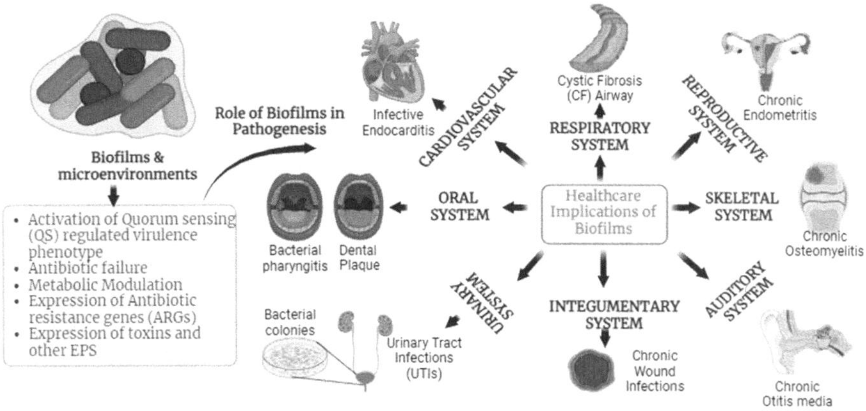

FIGURE 5.2 Schematic representation of the effect of biofilms on bacterial pathogenesis and healthcare implications.

composed of glycoproteins and polysaccharides are the integral component in biofilm architecture and support the survival under extreme environmental conditions (Hall and Mah, 2017; Singh et al., 2017).

The maintenance of metabolic heterogeneity within the biofilm matrices also impede the antimicrobial treatment regimens as the therapeutic procedures act upon the metabolically active bacterial cells thereby decreasing the antimicrobial susceptibility (Schiessl et al., 2019). The genetic adaptation within the biofilm matrix is also essential for reduced antibiotic susceptibility and concomitant increase in the resistance profile. The mutations of specific genes (e.g., *sbmA*: inner membrane transporter; *fusA*: essential elongation factor, etc.) during biofilm development instead of during planktonic stage eventually promote tolerance towards antibiotic treatment. Similarly, the relative expression of several multidrug-resistant regulatory operons within the biofilm matrix stimulates the improved antibiotic tolerance mechanism (Singh et al., 2017; Usui et al., 2023). In addition, the relative presence of specific antibiotic-modifying enzymes such as β-lactamases within the biofilm matrices reduce the antibiotic susceptibility by concomitant degradation of antibiotics and modification of antibiotic targets. Apart from these factors, the integral component of biofilm matrix, eDNA, also promotes substantial alteration in the extracellular environment thereby promoting antibiotic resistance (Hall and Mah, 2017). The integral components of biofilms (i.e. EPS, which includes rhamnolipids, alginates, eDNA, motility factors, etc.) are produced under the regulation of the highly controlled cell-cell signalling cascade known as QS. The QS communication between bacterial cells encased within the biofilm matrix results in the activation of integral components of biofilm, and thus QS signalling cascade-mediated biofilm dynamics play a pivotal role in conferring antibiotic tolerance and resistance (Singh et al., 2017; Zhao et al., 2020). The acquisition of antibiotic resistance in biofilms is often regulated by the active expression of multidrug-resistant efflux pump transporters that not only expel the antimicrobial agents out of the system but also reduce the antibiotic susceptibility (Soto, 2013; Huang et al., 2022).

5.2 CONCEPT OF BIOFOULING

Biofouling, or biological fouling, corresponds to the unwanted accumulation of living organisms (i.e., microorganisms, algae, plants, and small animals) onto different biotic and abiotic surfaces, affecting the functional attributes of such surfaces and/or objects (Ferreira-Vancato et al., 2020; Dobretsov and Rittschof, 2023). The incidence of biofouling is observed in ship and submarine hulls, biomedical devices, water distribution systems through pipework, and industrial setups like paper manufacturing and food processing industries (Bixler and Bhushan, 2012). The concept of biofouling has received considerable attention in the last few decades as approximately 0.17×10^4 number of species comprising more than 0.4×10^4 number of organisms are directly or indirectly associated with the occurrence of biofouling. A highly diverse group of bacteria (e.g., *Pseudomonas* sp., *Vibrio* sp., etc.) and diatoms (e.g., *Achnantes brevipes*, *Amphiprora paludosa*, *Licmophora abbreviata*, *Amphora coffeaeformis*, etc.) are regarded as the primary colonizers to colonize the surface and form biofilm. Spores of macroalgae (e.g., *Ulothrix zonata*, *Entermorpha intestinalis*, etc.) and protozoa (e.g., *Vorticella* sp., *Vaginicola* sp., *Zoolhamnium* sp., etc.) constitute the secondary colonizers that harbour the different chemicals secreted by primary colonizers. Several groups of invertebrate larvae (e.g., Tunicates, Molluscs, Coelenterates, Crustaceans, Polychaetes, Bryozoa, etc.) are considered as the tertiary macrofoulers responsible for the majority of macrofouling and associated consequences (Shan et al., 2011; Nakano and Strayer, 2014). These diverse taxa have the inherent ability to form a biofouling community in both freshwater as well as marine ecosystems causing widespread implications.

The biofouling occurs either at the micro level (i.e., formation and development of biofilms by microorganisms) or the macro level (i.e., involvement of macroorganisms) based on the involvement of living organisms in the biofouling event. Based on the distinct fouling properties, fouling organisms are either calcareous (hard) type (e.g., Barnacles, Bryozoans, Molluscs, Polychaetae, Zebra mussels) or soft type (e.g., Seaweed, algae, bacteria etc.). The biofilm formation on biomedical devices (i.e., termed as medical biofouling) is exclusively associated with microorganisms associated microfouling whereas the biofouling in industrial sectors and marine environment involves not only biofilm formation by microorganisms but also the combination of both macrofouling as well as inorganic fouling (Bixler and Bhushan, 2012; Gu, 2012). The process of biofouling begins with the formation of biofilms. However, not every biofilm results in biofouling. Biofilms can lead to industrial and biomedical biofouling only beyond the "threshold of interference" (Flemming, 2020). The development of biofouling involves a series of sequential events starting from (i) formation of conditioning film by adsorption of dissolved organic matters onto the surface; (ii) conditioning film further promotes the colonization of primary colonizers (including bacteria and diatoms) and encourages the formation of multispecies microbial biofilms (microfouling); (iii) pioneer microbial community subsequently promotes settlement of secondary colonizers by producing several classes of chemical cues; and (iv) eventual settlement of invertebrate larvae and macroalgal zoospores onto the organic surfaces to form macroscopic fouling community (macrofouling) (Martin-Rodriguez et al., 2015). Several abiotic and biotic factors such as temperature, geographical distribution, influx of solar radiation, species diversity, and environmental pollution explicitly drive the formation of the biofouling community. The manifestation of biofilms (microfouling) is considered as an influential step in the progress of biofouling as the biofilms protect the harboring microorganisms from hostile environments and allow them to continue the colonization process in order to cross the threshold interference for formation of biofouling community (Flemming, 2011).

5.2.1 Role of Biofilms in Biofouling

Among the several factors, bacterial biofilms remain the primary influencers of establishing biological fouling (biofouling) in the majority of industrial systems such as oil and gas distribution, drinking water distribution, nuclear power stations, geothermal systems, desalination plants, etc., by promoting mechanical blockages, inducing microbial corrosion, producing contamination, and causing economic loss (Barton et al., 2023). As evident from biofouling succession stages, the bacterial settlement followed by biofilm formation remains the basic step. Once the primary colonizers (i.e., bacteria over the surfaces) get settled, in the subsequent progression stages, other organisms starting from microalgae to marine invertebrate larvae undergo settlement over these surfaces. In the final stage, macrofouling by barnacles, ascidians, and mussels takes place (Rao, 2022). The success of biofouling with catastrophic influence largely depends upon the successful establishment of bacterial biofilms over various biotic and abiotic surfaces. Once the biofilms get established, they provide novel platforms for further colonization and settlement of other organisms leading to fouling with significant economic implications (de Carvalho, 2018).

5.2.2 Implications of Biofouling

The increased incidence of biofouling has a profound impact on socio-economics, healthcare systems, and environmental sustainability. Since aquaculture infrastructure involves the regular use of submerged components for substantial period, the development of biofouling on these surfaces has a detrimental effect on aquaculture operations, which invariably results in loss of production (approximately 10–30%) and increase in operational cost (i.e., approximately 30%) leading to significant economic implications (Fitridge et al., 2012; Lacoste and Gaertner-Mazouni, 2014). It has also been observed that the intensity of biofouling on different marine plastic surfaces lowers the natural photochemical degradation process thereby promoting environmental imbalance in the marine atmosphere (Nelson et al., 2021). In addition, the development of biofouling also alters the polarity of marine microplastics by minimizing their physical integrity. Once the physical integrity of the polymer is lost, the traditional bioremediation strategies fail to remediate the environmental pollutants leading to environmental pollution(Melkebeke et al., 2020). Biofouling also influences operational management of maritime shipping industries as the development of fouling over ship hulls results increased operational costs (Schultz et al., 2011; Davidson et al., 2021).

Increased operational and management costs also increase fuel consumption, which concomitantly increases the release of greenhouse gases leading to environmental deterioration (Song et al., 2020). The occurrence of biofouling also greatly affects the functional attributes and performance of the urbines that are used to convert the kinetic energy of the tidal currents into valuable mechanical power (Stringer and Polagye, 2020). The increased incidence of membrane fouling in the reverse osmosis (RO)-based water treatment sectors also influence shortening of membrane lifespan with increased operational cost (Jamieson and Leterme, 2021). Biofouling is also prominent in food and beverages industries. Biofouling in the food industries leads to food product contamination with decreased shelf-life, corrosion of food equipment, food spoilage, and most importantly the promotion of foodborne diseases (e.g., mild gastroenteritis, hemorrhagic colitis, severe liver abnormalities, meningitis, thrombotic thrombocytopenic purpura, etc.) (Samrot et al., 2021). Biofouling assemblages also have catastrophic influence on fish health as the abundance of Cnidarians taxa in the fouling congregation produce several types of toxins with subsequent health hazards (Fletcher et al., 2023). Thus, it is imperative to look for effective biofouling control measures as well as remedial strategies.

5.3 STRATEGIES TO CONTROL BACTERIAL BIOFILMS

Scientific research is in progress with the aim to eradicate biofouling by targeting different stages of biofilm progression starting from preventing the formation of biofilms to the elimination of mature biofilms (Bi et al., 2021; Razdan et al., 2022). Strategies include interference with QS signalling pathways, mechanism of adhesion, and disruption of various functional components of biofilm matrices such as eDNA, extracellular proteins, lipopolysaccharides, exopolysaccharides, and secondary messengers (e.g., cyclicGMP). Moreover, exploring the physical and chemical heterogeneity within the biofilm matrix and targeting the host-pathogen interactions are also instrumental in the therapeutic management of biofilms (Koo et al., 2017; Roy et al., 2018). In the process of development of potential therapeutic agents targeting biofilm mitigation, it is also important to consider the efficacy of the therapeutic approach by efficient management of limitations such as relative cytotoxicity, failed response towards resistance, and recurrent infections (Abdelhamid and Yousef, 2023).

5.3.1 Natural Product-Based Antibiofilm Strategies

In the last few decades, the scientific world has witnessed the pivotal reflection of natural products in the design and development of antibiofilm agents against the recalcitrant biofilm matrices. The natural products derived from plants, microbes, animals, etc., are being explored and scientifically investigated for their prospective role in the therapeutic management of the growth and development of biofilms and associated infections as they are considered as eco-friendly and relatively safe to use (Melander et al., 2020; Guo et al., 2023). Among the therapeutic targets, inhibition of biofilm matrix formation, inhibition of virulence phenotype production by interfering with QS signalling cascade, and the suppression of bacterial cell adhesion are regularly employed as potential antibiofilm agents derived from natural sources (Lu et al., 2019) (Figure 5.3).

Among the natural products, plant-derived bioactive components are widely employed as putative drug candidates as antibiofilm agents. With the rich diversity of natural products and the variety of sources from which they are obtained is imperative that the possible antibiofilm effects could also vary in terms of therapeutic targets (Mishra et al., 2020). Medicinal plants are a rich source of bioactive secondary metabolites of pharmacological relevance and have been shown to be efficient in management of biofilm-related infections. Bioactive compounds such as Ginkgloic acid, Cinnamaldehyde, Phloretin, Hordenine, Myricetin, Quercetin, etc., obtained from different parts of plants, are being investigated for their promising roles in biofilm mitigation. These natural compounds can also be used in combination with antimicrobial agents like antibiotics for improved therapeutics and decreased selective pressure (Lu et al., 2019; Ghosh et al., 2023). The marine ecosystem harbors a diverse array of microorganisms including actinobacteria, bacteria, and fungi with the ability to produce a wide spectrum of bioactive natural products such as polyketides (e.g., Flavipesin A), alkaloids (e.g., Penicillenols A2, B1), terpenoids (e.g., Aszonapyrone A), peptides (e.g., Nesfactin), etc., with diverse range of pharmacological properties with special reference to antimicrobial and antibiofilm activities (Deng et al., 2022; Wibowo et al., 2023). Table 5.1 lists the plant- and microbe-derived bioactive compounds reported for their biofilm inhibition potential against pathogenic microorganisms (Table 5.1).

5.3.2 Nanoformulations for Biofilm Mitigation

The intrinsic tolerance and resistance shown by recalcitrant biofilms substantially limits the therapeutic efficacy of antimicrobial agents. Further, the bioavailability and biodegradability issues also limit their widespread applications as therapeutic agents. In this context, novel drug molecules may be developed as potential alternative therapies to tackle the intricacies of biofilm-mediated infections. The intervention of nanotechnology could also improve the therapeutic efficacy of biofilm inhibition (Sahli et al., 2022). The development of functionalized nanostructured surfaces (e.g., liposomes, dendrimers, carbon nanotubes etc.), synthesis of nanoparticles (NPs) (e.g., polymeric NPs, metal-based NPs, solid-lipid NPs, etc.), preparation of nanocomposites (e.g., drug conjugated nanomaterials), etc., are currently being employed to overcome the limitations associated with current therapeutic approaches to tackling biofilms (Elbourne et al., 2019; Kumar et al., 2023). Novel physicochemical properties such as high surface area-to-volume ratio, efficient optical absorption, improved tissue penetration, reduced adverse effects, increased therapeutic propensity, etc., facilitate improved biomedical applications including biofilm-associated hospital-acquired infections with special emphasis on biofilm hindrance, its control measures and eradication (Asare et al., 2022; Kumar et al., 2023; Mohanta et al., 2023).

Green nanotechnology (i.e., synthesis of NPs using microbial and/or plant extracts) has helped the scientific community design green synthesized NPs and

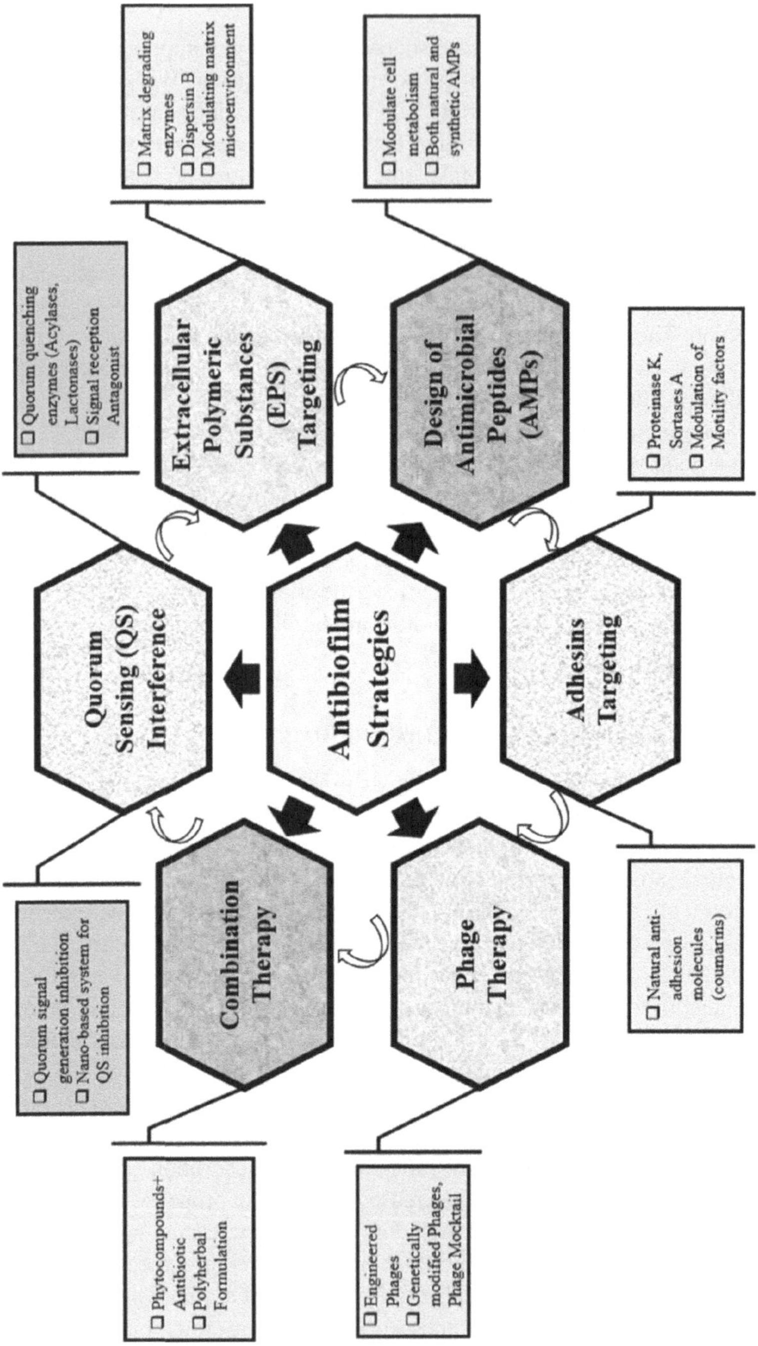

FIGURE 5.3 Schematic overview of the different therapeutic strategies employed for biofilm mitigation and its eradication.

TABLE 5.1
List of Plants, and Microbes Derived Bioactive Compounds Reported for Their Promising Biofilm Inhibition Potential against Pathogenic Microorganisms

A. Plants derived Phytocompounds as antibiofilm agents						
Sl. No.	**Chemical class**	**Phytocompound**	**Source of plants**	**Target microorganism**	**Biological effects**	**Reference**
1.	**Alkaloids**	Piperine	*Piper* sp.	*Pseudomonas aeruginosa*	Mitigates biofilm mechanics by modulating the quorum sensing (QS) signalling and surface hydrophobicity	Das et al., 2023
2.		Reserpine	*Rauwolfia serpentina*	*Staphylococcus aureus*	Significantly inhibits biofilm formation by altering exopolysachharides production, and down regulation of QS mediated virulence genes	Parai et al., 2020
3.	**Anthocyanins**	Carboxypyranocyanidin-3-O-glucoside	-	*S. aureus*, *P. aeruginosa*	Regulates QS mediated biofilm responses	Coelho et al., 2021
4.		Malvidin-3-glucoside	Anthocyanin rich extract of plants	*S. aureus*	Inhibits biofilm formation by modulating the adhesion properties	Silva et al., 2022
5.	**Anthraquinones**	2-ethylanthraquinone	Pure	*Candida albicans*	Mitigates biofilm dynamics by reduction in EPS production and down regulation of biofilm associated genes	Song et al., 2022

(*continued*)

TABLE 5.1 (Continued)
List of Plants, and Microbes Derived Bioactive Compounds Reported for Their Promising Biofilm Inhibition Potential against Pathogenic Microorganisms

A. Plants derived Phytocompounds as antibiofilm agents

Sl. No.	Chemical class	Phytocompound	Source of plants	Target microorganism	Biological effects	Reference
6.	**Carotenoids**	Lutein	-	*P. aeruginosa*	Inhibits biofilm forming capabilities without affecting growth kinetics	Mahavy et al., 2022
7.	**Coumarins**	4-Farnesyloxycoumarin	Pure	*P. aeruginosa* PAO1	Controls the biofilm formation by modulating the QS gene regulation	Tajani et al., 2023
8.		6-Methylcoumarin	Pure	*P. aeruginosa* PAO1	Shows promising antibiofilm potential	Bajire et al., 2021
9.	**Flavonoids**	Morin	*Psidium guajava*	*S. aureus*	Inhibits biofilm dynamics by targeting SarA	Chemmugil et al., 201921
10.		Methoxyisoflavan	*Trigonella stellata*	*P. aeruginosa*	Exhibits biofilm mitigation by regulation of QS behavior	Naga et al., 2022
11.		Quercetin	Pure	Methicillin resistant *S. aureus*	Regulates biofilm dynamics by modulating the expression profiles of SarA gene	Liu et al., 2023
12.	**Phenolic acids**	Salicylic acid	*Salix* sp.	*P. aeruginosa*	Intereference in biofilm mechanics by regulating the QS regulatory network	Ahmed et al., 2019
13.		Quinic acid	*Lonicerae Japonicae Flos*	*P. aeruginosa*	Inhibits biofilm formation by regulating the production of EPS	Lu et al., 2021

14.	**Phenylpropanoids**	Cinnamaldehyde	Cinnamol essential oil	*Enterococcus faecalis*	Regulates biofilm dynamics by down regulation of associated genes	Akshaya et al., 2023
15.		Curcumin	*Curcuma longa*	*S. aureus*	Regulates surface wound associated biofilm dynamics	Gao et al., 2022
16.	**Tannins**	Punicalagin (Ellagitannins)	*Punica granatum*	*S. aureus* ATCC 29213	Exhibits promising biofilm inhibitory activities without affecting the bacterial growth	Xu et al., 2023
17.	**Terpenes and Terpenoids**	Terpinen-4-ol	*Pandanus odorifer* essential oil	*S. aureus, and Klebsiella pneumoniae*	Regulates biofilm formation by critically affecting the EPS production	Cheruvanachari et al., 2022
18.		2-Phenyl ethyl methyl ether (PEME)	*P. odorifer* essential oil	*S. aureus*	Controls biofilm mechanics by reducing the EPS production	Cheruvanachari et al., 2023
19.		Nerolidol	*Myrocarpus frondosus*	*P. aeruginosa, K. pneumoniae, S. aureus*	Controls biofilm formation	de Moura et al., 2021
20.		Linalool	*Hedychium larsenii*	*Streptococcus pyogenes*	Destabilizes biofilm matrix formation	Praseetha et al., 2023

B. Microbes derived Bioactive compounds as antibiofilm agents

Sl. No.	Chemical class	Bioactive compounds	Source of Microoganism	Target microorganism	Biological effects	Reference
21.	**Alkaloids**	1,3,6-trihydroxy-7-methoxy-9H-xanthen-9-one	*Penicillium citrinum-314*	*P. aeruginosa*	Exhibit biofilm inhibition without affecting the bacterial growth	Abdel Razek et al., 2020
22.		Coniine/(S)-2-propylpiperidine	*Streptomyces* sp.	*Vibrio harveyi* and *A. hydrophila*	Regulates biofilm formation	Miller et al., 2022
23.		Macrocidin Z, Dihydromacrocidin Z	*Phoma* sp.	*S. aureus*	Inhibit the growth of biofilm matrix	Treiber et al., 2021

(*continued*)

TABLE 5.1 (Continued)
List of Plants, and Microbes Derived Bioactive Compounds Reported for Their Promising Biofilm Inhibition Potential against Pathogenic Microorganisms

B. Microbes derived Bioactive compounds as antibiofilm agents

Sl. No.	Chemical class	Bioactive compounds	Source of Microoganism	Target microorganism	Biological effects	Reference
24.	**Anthraquinones**	Anthraquinone-2-carboxlic acid, and Rhein	*Kitasatospora albolonga R62*	Methicillin resistant *S. aureus*	Biofilm disruption by promoting dispersion	Song et al., 2021
25.	**Carotenoids**	Lutein	*Chlorella pyrenoidosa*	*P. aeruginosa* PAO1	Regulates biofilm formation by modulating the QS regulatory mechanisms	Sampathkumar et al., 2019
26.	**Glycolipids**	Biosurfactant	*Lactobacillus rhamnosus*	*Bacillus subtilis, Escherichia coli, P. aeruginosa, S. aureus*	Regulates bacterial adhesion capabilities and biofilm eradication	Patel et al., 2021
27.	**Peptides**	Epicotripeptin	*Epicoccum nigrum* M13	*B. subtilis, S. aureus*	Shows moderate biofilm inhibition	Qader et al., 2021
28.		Phragamides A and B, Tenuazonic acid	*Alternaria alternata* 13A	*B. subtilis, S. aureus*	Exhibit promising biofilm inhibition	Qader et al., 2021
29.	**Terpenoids**	2,6-Di-tert-butyl, 1,4-benzoquinone	*Nocardiopsis synnemataformans* RMN 4	*Proteus mirabilis*	Exhibits biofilm inhibition by altering biofilm metabolic activities and reduction of EPS production	Govindan et al., 2023
30.	**Macrolide Antibiotic**	Emycin-E	*Streptomyces griseus* RG1011	*S. aureus*	Modulates biofilm forming ability of pathogenic microorganism	Sankhwar et al., 2023

investigate their biological activities. For example, the flower extract of Aparjita (*Clitoria ternatea*) was employed as reducing agent for the green synthesis of zinc oxide nanoparticles (ZnO NPs). The synthesized ZnO NPs exhibited promising influence on the eradication of oral biofilms associated with oral pathogenic bacteria, *Porphyromonas gingivalis*, and *Alcaligenes faecalis* by reducing the production of biofilm components (i.e., EPS) (Lahiri et al., 2022). Similarly, polymeric NPs such as carvacrol-loaded chitosan NPs also improved the biofilm inhibition efficacy of carvacrol by facilitating penetration through the biofilm matrices in multidrug resistant *P. aeruginosa*. Dextran-based polymeric NPs also improve the therapeutic efficacy of Tobramycin by neutralization of the ionic interactions associated with biofilm matrices (Bernal-Mercado et al., 2022; Blanco-Cabra et al., 2022). In addition, these metal-based and polymeric NPs also hinder the QS signalling cascade thereby modulating the phenotypic production of virulence factors linked with biofilm formation and maturation in the majority of pathogenic microorganisms. Interference in QS regulatory pathway alters biofilm dynamics and thus may influence biofilm mitigation. Moreover, NPs provide avenues to be used in combination with bioactive phytochemicals for the design of novel nanocomposites with enhanced photothermal and photodynamic properties and could be instrumental in the inhibition of biofilm matrices (Lin et al., 2021; Alavi et al., 2023).

In the design and development of novel nanoformulations, encapsulation of Tirapazamine into gold NP-hybrid liposomes aggravates biofilm mitigation by creating a suitable hypoxia-induced microenvironment (Zou et al., 2023). Similar to liposomes, niosomes are also categorized as prolific nanocarriers for the encapsulation of therapeutic drugs and improve their efficacy. In this context, niosome= encapsulated antibiotic Imipenem improved the efficacy of Imipenem in preventing biofilm formation in multi-drug resistant (MDR) *Staphylococcus epidermidis* (Piri-Gharaghie et al., 2022). In an earlier report, hybrid nanoformulations comprising Graphene oxide-silver NPs coated nickel-titanium alloy were designed for effective mitigation of biofilm dynamics in *Streptococcus mutans* (Pipattanachat et al., 2021). More recently, curcumin-ZnO nanocomposite and gold nanocluster-decorated graphitic carbon nitride nanocomposites were spatially designed to dismantle the biofilm matrices in MDR *P. aeruginosa* and *Escherichia coli,* respectively (Luo et al., 2023; Shome et al., 2023).

5.4 ANTIBIOFOULING STRATEGIES

In addition to developing the antibiofilm agents targeting the recalcitrant biofilms the scientific community should also focus on developing long-term antibiofouling approaches for benefits in terms of health, economy, and the environment. As discussed earlier, biofouling remains a challenging situation for the majority of implanted biomedical devices including biosensors. Antibiofouling approaches could be employed either by physical and/or chemical methods. The physical antibiofouling methods include the exploration of several physical forces with an aim to modify surface properties. Meanwhile, chemical mode of antibiofouling employ the use of antifouling paints with biocides including copper, silver, thiocyanate, Copper powder, Irgarol

1051, Zinc pyrithione, and Tributyltin (TBT) (Gizer et al., 2023). The non-toxic antifouling coatings include hydrophobic and hydrophilic coatings. The hydrophobic coatings are based on low surface and low friction energies, which provide a suitable environment for hydrophobic surfaces that in turn prohibit the growth of larger organisms of biofouling importance. The hydrophilic coatings rely on the amount of hydration, which in turn catastrophically increases the energy needed for the organisms to attach to the surface. Among the different hydrophobic coatings used for antifouling approaches, polydimethylsiloxane (PDMS) is considered to be industrially relevant and is widely used in biofouling management (Verma et al., 2020). Zwitterionic hydrophilic coatings such as glycine betaine and sulfobetaine are also regularly used in benzoxazine resins in for antibiofouling strategies (Zhao J et al., 2023). Several mechanisms of action such as modification of surface topography, use of QS inhibitors, alteration of metabolic pathways, and prevention of adsorption to prevent biofilm formation are employed by the majority of biobased antifouling coatings (Kyei et al., 2020). In recent years, an encapsulated dichlorooctylisothiazolinone (DCOIT) coating has been developed and is considered to be influential in characteristic reduction of algal fouling over different marine surfaces without affecting the coral settlement (Roepke et al., 2022). The marine antifouling coating market has taken a large share in the marine market in the last few years (with the market value of $5.91 billion in 2021) with the Asia-Pacific region considered to be the fastest growing market in the growth of antifouling coatings of industrial relevance. The antifouling market is expected to reach $13.18 billion by 2031, with a steady compound annual growth rate (CAGR) of 8.48%.

The prevention and mitigation of biofouling could be achieved either through active and/or passive approaches. Active strategies include the use of external stimuli-responsive approaches including temperature, pH, electrical, and magnetic field for antibiofouling applications. The use of biomimicking natural materials (e.g., chitosan, hyaluronic acid, gelatin, collagen etc.), hydrophilic constituents (e.g., polyethylene glycol, hydrogel), superhydrophobic materials, drug-eluting materials (e.g., nitric oxide), and micropatterning of substances (for both prevention of adhesion as well as removal of attached cells from surface) are considered as passive approaches to tackling biofouling (Damodaran and Murthy, 2016; Xu and Lee, 2020). Since physicochemical properties such as surface hydrophilicity, roughness, and surface charge are impeccable in biofouling development, modification of these surface properties by coating or grafting of hydrophilic monomers (e.g., polyethylene glycol (PEG)ylated materials, polyethylenimine (PEI), sericin, branched poly(amido amine) (PAMAM), aminoacids, etc.) could provide strategic solution to surface biofouling by resisting the initial attachment of biofoulants (Zhang et al., 2022). Apart from hydrophilic surface modifications, hydrophobic as well as amphiphilic surfaces could also be employed to discourage the settlement of biofoulant organisms. Recent research on controlling surface topography by employing 3D surfaces could also influence biofouling mechanisms. More recently, the use of engineered polymer coatings for their antiadhesion properties and spatial use of antimicrobial surface-coated materials for prevention and mitigation of biofouling organisms have been explored (Krishnan et al., 2008; Joardar and Dutta, 2023).

Due to the catastrophic consequences Schiff-based compounds through naturally derived Tobramycin and protocatechualdehyde (obtained from *Streptomyces* and black fern, respectively) were synthesized and a self-renewal Schiff-based metal composite coating was designed. These naturally derived Schiff-based metal composites exhibited promising antiadhesion properties by generation of antimicrobial substances thereby affecting the colonization of bacterial and algal biofoulants (Liu et al., 2023). Among the explored antibiofouling techniques, enzymatic degradation of exopolymeric materials could provide an alternative approach. A consortium of carbohydrases could be employed for specific targeting of heterogenous polysaccharides associated with matrix formation (Kaur et al., 2022).

5.5 RECENT TRENDS AND NOVEL AVENUES IN BIOFILM MITIGATION AND ANTIBIOFOULING STRATEGIES

As discussed earlier, the tendency of microbes to form biofilms and the subsequent assemblage of other biofoulants over the surfaces to confer biofouling have resulted in severe consequences (Shrestha et al., 2022). Due to the widespread negative effects of biofilms and biofouling, it is imperative to look for alternative approaches to prevent the formation of and successful removal of biofilm matrices.

The therapeutic and industrial management of biofilms needs to be explored in the postantibiotic era. As alternatives to conventional antibiotics and synthetic derivatives, plant-derived natural products (e.g., phytochemicals, essential oils, etc.) may be useful for effective remediation of biofilms on a large-scale basis (Pattnaik et al., 2023). In the therapeutic regulation of bacterial biofilms, several aspects such as degradation of EPS, regulation of QS cascade, and modulation of biofilm metabolism are taken into consideration (Abdelhamid and Yousef, 2023). In recent times, there has been more evidence reported on the successful use of a combination of several antibiotics together or pharmacologically relevant phytochemicals with antibiotics for effective remediation of biofilm dynamics depicting their potential synergistic activities (Krukiewicz et al., 2022). While the majority of research is oriented towards engineered surface properties with an aim to provide several surface modification strategies to control the development of biofilms over biomedical devices (Dziadek et al., 2021; Krukiewicz et al., 2022), alternative therapies employing enzymes, QS inhibitors, antimicrobial peptides (AMPs), and other bioactive phytochemicals for targeting bacterial biofilms have shown therapeutic efficacy. With this aim, smart delivery platforms could be designed for effective delivery of these agents for improving therapeutic efficacy. Among the described drug delivery platforms, novel nanoplatforms (metal-based NPs, nanocomposites, polymeric NPs, carbon-based NPs, etc.) could be spatially designed for target-based drug delivery with long-term therapeutic effects (Bi et al., 2021).

In the development of potent antibiofilm strategies, the electrochemical approach aims to mitigate the biofilm matrices by spatially altering the biofilm permeability through destabilization of electron transport chain on bacterial biofilm membrane by administration of lower dose of antimicrobial agents in a weak electric field

(Srinivasan et al., 2021; Zhao et al., 2023). In addition, the implementation of antimicrobial photodynamic therapy (aPDT) to control biofilm dynamics by producing highly reactive oxygen species (ROS), which concomitantly produce lethal effects on biomolecular targets including biofilm components such as proteins, lipids, and nucleic acids (Martins Antunes de Melo et al., 2021; Srinivasan et al., 2021), may be used. Photothermal therapy (PTT) has also gained attention as a potent non-invasive approach to tackling biofilms by promoting photosensitizers (e.g., toluidine blue O, indocyanine green, etc.) and/or photothermally active nanomaterials to absorb near infrared (NIR) light leading to generation of heat for biofilm eradication (Beytollahi et al., 2017; Kaur et al., 2021).

Apart from biofilm eradication strategies, more environmentally friendly approaches need to be explored to control biological fouling in industry. Since surface properties and particle charges play a pivotal role in biofouling, electrokinetic studies using external electrical fields can be used to modulate surface properties and thereby control biofouling (Gule et al., 2016). Since marine biofouling has the greatest role to play in the deterioration of aquatic ecosystem across the globe, strategic action plans should be undertaken to control biofouling hazards. These include the active exploration of natural compounds (e.g., antimicrobial peptides, QS inhibitory compounds, industrially relevant enzymes), use of disinfectants, routine exposure to radiation therapies, employment of advanced cleaning technologies, surface engineering technologies for modification of surfaces using antimicrobial surface coatings, and use of spatial hydrodynamic and acoustic cavitation (Romeu and Mergulhão, 2023). Among the surface modification strategies using surface coatings, polymer-based sustainable coatings with respect to their self-renewability, switchability, and biomimetic properties could be considered for biofouling control (Qiu et al., 2022).

In the development of novel strategies targeting biofilms and biofouling, the spatial use of AMPs could also provide novel avenues in controlling biofouling hazards due to their unique physicochemical properties and structural diversity (Patra et al., 2022). The active exploration of antimicrobial nanomaterials in the management of bacterial biofilms and biofouling also add a new dimension to current inhibition approaches. Since the antimicrobial-loaded nanomaterials prevent the initial adhesion of colonizing biofoulant organisms on different surfaces, they are an eco-friendly approach to tackling the nuisances of biofilm and biofouling (Samal et al., 2023).

5.6 CONCLUSION

The catastrophic role of biofilms and biological fouling in animal health, aquatic health, industrial progress, economies, pharmaceutical productivity, and environmental sustainability have encouraged the scientific community to take necessary measures to control the hazardous effects of these processes. While strategic plans are being considered for biofilm eradication and biofouling control, more intensive and target-oriented approaches need to be discovered for complete control of biofilm-associated hazards. With the use of advanced bioengineering methodologies, potential control measures could be developed with the cooperation of the scientific community, funding agencies, government bodies, international policy makers, and environmental agencies.

ACKNOWLEDGMENTS

The author would like to acknowledge DST-SERB, Govt. of India for providing financial support through the N-PDF programme (Ref. PDF/2021/001260).

REFERENCES

Abdel Razek, M.M.M., Moussa, A.Y., El-Shanawany, M.A., and Singab, A.N.B. 2020. A new phenolic alkaloid from *Halocnemum strobilaceum* endophytes: Antimicrobial, antioxidant and biofilm inhibitory activities. *Chem. Biodivers.*, 17(10), e2000496. https://doi.org/10.1002/cbdv.202000496

Abdelhamid, A.G., and Yousef, A.E. 2023. Combating bacterial biofilms: Current and emerging antibiofilm strategies for treating persistent infections. *Antibiotics.*, 12(6), 1005. https://doi.org/10.3390/antibiotics12061005

Achinas, S., Charalampogiannis, N., and Euverink, G.J.W. 2019. A brief recap of microbial adhesion and biofilms. *Appl. Sci.*, 9(14), 2801. https://doi.org/10.3390/app9142801

Ahmed, S.A.K.S., Rudden, M., Smyth, T.J., Dooley, J.S.G., Marchant, R., and Banat, I.M. 2019. Natural quorum sensing inhibitors effectively down regulate gene expression of *Pseudomonas aeruginosa* virulence factors. *Appl. Microbiol. Biotechnol.*, 103, 3521–3535. https://doi.org/10.1007/s00253-019-09618-0

Akshaya, B.S., Premraj, K., Iswarya, C., Muthusamy, S., Ibrahim, H.M., Khalil, H.E., et al. 2023. Cinnamaldehyde inhibits *Enterococcus faecalis* biofilm formation and promotes clearance of its colonization by modulation of phagocytes *in vitro*. *Microb. Pathog.*, 181, 106157. https://doi.org/10.1016/j.micpath.2023

Alavi, M., Li, L., and Nokhodchi, A. 2023. Metal, metal oxide and polymeric nanoformulations for the inhibition of bacterial quorum sensing. *Drug. Dis. Today.*, 28(1), 103392. https://doi.org/10.1016/j.drudis.2022.103392

Allewell, N.M. 2016. Introduction to biofilms thematic minireview series. *J. Biol. Chem.*, 291(24), 12527-12528. https://doi.org/10.1074/jbc.R116.734103

Asare, E.O., Mun, E.A., Marsili, E., and Paunov, V.N. 2022. Nanotechnologies for control of pathogenic microbial biofilms. *J. Mater. Chem. B.*, 10, 5129–5153. https://doi.org/10.1039/D2TB00233G

Bai, X., Nakatsu, C.H., and Bhunia, A.K. 2021. Bacterial biofilms and their implications in pathogenesis and food safety. *Foods.*, 10(9), 2117. https://doi.org/10.3390/foods10092117

Bajire, S.K., Jain, S., Johnson, R.P., and Shastry, R.P. 2021. 6-Methylcoumarin attenuates quorum sensing and biofilm formation in *Pseudomonas aeruginosa* PAO1 and its applications on solid surface coatings with polyurethane. *Appl. Microbiol. Biotechnol.*, 105, 8647–8661. https://doi.org/10.1007/s00253-021-11637-9

Bamford, N.C., MacPhee, C.E., and Stanley-Wall, N.R. 2023. Microbial Primer: An introduction to biofilms – what they are, why they form and their impact on built and natural environments. *Microbiol.*, 169(8), 001338. https://doi.org/10.1099/mic.0.001338

Barton, F., Spencer, B.F., Tartèse, R., Graham, J., Shaw, S., Morris, K., et al. 2023. The potential role of biofilms in promoting fouling formation in radioactive discharge pipelines. *Biofouling*, 39(8), 785–799. https://doi.org/10.1080/08927014.2023.2269532

Berlanga, M., and Guerrero, R. 2016. Living together in biofilms: The microbial cell factory and its biotechnological implications. *Microb. Cell. Fact.*, 15, 165. https://doi.org/10.1186/s12934-016-0569-5

Bernal-Mercado, A.T., Juarez, J., Valdez, M.A., Ayala-Zavala, J.F., Del-Toro-Sánchez, C.L., and Encinas-Basurto, D. 2022. Hydrophobic chitosan nanoparticles loaded with carvacrol against *Pseudomonas aeruginosa* biofilms. *Molecules*, 27(3), 699. https://doi.org/10.3390/molecules27030699

Beytollahi, L., Pourhajibagher, M., Chiniforush, N., Ghorbanzadeh, R., Raoofian, R., Pourakbari, B., et al. 2017. The efficacy of photodynamic and photothermal therapy on biofilm formation of *Streptococcus mutans*: An *in vitro* study. *Photodiagnosis. Photodyn. Ther.*, 17, 56–60. https://doi.org/10.1016/j.pdpdt.2016.10.006

Bi, Y., Xia, G., Shi, C., Wan, J., Liu, L., Chen, Y., et al. 2021. Therapeutic strategies against bacterial biofilms. *Fundamental. Res.*, 1(2), 193–212. https://doi.org/10.1016/j.fmre.2021.02.003

Bixler, G.D., and Bhushan, B. 2012. Biofouling: Lessons from nature. *Phil. Trans. R. Soc. A.*, 370, 2381–2417. https://doi.org/10.1098/rsta.2011.0502

Blanco-Cabra, N., Movellan, J., Marradi, M., Gracia, R., Salvador, C., Dupin, D., et al. 2022. Neutralization of ionic interactions by dextran-based single-chain nanoparticles improves tobramycin diffusion into a mature biofilm. *npj Biofilms. Microbiomes.*, 8, 52. https://doi.org/10.1038/s41522-022-00317-9

Caldara, M., Belgiovine, C., Secchi, E., and Rusconi, R. 2022. Environmental, microbiological, and immunological features of bacterial biofilms associated with implanted medical devices. *Clin. Microbiol. Rev.*, 35(2), e0022120. https://doi.org/10.1128/cmr.00221-20

Chandki, R., Banthia, P., and Banthia, R. 2011. Biofilms: A microbial home. *J. Indian. Soc. Periodontol.*, 15(2), 111–114. https://doi.org/10.4103/0972-124X.84377

Chemmugil, P., Lakshmi, P.T.V., and Annamalai, A. 2019. Exploring Morin as an anti-quorum sensing agent (anti-QSA) against resistant strains of *Staphylococcus aureus*. *Microb. Pathog.*, 127, 304–315. https://doi.org/10.1016/j.micpath.2018.12.007

Cheruvanachari, P., Pattnaik, S., Mishra, M., Pragyandipta, P., and Naik, P.K. 2022. Terpinen-4-ol, An active constituent of kewda essential oil, mitigates biofilm forming ability of multidrug resistant *Staphylococcus aureus* and *Klebsiella pneumonia*. *J. Biol. Active. Prod. Nat.*, 12(5), 406–420, https://doi.org/10.1080/22311866.2022.2154264

Cheruvanachari, P., Pattnaik, S., Mishra, M., Pragyandipta, P., Pattnaik, A., and Naik, P.K. 2023. Deciphering the antibiofilm potential of 2-Phenylethyl methyl ether (PEME), a bioactive compound of Kewda essential oil against *Staphylococcus aureus*. *Microb. Pathog.*, 179, 106093. https://doi.org/10.1016/j.micpath.2023.106093

Coelho, P., Oliveira, J., Fernandes, I., Araujo, A., Pereira, A.R., Gameiro, P., and Bessa, L.J. 2021. Pyranoanthocyanins interfering with the quorum sensing of *Pseudomonas aeruginosa* and *Staphylococcus aureus*. *Int. J. Mol. Sci.*, 22(16), 8559. https://doi.org/10.3390/ijms.22168559

Costa, R.C., Bertolini, M., Costa Oliveira, B.E., Nagay, B.E., Dini, C., Benso, B., et al. 2023. Polymicrobial biofilms related to dental implant diseases: Unravelling the critical role of extracellular biofilm matrix. *Crit. Rev. Microbiol.*, 49(3), 370–390. https://doi.org/10.1080/1040841X.2022.2062219

Damodaran, V.B., and Murthy, N.S. 2016. Bio-inspired strategies for designing antifouling biomaterials. *Biomater. Res.*, 20, 18. https://doi.org/10.1186/s40824-016-0064-4

Das, S., Paul, P., Dastidar, D.G., Chakraborty, P., Chatterjee, S., Sarkar, S., et al. 2023. Piperine exhibits potential antibiofilm activity against *Pseudomonas aeruginosa* by accumulating reactive oxygen species, affecting cell surface hydrophobicity and quorum sensing. *Appl. Biochem. Biotechnol.*, 195(5), 3229–3256. https://doi.org/10.1007/s12010-022-04280-1

Davidson, I., Cahill, P., Hinz, A., Kluza, D., Scianni, C., and Georgiades, E. 2021. A review of biofouling of ships' internal seawater systems. *Front. Mar. Sci.*, 8, 761531. https://doi.org/10.3389/fmars.2021.761531

de Carvalho, C.C.C.R. 2018. Marine biofilms: A successful microbial strategy with economic implications. *Front. Mar. Sci.*, 5, 126. https://doi.org/10.3389/fmars.2018.00126

de Moura, D.F., Rocha, T.A., de Mello Barros, D., da Silva, M.M., dos Santos Santana, M., Neta, B.M., et al. 2021. Evaluation of the antioxidant, antibacterial, and antibiofilm activity of the sesquiterpene nerolidol. *Arch. Microbiol.*, 203(7), 4303–4311. https://doi.org/10.1007/s00203-021-02377-5

Deng, Y., Liu, Y., Li, J., Wang, X., He, S., Yan, X., Shi, Y., Zhang, W., and Ding, L. 2022. Marine natural products and their synthetic analogs as promising antibiofilm agents for antibiotics discovery and development. *Eur. J. Med. Chem.*, 239, 114513. https://doi.org/10.1016/j.ejmech.2022.114513

Dobretsov, S., and Rittschof, D. 2023. Biofouling and antifouling: Interactions between microbes and larvae of invertebrates. *Int. J. Mol. Sci.*, 24(7), 6531. https://doi.org/10.3390/ijms24076531

Donlan, R.M. 2002. Biofilms: microbial life on surfaces. *Emerg. Infect. Dis.*, 8(9), 881–890. https://doi.org/10.3201/eid0809.020063

Dziadek M., Dziadek K., Checinska K., Zagrajczuk B., Golda-Cepa M., Brzychczy-Wloch M., et al. 2021. PCL and PCL/bioactive glass biomaterials as carriers for biologically active polyphenolic compounds: Comprehensive physicochemical and biological evaluation. *Bioact. Mater.*, 6, 1811–1826. https://doi.org/10.1016/j.bioactmat.2020.11.025

Elbourne, A., Truong, V.K., Cheeseman, S., Rajapaksha, P., Gangadoo, S., Chapman, J., and Crawford, R.J. 2019. The use of nanomaterials for the mitigation of pathogenic biofilm formation. In: *Methods in Microbiology. Nanotechnology*. Gurtler, V., Ball, A.S., Soni, S. (Eds.). Elsevier Inc. pp. 61-92. https://doi.org/10.1016/bs.mim.2019.04.002

Ferreira-Vancato, Y.C.S., Dantas, F.M.L., and Fleury, B.G. 2020. Nanobiocides against marine biofouling. In: *Studies in Natural Products Chemistry. Bioactive Natural Products*. Rahman, A (Ed.). Elsevier Inc. pp. 463–514. https://doi.org/10.1016/B978-0-12-819483-6.00013-8

Fitridge, I., Dempster, T., Guenther, J., and de Nys, R. 2012. The impact and control of biofouling in marine aquaculture: A review. *Biofouling.*, 28(7), 649–669. https://doi.org/10.1080/08927014.2012.700478

Flemming, H.C. 2011. Microbial biofouling: Unsolved problems, insufficient approaches, and possible solutions. In: Biofilm Highlights. Vol. 5, Flemming, H.C., Wingender, J., Szewzyk, U. (Eds.). Springer Nature. pp. 81–109. https://doi.org/10.1007/978-3-642-19940-0_5

Flemming, H.C. 2020. Biofouling and Me: My Stockholm syndrome with biofilms. *Water. Res.*, 173, 115576. https://doi.org/10.1016/j.watres.2020.115576

Flemming, H.C., van Hullebusch, E.D., Neu, T.R., Nielsen, P.H., Seviour, T., Stoodley, P., et al. 2023. The biofilm matrix: Multitasking in a shared space. *Nat. Rev. Microbiol.*, 21(2), 70–86. https://10.1038/s41579-022-00791-0

Fletcher, L.M., Davidson, I.C., Bucknall, B.G., and Atalah, J. 2023. Salmon farm biofouling and potential health impacts to fish from stinging cnidarians. *Aquaculture.*, 568, 739315. https://doi.org/10.1016/j.aquaculture.2023.739315

Gao, K., Zhang, B., and Zhao, F. 2023. Antibiofilm effect of curcumin against *Staphylococcus aureus* surface wound biofilm–associated infection: *In vitro* and *in silico*. *Appl. Biochem. Biotechnol.*, 195, 5329–5337. https://doi.org/10.1007/s12010-022-03844-5

Ghosh, S., Lahiri, D., Nag, M., Dey, A., Sarkar, T., Biswas, R., et al. 2023. Analysis of antibiofilm activities of bioactive compounds from honeyweed (*Leonurus sibiricus*) against *P. aeruginosa*: an *in vitro* and *in silico* approach. *Appl. Biochem. Biotechnol.*, 195, 5312–5328. https://doi.org/10.1007/s12010-021-03797-1

Gizer, G., Önal, U., Ram, M., and Sahiner, N. 2023. Biofouling and mitigation methods: A review. *Biointerface. Res. Appl. Chem.*, 13(2), 185. https://doi.org/10.33263/BRIAC132.185

Govindan, R., Gnanasekaran, C., Govindan, R., Muthuchamy, M., Quero, F., Jothi, A., et al. 2023. Anti-quorum sensing and anti-biofilm effect of *Nocardiopsis synnemataformans* RMN 4 (MN061002) compound 2,6-Di-tert-butyl, 1,4-Benzoquinone against biofilm-producing bacteria. *Appl. Biochem. Biotechnol.*, https://doi.org/10.1007/s12010-023-04738-w

Gu, J.D. 2012. Biofouling and prevention: Corrosion, biodeterioration and biodegradation of materials. In: *Handbook of Environmental Degradation of Materials*. Vol. 2, Kutz, M. (Ed.). Elsevier Inc. pp. 243–282. https://doi.org/10.1016/B978-081551500-5.50011-2

Gule, N.P., Begum, N.M., and Klumperman, B. 2016. Advances in biofouling mitigation: A review. *Crit. Rev. Environ. Sci. Technol.*, 46(6), 535–555. https://doi.org/10.1080/10643389.2015.1114444

Guo, N., Bai, X., Shen, Y., and Zhang, T. 2023. Target-based screening for natural products against *Staphylococcus aureus* biofilms. *Crit. Rev. Food. Sci. Nutrition.*, 63(14), 2216–2230. https://doi.org/10.1080/10408398.2021.1972280

Hall, C.W., and Mah, T.F. 2017. Molecular mechanisms of biofilm-based antibiotic resistance and tolerance in pathogenic bacteria. *FEMS Microbiol. Rev.*, 41(3), 276–301. https://doi.org/10.1093/femsre/fux010

Huang, L., Wu, C., Gao, H., Xu, C., Dai, M., Huang, L., et al. 2022. Bacterial multidrug efflux pumps at the frontline of antimicrobial resistance: An overview. *Antibiotics (Basel).*, 11(4), 520. https://doi.org/10.3390/antibiotics11040520

Huemer, M., Shambat, S.M., Brugger, S.D., and Zinkernagel, A.S. 2020. Antibiotic resistance and persistence—Implications for human health and treatment perspectives. *EMBO Rep.*, 21, e51034. https://doi.org/10.15252/embr.202051034

Jamieson, T., and Leterme, S.C. 2021. Influences and impacts of biofouling in SWRO desalination plants. *Crit. Rev. Environ. Sci. Technol.*, 51(12), 1281–1301. https://doi.org/10.1080/10643389.2020.1757937

Jana, S., Charlton, S.G., Eland, L.E., Burgess, J.G., Wipat, A., Curtis, T.P., and Chen, J. 2020. Nonlinear rheological characteristics of single species bacterial biofilms. *npj Biofilms. Microbiomes.*, 6, 19. https://doi.org/10.1038/s41522-020-0126-1

Joardar, I., and Dutta, S.A. 2023. Selective review on the novel approaches and potential control agents of anti-biofouling and anti-biofilming. *Appl. Biochem. Biotechnol.*, 195, 5605–5617. https://doi.org/10.1007/s12010-022-04160-8

Johnston, E.L., Zavan, L., Bitto, N.J., Petrovski, S., Hill, A.F., and Kaparakis-Liaskos, M. 2023. Planktonic and biofilm-derived *Pseudomonas aeruginosa* outer membrane vesicles facilitate horizontal gene transfer of plasmid DNA. *Microbiol. Spectr.* 11(2), e0517922. https://doi.org/10.1128/spectrum.05179-22

Kaur, H., Kaur, A., Soni, S.K., and Rishi, P. 2022. Microbially-derived cocktail of carbohydrases as an anti-biofouling agents: A 'green approach'. *Biofouling.*, 38(5), 455–481. https://doi.org/10.1080/08927014.2022.2085566

Kaur, K., Reddy, S., Barathe, P., Shriram, V., Anand, U., Proćków, J., and Kumar V. 2021. Combating drug-resistant bacteria using photothermally active nanomaterials: A perspective review. *Front. Microbiol.*, 12, 747019. https://doi.org/10.3389/fmicb.2021

Koo, H., Allan, R., Howlin, R., Stoodley, P., and Hall-Stoodley, L. 2017. Targeting microbial biofilms: Current and prospective therapeutic strategies. *Nat. Rev. Microbiol.*, 15, 740–755. https://doi.org/10.1038/nrmicro.2017.99

Krishnan, S., Weinman, C.J., and Ober, C.K. 2008. Advances in polymers for anti-biofouling surfaces. *J. Mater. Chem.*, 18, 3405–3413. https://doi.org/10.1039/b801491d

Krukiewicz, K., Kazek-Kęsik, A., Brzychczy-Włoch, M., Łos, M.J., Ateba, C.N., Mehrbod, P., et al. 2022. Recent advances in the control of clinically important biofilms. *Int. J. Mol. Sci.*, 23(17), 9526. https://doi.org/10.3390/ijms23179526

Kumar, L., Bisen, M., Harjai, K., Chhibber, S., Azizov, S., Lalhlenmawia, H., and Kumar, D. 2023. Advances in nanotechnology for biofilm inhibition. *ACS Omega.*, 8(24), 21391-21409. https://doi.org/10.1021/acsomega.3c02239

Kyei, S.K., Darko, G., and Akaranta, O. 2020. Chemistry and application of emerging eco-friendly antifouling paints: A review. *J. Coatings. Technol. Res.*, 17(4), 1–18. https://doi.org/10.1007/s11998-019-00294-3

Lacoste, E., and Gaertner-Mazouni, N. 2014. Biofouling impact on production and ecosystem functioning: A review for bivalve aquaculture. *Rev. Aquaculture.*, 6, 1–10. https://doi.org/10.1111/raq.12063

Lahiri, D., Ray, R.R., Sarkar, T., Upadhye, V.J., Ghosh, S., Pandit, S., et al. 2022. Anti-biofilm efficacy of green-synthesized ZnO nanoparticles on oral biofilm: *In vitro and in silico* study. *Front. Microbiol.*, 13, 939390. https://doi.org/10.3389/fmicb.2022.939390

Larsson, D.G.J., and Flach, C.F. 2022. Antibiotic resistance in the environment. *Nat. Rev. Microbiol.*, 20, 257–269. https://doi.org/10.1038/s41579-021-00649-x

Li, W., Yang, Z., Hu, J., Wang, B., Rong, H., Li, Z., et al. 2022. Evaluation of culturable 'last-resort' antibiotic resistant pathogens in hospital wastewater and implications on the risks of nosocomial antimicrobial resistance prevalence. *J. Hazard. Mater.*, 438, 129477. https://doi.org/10.1016/j.jhazmat.2022.129477

Lichtenberg, M., Coenye, T., Parsek, M.R., Bjarnsholt, T., and Jakobsen, T.H. 2023. What's in a name? Characteristics of clinical biofilms. *FEMS Microbiol. Rev.*, 47(5), fuad050. https://doi.org/10.1093/femsre/fuad050

Lin, Y.K., Yang, S.C., Hsu, C.Y., Sung, J.T., and Fang, J.Y. 2021. The antibiofilm nanosystems for improved infection inhibition of microbes in skin. *Molecules.*, 26(21), 6392. https://doi.org/10.3390/molecules26216392

Liu, P., Kang, X., Chen, X., Luo, X., Li, C., and Wang, G. 2023. Quercetin targets SarA of methicillin-resistant *Staphylococcus aureus* to mitigate biofilm formation. *Microbiol. Spectr.*, 29, e0272223. https://doi.org/10.1128/spectrum.02722-23

Liu, W., Yan, M., and Zhao, W. 2023. Antibacterial-renew dual-function anti-biofouling strategy: Self-assembled Schiff-base metal complex coatings built from natural products. *J. Colloid. Interface. Sci.*, 629 (Part A), 496–507. https://doi.org/10.1016/j.jcis.2022.08.178

Lu, L., Hu, W., Tian, Z., Yuan, D., Yi, G., Zhou, Y., et al. 2019. Developing natural products as potential anti-biofilm agents. *Chin. Med.*, 14, 11. https://doi.org/10.1186/s13020-019-0232-2

Lu, L., Zhao, Y., Yi, G., Li, M., Liao, L., Yang, C., et al. 2021. Quinic acid: A potential antibiofilm agent against clinical resistant *Pseudomonas aeruginosa. Chin. Med.*, 16, 72. https://doi.org/10.1186/s13020-021-00481-8

Luo, Y., Zhang, H., Cui, Q., Li, H., and Li, L. 2023. Gold nanoclusters-decorated graphitic carbon nitride nanocomposites with antibacterial and anti-biofilm activities. *Colloid. Surf. A Physicochem. Eng. Aspect.*, 676 (Part B), 132189. https://doi.org/10.1016/j.colsurfa.2023.132189

Mahavy, C.E., Mol, A., Andrianarisoa, B., Duez, P., El Jaziri, M., Baucher, M., et al. 2022. The xanthophyll carotenoid lutein reduces the invasive potential of *Pseudomonas aeruginosa* and increases its susceptibility to Tobramycin. *Int. J. Mol. Sci.*, 23, 7199. https://doi.org/10.3390/ijms23137199

Martin-Rodriguez, A.J., Babbaro, J.M.F., Lahoz, F., Sanson, M., Martin, V.S., Norte, M., and Fernandez, J.J. 2015. From broad-spectrum biocides to quorum sensing disruptors and mussel repellents: Antifouling profile of alkyl triphenylphosphonium salts. *PLoS ONE*, 10(4), e0123652. https://doi.org/10.1371/journal.pone.0123652

Martins Antunes de Melo, W.C., Celiešiūtė-Germanienė, R., Šimonis, P., and Stirkė, A. 2021. Antimicrobial photodynamic therapy (aPDT) for biofilm treatments. Possible synergy between aPDT and pulsed electric fields. *Virulence.*, 12(1), 2247–2272. https://doi.org/10.1080/21505594.2021

Melander, R.J., Basak, A.K., and Melander, C. 2020. Natural products as inspiration for the development of bacterial antibiofilm agents. *Nat. Prod. Rep.*, 37, 1454–1477. https://doi.org/10.1039/D0NP000222A

Melkebeke, M.V., Janssen, C., and De Meester, S. 2020. Characteristics and sinking behavior of typical microplastics including the potential effect of biofouling: Implications for remediation. *Environ. Sci. Technol.*, 54(14), 8668–8680. https://doi.org/10.1021/acs.est.9b07378

Miller, T., Waturangi, D.E., Yogiara. 2022. Antibiofilm properties of bioactive compounds from *Actinomycetes* against foodborne and fish pathogens. *Sci. Rep.*, 12(1), 18614. https://doi.org/10.1038/s41598-022-23455-8

Mishra, R., Panda, A.K., De Mandal, S., Shakeel, M., Bisht, S.S., and Khan, J. 2020. Natural anti-biofilm agents: Strategies to control biofilm-forming pathogens. *Front. Microbiol.*, 11, 566325. https://doi.org/10.3389/fmicb.2020.566325

Mohanta, Y.K., Chakrabartty, I., Mishra, A.K., Chopra, H., Mahanta, S., Avula, S.K., et al. 2023. Nanotechnology in combating biofilm: A smart and promising therapeutic strategy. *Front. Microbiol.*, 13, 1028086. https://doi.org/10.3389/fmicb.2022.1028086

Naga, N.G., Zaki, A.A., El-Badan, D.E., Rateb, H.S., Ghanem, K.M., and Shaaban, M.I. 2022. Methoxyisoflavan derivative from *Trigonella stellata* inhibited quorum sensing and virulence factors of *Pseudomonas aeruginosa*. *World. J. Microbiol. Biotechnol.*, 38, 156. https://doi.org/10.1007/s11274-022-03337-x

Nakano, D., and Strayer, D.L. 2014. Biofouling animals in fresh water: Biology, impacts, and ecosystem engineering. *Front. Ecol. Environ.*, 12(3), 167–175. https://doi.org/10.1890/130071

Nandakumar, V., Chittaranjan, S., Kurian, V.M., and Doble, M. 2013. Characteristics of bacterial biofilm associated with implant material in clinical practice. *Polym. J.*, 45, 137–152. https://doi.org/10.1038/pj.2012.130

Nelson, T.F., Reddy, C.M., and Ward, C.P. 2021. Product formulation controls the impact of biofouling on consumer plastic photochemical fate in the ocean. *Environ. Sci. Technol.*, 55, 8898–8907. https://doi.org/10.1021/acs.est.1c02079

Nesse, L.L., Osland, A.M., and Vestby, L.K. 2023. The role of biofilms in the pathogenesis of animal bacterial infections. *Microorganisms*, 11(3), 608. https://doi.org/10.3390/microorganisms11030608

Parai, D., Banerjee, M., Dey, P., and Mukherjee, S.K. 2020. Reserpine attenuates biofilm formation and virulence of *Staphylococcus aureus*. *Microb. Pathog.*, 138, 103790. https://doi.org/10.1016/j.micpath.2019.103790

Parrilli, E., Tutino, M.L., and Marino, G. 2022. Biofilm as an adaptation strategy to extreme conditions. *Rend. Fis. Acc. Lincei.*, 33, 527–536. https://doi.org/10.1007/s12210-022-01083-8

Patel, M., Siddiqui, A.J., Hamadou, W.S., Surti, M., Awadelkareem, A.M., Ashraf, S.A., et al. 2021. Inhibition of bacterial adhesion and antibiofilm activities of a glycolipid biosurfactant from *Lactobacillus rhamnosus* with its physicochemical and functional properties. *Antibiotics*, 10(12), 1546. https://doi.org/10.3390/antibiotics10121546

Patra, A., Das, J., Agrawal, N.R., Kushwaha, G.S., Ghosh, M., and Son, Y-O. 2022. Marine antimicrobial peptides-based strategies for tackling bacterial biofilm and biofouling challenges. *Molecules*, 27(21), 7546. https://doi.org/10.3390/molecules27217546

Pattnaik, S., Mishra, M., Singh, H., and Naik, P.K. 2023. Novel perspectives on phytochemicals-based approaches for mitigation of biofilms in ESKAPE pathogens: recent trends and future avenues. In: *Recent Frontiers of Phytochemicals: Applications in Food, Pharmacy, Cosmetics, and Biotechnology*. Pati, S., Sarkar, T., Lahiri, D. (Eds.). Elsevier Inc. pp. 433–454. https://doi.org/10.1016/B978-0-443-19143-5.00005-0

Patwardhan, S.B., Pandit, C., Pandit, S., Verma, D., Lahiri, D., Nag, M., et al. 2023. Illuminating the signalomics of microbial biofilm on plant surfaces. *Biocatalysis. Agric. Biotechnol.*, 47, 102537. https://doi.org/10.1016/j.bcab.2022.102537

Penesyan, A., Paulsen, I.T., Kjelleberg, S., and Gillings, M.R. 2021. Three faces of biofilms: A microbial lifestyle, a nascent multicellular organism, and an incubator for diversity. *npj Biofilms. Microbiomes.*, 7, 80. https://doi.org/10.1038/s41522-021-00251-2

Pipattanachat, S., Qin, J., Rokaya, D., Thanyasrisung, P., and Srimaneepong, V. 2021. Biofilm inhibition and bactericidal activity of NiTi alloy coated with graphene oxide/silver nanoparticles via electrophoretic deposition. *Sci. Rep.*, 11, 14008. https://doi.org/10.1038/s41598-021-92340-7

Piri-Gharaghie, T., Jegargoshe-Shirin, N., Saremi-Nouri, S., Khademhosseini, S., Hoseinnezhad-lazarjani, E., Mousaviet, A., et al. 2022. Effects of Imipenem-containing Niosome nanoparticles against high prevalence methicillin-resistant *Staphylococcus epidermidis* biofilm formed. *Sci. Rep.*, 12, 5140. https://doi.org/10.1038/s41598-022-09195-9

Praseetha, S., Sukumaran, S.T., Dan, M., Augustus, A.R., Pandian, S.K., and Sugathan, S. 2023. The anti-biofilm potential of Linalool, a major compound from *Hedychium larsenii*, against *Streptococcus pyogenes* and its toxicity assessment in *Danio rerio*. *Antibiotics*, 12(3), 545. https://doi.org/10.3390/antibiotics12030545

Qader, M.M., Hamed, A.A., Soldatou, S., Abdelraof, M., Elawady, M.E., Hassane, A.S.I., et al. 2021. Antimicrobial and antibiofilm activities of the fungal metabolites isolated from the marine endophytes *Epicoccum nigrum* M13 and *Alternaria alternata* 13A. *Marine. Drugs.*, 19(4), 232. https://doi.org/10.3390/md19040232

Qiu, H., Feng, K., Gapeeva, A., Meurisch, K., Kaps, S., Li, X., et al. 2022. Functional polymer materials for modern marine biofouling control. *Prog. Polymer. Sci.*, 127, 101516. https://doi.org/10.1016/j.progpolymsci.2022.101516

Rao, T.S. 2022. Biofouling (macro-fouling) in sea water intake systems. In: *Water-Formed Deposits: Fundamentals and Mitigation Strategies*. Amjad, Z., Demadis, K.D. (Eds.). Elsevier Inc. pp. 565–587. https://doi.org/10.1016/B978-0-12-822896-8.00016-9

Razdan, K., Garcia-Lara, J., Sinha, V.R., and Singh, K.K. 2022. Pharmaceutical strategies for the treatment of bacterial biofilms in chronic wounds. *Drug. Dis. Today.*, 27(8), 2137–2150. https://doi.org/10.1016/j.drudis.2022.04.020

Roepke, L.K., Brefeld, D., Soltmann, U., Randall, C.J., Negri, A.P., and Kunzmann, A. 2022. Antifouling coatings can reduce algal growth while preserving coral settlement. *Sci. Rep.*, 12, 15935. https://doi.org/10.1038/s41598-022-19997-6

Romeu, M.J., and Mergulhão, F. 2023. Development of antifouling strategies for marine applications. *Microorganisms*, 11(6), 1568. https://doi.org/10.3390/microorganisms11061568

Roy, R., Tiwari, M., Donelli, G., and Tiwari, V. 2018. Strategies for combating bacterial biofilms: A focus on anti-biofilm agents and their mechanisms of action. *Virulence*, 9(1), 522–554. https://doi.org/10.1080/21505594.2017.1313372

Roy, S., Chowdhury, G., Mukhopadhyay, A.K., Dutta, S., Basu, S. 2022. Convergence of biofilm formation and antibiotic resistance in *Acinetobacter baumannii* infection. *Front. Med. (Lausanne)*, 9, 793615. https://doi.org/10.3389/fmed.2022.793615

Ruhal, R., and Kataria, R. 2021. Biofilm patterns in gram-positive and gram-negative bacteria. *Microbiol. Res.*, 251, 126829. https://doi.org/10.1016/j.micres.2021.126829

Sahli, C., Moya, S.E., Lomas, J.S., Gravier-Pelletier, C., Briandet, R., and Hémadi, M. 2022. Recent advances in nanotechnology for eradicating bacterial biofilm. *Theranostics.*, 12(5), 2383–2405. https://doi.org/10.7150/thno.67296

Samal, S., Misra, M., Rangarajan, V., and Chattopadhyay, S. 2023. Antimicrobial nanoparticles mediated prevention and control of membrane biofouling in water and wastewater treatment: Current trends and future perspectives. *Appl. Biochem. Biotechnol.*, 195, 5458–5477. https://doi.org/10.1007/s12010-023-04497-8

Sampathkumar, S.J., Srivastava, P., Ramachandran, S., Sivashanmugam, K., and Gothandam, K.M. 2019. Lutein: A potential antibiofilm and antiquorum sensing molecule from green microalga *Chlorella pyrenoidosa*. *Microb. Pathog.*, 135, 103658. https://doi.org/10.1016/j.micpath.2019.103658

Samrot, A.V., Mohamed, A.A., Faradjeva, E., Jie, L.S., Sze, C.H., Arif A., et al. 2021. Mechanisms and impact of biofilms and targeting of biofilms using bioactive compounds – A review. *Medicina (Kaunas).*, 57(8), 839. https://doi.org/10.3390/medicina57080839

Sankhwar, R., Kumar, A., Yadav, S., Singh, V., and Gupta, R.K. 2023. Emycin-E purified from *Streptomyces* sp. RG1011 from Himalayan soil has antibiofilm activity against *Staphylococcus aureus*. *Microb. Pathog.*, 182, 106256. https://doi.org/10.1016/j.micpath.2023.106256

Schiessl, K.T., Hu, F., Jo, J., Nazia, S.Z., Wang, B., Price-Whelan, A., et al. 2019. Phenazine production promotes antibiotic tolerance and metabolic heterogeneity in *Pseudomonas aeruginosa* biofilms. *Nat. Commun.*, 10, 762. https://doi.org/10.1038/s41467-019-08733-w

Schultz, M.P., Bendick, J.A., Holm, E.R., and Hertel, W.M. 2011. Economic impact of biofouling on a naval surface ship. *Biofouling*, 27(1), 87–98. https://doi.org/10.1080/08927014.2010.542809

Sentenac, H., Loyau, A., Leflaive, J., and Schmeller, D.S. 2022. The significance of biofilms to human, animal, plant and ecosystem health. *Funct. Ecol.*, 36(2), 294–33. https://doi.org/10.1111/1365-2435.13947

Shan, C., JiaDao, W., HaoSheng, C., and DaRong, C. 2011. Progress of marine biofouling and antifouling technologies. *Chinese Sci. Bull.*, 56, 598–612. https://doi.org/10.1007/s11434-010-4158-4

Sharma, S., Mohler, J., Mahajan, S.D., Schwartz, S.A., Bruggemann, L., and Aalinkeel, R. 2023. Microbial biofilm: A review on formation, infection, antibiotic resistance, control measures, and innovative treatment. *Microorganisms*, 11(6), 1614. https://doi.org/10.3390/microorganisms11061614

Shome, S., Talukdar, A.D., Nath, R., and Tewari, S. 2023. Curcumin-ZnO nanocomposite mediated inhibition of *Pseudomonas aeruginosa* biofilm and its mechanism of action. *J. Drug. Del. Sci. Technol.*, 81, 104301. https://doi.org/10.1016/j.jddst.2023.104301

Shrestha, L., Fan, H.M., Tao, H.R., and Huang, J.D. 2022. Recent strategies to combat biofilms using antimicrobial agents and therapeutic approaches. *Pathogens*, 11(3), 292. https://doi.org/10.3390/pathogens11030292

Silva, S., Costa, E.M., Machado, M., Morais, R., Calhau, C., and Pintado, M. 2022. Antiadhesive and antibiofilm effect of Malvidin-3-Glucoside and Malvidin-3-Glucoside/

Neochlorogenic acid mixtures upon *Staphylococcus*. *Metabolites*., 12(11), 1062. https://doi.org/10.3390/metabo12111062

Singh, S., Singh, S.K., Chowdhury, I., and Singh, R. 2017. Understanding the mechanism of bacterial biofilms resistance to antimicrobial agents. *Open. Microbiol. J*., 11, 53–62. https://doi.org/10.2174/1874285801711010053

Song, S., Demirel, Y.K., Muscat-Fenech, C.D.M., Tezdogan, T., and Atlar, M. 2020. Fouling effect on the resistance of different ship types. *Ocean Eng*., 216, 107736. https://doi.org/10.1016/j.oceaneng.2020.107736

Song, Y., Wang, Z., Long, Y., Mao, Y., Jiang, F., and Lu, Y. 2022. 2-Alkyl-anthraquinones inhibit *Candida albicans* biofilm via inhibiting the formation of matrix and hyphae. *Res. Microbiol*., 173(6–7), 103955. https://doi.org/10.1016/j.resmic.2022.103955

Song, Z.M., Zhang, J.L., Zhou, K., Yue, L.M., Zhang, Y., Wang, C.Y., et al. 2021. Anthraquinones as potential antibiofilm agents against Methicillin-resistant *Staphylococcus aureus*. *Front. Microbiol*., 12, 709826. https://doi.org/10.3389/fmicb.2021.709826

Soto, S.M. 2013. Role of efflux pumps in the antibiotic resistance of bacteria embedded in a biofilm. *Virulence*, 4(3), 223–229. https://doi.org/10.4161/viru.23724

Srinivasan, R., Santhakumari, S., Poonguzhali, P., Geetha, M., Dyavaiah, M., and Xiangmin, L. 2021. Bacterial biofilm inhibition: A focused review on recent therapeutic strategies for combating the biofilm mediated infections. *Front. Microbiol*., 12, 676458. https://doi.org/10.3389/fmicb.2021.676458

Stringer, C.C., and Polagye, B.L. 2020. Implications of biofouling on cross-fow turbine performance. *SN Appl Sci*., 2, 464. https://doi.org/10.1007/s42452-020-2286-2

Tajani, A.S., Tehranizadeh, Z.A., Pourmohammad, A., Pourmohammad, A., Iranshahi, M., Farhadi, F., et al. 2023. Anti-quorum sensing and antibiofilm activity of coumarin derivatives against *Pseudomonas aeruginosa* PAO1: Insights from in vitro and in silico studies. *Iran. J. Basic. Med. Sci*., 26(4), 445–452. https://doi.org/10.22038/IJBMS.2023.69016.15047

Thite, V., Kavi Bharathi, R., and Srinandan, C.S. 2023. Biofilm formation: A well-played game in bacterial pathogenesis. In: *Understanding Microbial Biofilms. Fundamentals to Applications*. Das, S., and Kungwani, N.A. (Eds.). Elsevier Inc. pp. 605–625. https://doi.org/10.1016/B978-0-323-99977-9.00009-0

Treiber, L., Pezolt, C., Zeng, H., Schrey, H., Jungwirth, S., Shekhar, A., et al. 2021. Dual agents: Fungal macrocidins and synthetic analogues with herbicidal and antibiofilm activities. *Antibiotics*, 10(8), 1022. https://doi.org/10.3390/antibiotics10081022

Uruén, C., Chopo-Escuin, G., Tommassen, J., Mainar-Jaime, R.C., and Arenas, J. 2021. Biofilms as promoters of bacterial antibiotic resistance and tolerance. *Antibiotics (Basel)*, 10(1), 3. https://doi.org/10.3390/antibiotics10010003

Usui, M., Yoshii, Y., Thiriet-Rupert, S., Ghigo, J.M., and Beloin, C. 2023. Intermittent antibiotic treatment of bacterial biofilms favors the rapid evolution of resistance. *Commun. Biol*., 6, 275. https://doi.org/10.1038/s42003-023-04601-y

Valen, H., and Scheie, A.A. 2018. Biofilms and their properties. *Eur. J. Oral. Sci*., 126(1), 13–18. https://doi.org/10.1111/eos.12425

Vani, S., Vadakkan, K., and Mani, B. 2023. A narrative review on bacterial biofilm: its formation, clinical aspects and inhibition strategies. *Futur. J. Pharm. Sci*., 9, 50. https://doi.org/10.1186/s43094-023-00499-9

Verma, S., Mohanty, S., and Nayak, S.K. 2020. Preparation of hydrophobic epoxy–polydimethylsiloxane–graphene oxide nanocomposite coatings for antifouling application. *Soft Matter*., 16(5), 1211–1226. https://doi.org/10.1039/C9SM01952A

Vestby, L.K., Grønseth, T., Simm, R., and Nesse, L.L. 2021. Bacterial biofilm and its role in the pathogenesis of disease. *Antibiotics*, 9(2), 59. https://doi.org/10.3390/antibiotics9020059

Wells, M., Schneider, R., Bhattarai, B., Currie, H., Chavez, B., Christopher, G., et al. 2023. Perspective: The viscoelastic properties of biofilm infections and mechanical interactions with phagocytic immune cells. *Front. Cell. Infect. Biol.*, 13, 1102199. https://doi.org/10.3389/fcimb.2023.1102199

Wibowo, J.T., Bayu, A., Aryati, W.D., Fernandes, C., Yanuar, A., Kijjoa, A., et al. 2023. Secondary metabolites from marine-derived bacteria with antibiotic and antibiofilm activities against drug-resistant pathogens. *Mar. Drug.*, 21(1), 50. https://doi.org/10.3390/md21010050

Xu, J., and Lee, H. 2020. Anti-biofouling strategies for long-term continuous use of implantable biosensors. *Chemosensors*, 8(3), 66. https://doi.org/10.3390/chemosensors8030066

Xu, Y., Guo, W., Luo, D., Li, P., Xiang, J., Chen, J., et al. 2023. Antibiofilm effects of punicalagin against *Staphylococcus aureus in vitro*. *Front. Microbiol.*, 14, 1175912. https://doi.org/10.3389/fmicb.2023.1175912

Zhang, H., Zhu, S., Yang, J., and Ma, A. 2022. Advancing strategies of biofouling control in water-treated polymeric membranes. *Polymers (Basel)*, 14(6), 1167. https://doi.org/10.3390/polym14061167

Zhao, A., Sun, J., and Liu, Y. 2023. Understanding bacterial biofilms: From definition to treatment strategies. *Front. Cell. Infect. Microbiol.*, 13, 1137947. https://doi.org/10.3389/fcimb.2023

Zhao, J., Chen, J., Zheng, X., Lin, Q., Zheng, G., Xu, Y., and Lin, F. 2023. Urushiol-based benzoxazine containing sulfobetaine groups for sustainable marine antifouling applications. *Polymers*, 15(10), 2383. https://doi.org/10.3390/polym15102383

Zhao, X., Wang, L.Y., Tang, C.Y., Li, K., Huang, Y.H., Duan, Y.R., et al. 2023. Electro-microenvironment modulated inhibition of endogenous biofilms by piezo implants for ultrasound-localized intestinal perforation disinfection. *Biomaterials.*, 295, 122055. https://doi.org/10.1016/j.biomaterials.2023.122055

Zhao, X., Yu, Z., and Ding, T. 2020. Quorum-sensing regulation of antimicrobial resistance in bacteria. *Microorganisms*, 8(3), 425. https://doi.org/10.3390/microorganisms8030425

Zheng, S., Bawazir, M., Dhall, A., Kim, H.E., He, L., Heo, J., and Hwang, G. 2021. Implication of surface properties, bacterial motility, and hydrodynamic conditions on bacterial surface sensing and their initial adhesion. *Front. Bioeng. Biotechnol.*, 9, 643722. https://doi.org/10.3389/fbioe.2021.643722

Zou, L., Li, X., Huang, Y., Wang, C., Fang, Y., Zhao, J., et al. 2023. Raspberry-like gold nanozyme-hybrid liposomes for hypoxia-enhanced biofilm eradication. *Nanotoday.*, 50, 101828. https://doi.org/10.1016/j.nantod.2023.101828

6 Microbial Biofilms and Corrosion

V.T. Anju, Siddhardha Busi, and Madhu Dyavaiah

6.1 WHAT ARE MICROBIAL BIOFILMS?

Biofilms are a population of cells enclosed in self-produced exopolymer substance (EPS) that allow survival in harsh microenvironments (Khan et al., 2022). EPS of biofilm constituting a microenvironment are characterized by various molecules such as nucleic acids, proteins, lipids, and other vital biomolecules. These biomolecules play a key role in the aggregation of microbial cells and form stable interactions. EPS form a watery gel-like matrix that helps microbes act synergistically in interspecies or intraspecies interactions and escape from the stress of unfavourable conditions (Liu et al., 2018). The biofilm EPS matrix is protective to cells and is involved in the transport of nutrients and signals among them. The formation of biofilm is a complex and dynamic process influenced by several biological and environmental factors (Y. G. and Mulky, 2023).

Biofilms are ubiquitous in various environments such as nature, industries, hospitals, etc., and can cause diseases, biofouling, and biocorrosion as well as affect health and environment (Jia et al., 2017). In comparison to the monospecies biofilms, mixed biofilms are more challenging and cause serious corrosion in various sectors. The presence of more than one species of bacteria in biofilms can cause different interactions and metabolic reactions in the aerobic and anaerobic region of biofilms (Videla and Herrera, 2009; Kip and Van Veen, 2015).

6.1.1 Biofilm Developmental Steps

Biofilms are formed on metal surfaces by planktonic cells and are transported to the metal surface by sedimentation, motility, cell surface hydrophobicity, and fluid dynamic forces. Conditioning films, metal composition, surface active molecules, and roughness favor the biofilm adhesion process. To prevent microbial attachment on surface, metal surface modification methods are studied (Eckert, 2015). Adhesion of bacteria to the metal surfaces is one of the important steps in the formation of mature biofilms. There are several phases in the adhesion process. The first process involves the transport of cells to the metal surface either by diffusion or convection displacement or active transport, followed by the reversible binding and then by irreversible binding to the material surface. Finally, cells colonize to the surface

DOI: 10.1201/9781003487203-6

and proliferate to form mature biofilms. The reversible binding is facilitated by the specialized cell structures such as pili, flagella, and fimbriae. The irreversible binding is mediated by the EPS produced by the cells. EPS production provides structural stability to biofilm cells (Pal and Lavanya, 2022). Sulfur-reducing bacteria (SRB) are commonly involved in the biocorrosion of steel and end up with localized corrosion (Y. G. and Mulky, 2023). An example of marine metal corrosion-causing biofilms is *Pseudomonas aeruginosa*. This bacterium is widely reported in the biocorrosion of marine and industrial environments (Liu et al., 2023).

6.2 METAL CORROSION: OVERVIEW

Metal corrosion is a big concern in industries due to the economic loss. Corrosion affects various industries and their production. There are abiotic and biotic mechanisms contributing to the metal corrosion. Corrosion occurs naturally when the metal surface encounters the water present in the surrounding area. This water acts as an electrolyte, which helps in the conductions of ions to case corrosion. Other corrosion- influenced factors such as seawater, industrially produced chemicals, soil resistivity, physical stress, and moisture levels are also involved in the induction or acceleration of corrosion circumstances (Procópio, 2020). Metal corrosion is driven by a electrochemical process in which electrons are transferred from a zero valent metal state (present at anode) to external acceptor of electron (at cathodic site) (Revie, 2011; Beech and Sunner, 2004).

Corrosion is a complex process contributed by several factors. Microorganisms are found to be involved in mechanisms of corrosion. Microbes alter conditions such as pH, oxidation-reduction potential, and release of corrosive metabolites or extracellular enzymes and thereby induce or speed up the corrosion process. Microbially influenced corrosion (MIC) is one of the key players in corrosion (Little and Lee, 2009). Metals release electrons and get oxidized under specific conditions in the presence of certain non-metals. Oxygen-like non-metals tend to gain electrons and turn into anions. These metals in their oxidized state are corroded. Thus, corrosion is a chemically favored oxidation process. For instance, iron in the presence of oxygen donates electrons and forms iron cations and oxygen anions. Thus, the corrosion product is the oxidized product of iron (i.e., iron oxide (FeO)) (Groysman, 2010).

This is the equation describing iron oxidation (Groysman, 2010):

$$\mathrm{Fe} \rightarrow \mathrm{Fe}^{2+} + 2e^{-} \tag{1}$$

$$\mathrm{O} + 2e^{-} \rightarrow \mathrm{O}^{2-} \tag{2}$$

The different factors affecting the metallic corrosion are temperature, pH, metal surface state, immersed solution, immersion time, chemical nature of metal, aeration state of solution, and hydrodynamic condition. An elevation in temperature and movement of corrosive fluids amplifies the rate of corrosion. The contact time of metals with aqueous solution determines the rate of corrosion. According to metal surface characteristics such as depositions or films formed on metal surface, biological growth on surface, microstructure, grain size, secondary phases. and rough surface or tight gaps can also affect rate of corrosion (Ibrahimi et al., 2021).

Metal corrosions are observed in different forms such as uniform/general, pitting, galvanic, intergranular, crevice, parting, stress corrosion cracking, and erosion corrosion. The simplest form of corrosion is uniform where the metal surface is homogenously attacked by a corrosive environment. Galvanic is a type of dissimilar corrosion where metals are immersed in an electrolyte solution and corrosion occurs in it. The most damaging corrosion is pitting corrosion where metals are corroded by a localized attack directing localized loss of metal. Mechanical corrosion is an erosion type where corrosion fatigue, fretting corrosion, and cavitation damage occurs. This type of corrosion affects piping works available in equipment such as pumps, turbine blades, and centrifuges that are continuously exposed to moving fluids. Here, rapidly flowing fluids cause mechanical wear on the metals and can lead to the separation of metal from the instrument. In stress corrosion cracking, metals develop cracks instead of noticeable damage due to the tensile stress exerted by the rapidly flowing media (Ibrahimi et al., 2021). According to the nature of metals, less noble metals are more prone to corrosion than the more noble metals (Harsimran et al., 2021).

In industries, metal instruments are easily corroded when chemicals, or other products produced in the industries such as food, paints, medicines, fuels, etc., are encountered. These rusted instruments can get deteriorated easily causing cracks or pinhole formation on metal surfaces. Also, the release of dangerous chemicals through corroded metal equipment impacts the environment in a negative manner and threatens humans. Overall, corrosion damages industrial machines and metallic equipment and reduces production turn over. Approximately US$6500 million/year economic loss is observed in India due to overall corrosion. Around 3.2% of waste products are generated among the domestic wastes due to corrosion in the United States. Corrosion control is enabled by proper awareness and by providing control methods (Harsimran et al., 2021).

6.3 MICROBIAL-MEDIATED CORROSION (MIC)

MIC is prevalent in many environments caused by different bacteria. According to the National Association of Corrosion Engineers (NACE) and American Society for Testing and Materials (ASTM), corrosion is defined as the impact of microorganisms/their activity or by both factors. It is known by different names such as biocorrosion, biodeterioration, and MIC (ASTM G193 2022). There are different definitions for MIC according to organizations such as ISO, IUPAC, etc. For example, metal corrosion is accepted in Europe and some international standards such as ISO 8044–2020. The International Union of Pure and Applied Chemistry (IUPAC) defined corrosion as an irreversible reaction of metals, ceramics, or polymer with the external environment (Knisz et al., 2023). Microorganisms can accelerate the process of corrosion 100 to 1000 times. Other than bacteria, fungi and archaea are also involved in the degradation of metals. Sulfur-oxidizing, sulfur-reducing bacteria, sulfur-reducing archaea, methanogens, iron-reducing, and oxidizing bacteria are a few examples of bacteria involved in metal corrosion (Procópio, 2020). Examples of iron-oxidizing bacteria are *Siderocapsa* sp, *Gallionella* sp, *Sphaerotilus* sp, and *Hyphomicrobium* sp. Slime formers involved in biocorrosion of metals include the *Pseudomonadaceae, Enterobacteriaceae, Micrococcaceae,* and *Bacillaceae* families. *Thiobacillus* and *Beggiatoa* sp. are examples of sulfur bacteria (Loto, 2017).

MIC or biocorrosion was first studied a hundred years ago where organisms were seen to act on metal surface and enable or hasten the corrosion process. Some microorganisms are involved in the inhibition of corrosion. MIC is caused by the cumulative action of medium, metal, and bacteria. During the process of MIC, microbes are stimulated to cathode or anode processes and the development of differential aeration conditions in the electrolyte medium. Metal corrosion induced by the growth and activities of biofilms on metal are related to MIC (Alasvand Zarasvand and Rai, 2014; Dao et al., 2021; Lavanya, 2021). MIC is recognized as a major challenge in different areas such as oil sectors, gas industries, medical implants, and ocean habitats and other water systems. In many places (ocean, dairy, water oil, food, cooling and paper systems, hospitals, power transmissions), biofilm formation is a major concern leading to huge expenses related to cleaning and maintenance activities (Omran and Abdel-Salam, 2020; Fu et al., 2021).

There are several indicators of the presence of MIC. For example, deep pits and SRB correspond to MIC. There are other indicators used in the confirmation of MIC. To validate MIC, microbiological analysis, and tests such as culture, molecular-based measures, optical microscopy, and DNA analysis are used. Stagnant conditions, short or intermittent flow rates in pipelines, presence of deep flask-shaped pits, and high sulfur compounds are other indicators of MIC (Keith et al., 2021).

6.3.1 Mechanisms of MIC

Microorganims growing as biofilms on metal surface accelerate the corrosion process. Both aerobic and anaerobic microbes are implicated in corrosion where the latter is most prevalent in metal corrosion. Microbial products such as organic acids, extracellular hydrogenases, and sulphides speed up the process of iron oxidation leading to the production of hydrogen. The products of metal corrosion such as hydrogen (H_2), nitrate, ferric iron, sulphate, or CO_2 are utilized by respiratory anaerobic microbes as terminal electron donors (Xu et al., 2023). Anaerobic bacteria-mediated corrosion is grouped into type I and II based on its respiration and fermentation processes. Type I is based on electron transfer during respiration. Extracellular electron transfer (EET) is considered as type I MIC. In this process, electrons are transferred from organic carbon oxidation to anode and externally generated electrons are transported to cytoplasm for biocorrosion mediated by oxidation-reduction mechanisms (Victoria et al., 2021; Xu et al., 2016; Du et al., 2007). Type II is based on the microbially produced metabolites causing corrosion. M-MIC or metabolite MIC is a type II process. It is based on the production of microbial metabolite rather than the chemical process. There is another type of biocorrosion (i.e., type III) where microbial cells extract organic carbon and energy by attacking extracellular organic compounds such as polyurethane and its organic plasticizer (Li et al., 2018; Gu, 2003). The classical theory describing metal corrosion is the cathodic depolarization hypothesis explained by Kuhr et al. (1934) when he discussed the high prevalence of metal corrosion consequences occuring in subterranean cast iron pipelines in the Dutch countryside (Li et al., 2013b).

In the corrosion process, alteration of metal surroundings and environment occurs by the growth of biofilms and their metabolism. For example, exopolymeric substance (EPS), alginate production and its accumulation on mild steel can enhance corrosion. Here, EPS prevents local oxygen transport to the surface. Biofilm growth is dynamic where cells grow in number as it matures. During the initial attachment period, some of the metal surface is observed with thick biofilm growth whereas adjoining areas covered by low bacterial numbers. These areas with low cell numbers are constantly exposed to surrounding medium. A differential aeration condition is observed on metal substratum where metabolic activity maximizes oxygen utilization at thick biofilm growth resulting in local anode. Cathodic area corresponds to the metal surface less colonized by bacteria and observed with minimum oxygen utilization. In the direct mechanism of corrosion, electrons are transferred from anode to cathode where oxygen acts as terminal electron acceptor (Hamilton, 2003).

Biofilm accelerates corrosion in different ways including modifying chemical agent transport to the surface of metal, biofilm separation resulting in relaxed removal of protective films from metals, differential aeration conditions, oxidation-reduction potential changes, and removal of inorganic passive layers form metal. The kind of microbial species that form biofilms and are involved in corrosion are influenced by different factors such as pH, temperature, salt concentration, activities of new species, nutrient availability, and metal composition (Pal and Lavanya, 2022). The steps of biofilm-mediated metal corrosion are described in Figure 6.1.

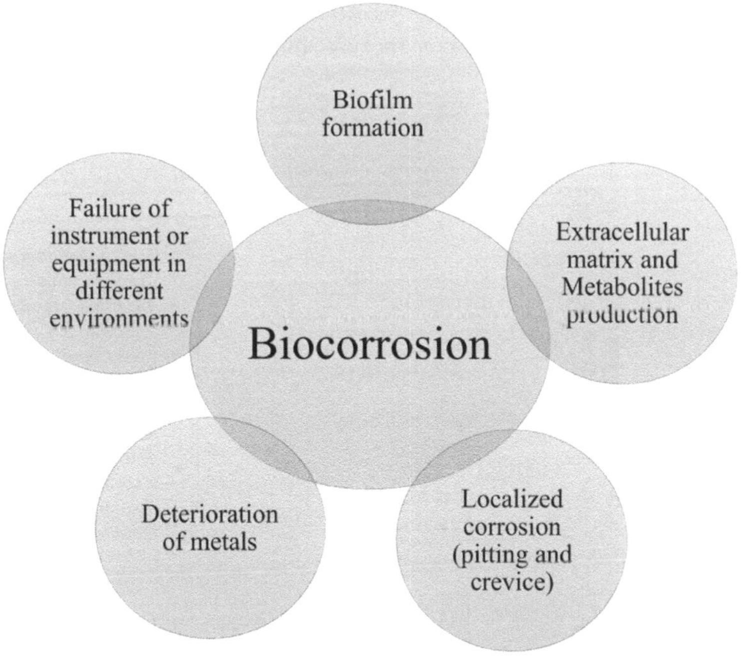

FIGURE 6.1 Different steps involved in the biocorrosion of metals mediated by biofilms.

6.3.2 Biocorrosion of Metals in Different Environments

SRB are the most observed and explored microorganisms in biocorrosion. Though they are obligate anaerobes, SRB can live in aerobic conditions (Figure 6.2). SRB use oxygen as final electron acceptor in aerobic respiration phase and begin to use sulfate respiration when oxygen is in low concentrations (Thauer et al., 2007). SRB-mediated corrosion is favored by the presence of hydrogenase enzyme, which uses hydrogen from cathodic site and leads to cathode depolarization. Iron sulfide produced because of corrosion acts as cathode causing elevation in iron dissolution. The rate of corrosion is dependent on the increased production of iron sulfide (Tripathi et al., 2021; Wang et al., 2023).

SRB-mediated corrosion is prevalent in the oil and gas industries leading to carbon steel pipeline failures. Transfer of electrons extracellularly is the most thermodynamically favorable reaction behind the SRB-induced corrosion of carbon steel in the presence of neutral medium. Here, electron is transferred from elemental iron for the reduction of sulfur causing the generation of energy (Jia et al., 2019). There are reports of SRB that causes corrosion of hot water tanks as some can thrive at temperatures up to 120°C. Thus, thermophilic SRB cause hot water tank corrosion through extracellular electron transfer mechanism. A rotten egg smell is observed in the carbon steel corrosion of water tanks (Jia et al., 2018). Methane (CH_4)-producing microbes are also involved in the MIC of metals. A study conducted reported high rates of carbon steel corrosion of water pipelines in water separator equipment at Halfdan field in 2008. MIC causes major problems like metal pitting and scaling in equipment. Organic acid producing bacteria are involved in the corrosion mechanism, which can decrease pH making a favorable condition for MIC (Augustinovic et al., 2012).

Copper metal corrosion is enabled by SRB despite the fact that copper ions (Cu^+ and Cu^{2+}) can kill microbes. Copper ions are precipitated by SRB utilizing HS^-. Copper

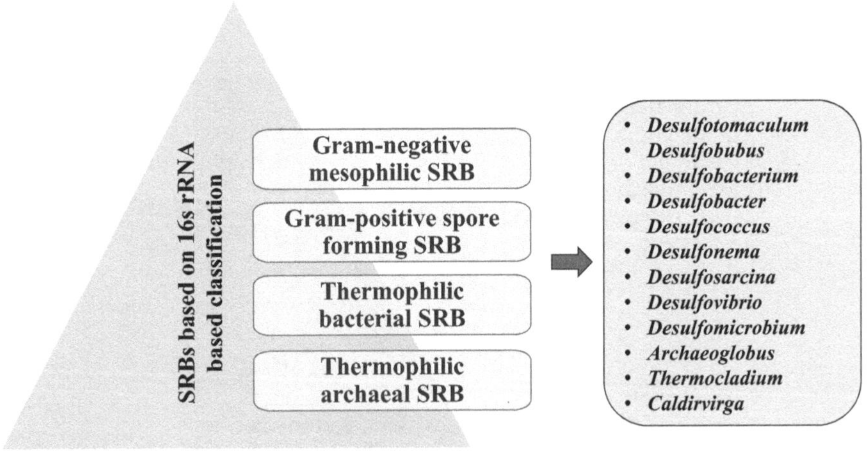

FIGURE 6.2 Grouping of SRBs based on 16s rRNA sequence and examples of genera associated with MIC.

Source: Diao et al., 2023.

ions are precipitated to insoluble chalcocite by metabolites produced by SRB (Wang et al., 2020). According to reports copper corrosion by SRB is stronger (10x) than that of carbon steel. Copper corrosion is also reported by nitrate-reducing bacteria that utilizes nitrate as final electron acceptor (Pu et al., 2020). Copper corrosion leading to piping failures has been observed in AC coils, drinking water, fire sprinklers, and heat exchanger pipes (Hooper et al., 2010) (Dou et al., 2021a). Nitrate-reducing marine bacterium, *Pseudomonas aeruginosa,* has been reported in the corrosion of Cu. Extracellular electrons and ammonia produced by the sessile cells and biofilms cause pitting corrosion of copper (Pu et al., 2020).

Some metals can form corrosion-resistant dense passive film by interacting with oxygen. Still, there are microorganisms that can breach these films and cause corrosion. For instance, industrial and consumer field stainless steels are used widely due to their better corrosion-resistance chromium oxide films (Li et al., 2017a). Nevertheless, sulfate- and nitrate-reducing bacterial biofilms can attack these metals. A study conducted on 304 stainless steel showed accelerated corrosion and pitting mechanism by nitrate-reducing *Bacillus cereus*. In this study, passive film integrity was destroyed by biofilms, which accelerated the corrosion by enhancing the dissolution of anode (Yu et al., 2020). The highly resistant titanium metals are also prone to corrosion attacks by bacteria. *P. aeruginosa* biofilms are reported for their ability to enhance corrosion rate of various metals in marine environment. Saleem Khan et al. (2019) studied the ability of *P. aeruginosa* to speed up the process of titanium corrosion in a simulate marine system. The aerobic and marine bacterium was able to form biofilms on titanium metals, and their metabolic products initiated the localized corrosion process by destroying the passive film (Saleem Khan et al., 2019).

The dominancy of biofilm-associated corrosion in water environments were observed by Ogawa and colleagues. In this study, SS400 carbon steel coupons were observed to be corroded by biofilms in water environment, and different coatings of zinc and copper affected the formation of biofilms. The copper coatings accelerated MIC corrosion compared to uncoated steel whereas the zinc-coated steel coupons were not corroded. The production of metabolites by biofilms accelerated the corrosion rate. Herein, biofilms affected corrosion rate rather than planktonic cells (Ogawa et al., 2020). Corrosion of magnesium alloys in marine environments were observed by the action of sulfate-reducing bacteria. The prolific growth of microbes in mudflat environment causes biocorrosion and failure of metal instruments and buried pipelines. To study the corrosion of magnesium alloys used as cathodic protection, *Desulfovibrio* sp. HQM3 was isolated from pipe-laying soil. In the study, isolated strain caused corrosion of AZ31B magnesium alloy. Pitting of metal also occurred accelerated by the isolated strain (Lan et al., 2022). Table 6.1 provides examples of metal corrosion/inhibition by different bacterial biofilms.

6.4 IMPACT OF BIOFILM-INFLUENCED CORROSION

MIC accounts for 10–20% of total observed corrosion, and according to NACE in 2013, the cost of MIC is responsible for 3.4% global GDP) The estimate shows loss of billions of US dollars/year due to corrosion. Economic problems associated with MIC include metal pipeline failures, instrument damage, safety hazards, and costs associated with the

TABLE 6.1
Role of Bacterial Biofilms on the Acceleration or Inhibition of Corrosion on Different Metals and Alloys

Microorganism	Corrosion acceleration/ inhibition	Type of metals	Reference
Mariprofundus sp. DIS-1	Acceleration	Carbon steel	Mumford et al. (2016)
Tenacibaculum mesophilum D-6 and *Bacillus* sp. Y-6	Inhibition	Q235 Carbon Steel	Ruans et al. (2023)
Enterococcus faecalis	Acceleration	Titanium Grade 4	Conforte et al. (2023)
B. safensis	Inhibition	Steel bars	Kanwal et al. (2023)
Thermacetogenium sp and *Thermodesulfobacterium commune*	Acceleration	Carbon steel	Lenhart et al. (2014)
Desulfovibrio ferrophilus and *Desulfopila corrodens*	Acceleration	Iron coupon	Enning et al. (2012)
Bacillus subtilis	Inhibition	Steel	Guo et al. (2017)
Pseudoalteromonas lipolytica	Acceleration	Steel	Guo et al. (2017)
Prolixibacter sp. MIC1-1	Acceleration	Iron	Iino et al. (2015)
Desulfopila inferna	Acceleration	Carbon steel	Rajala et al. (2015)
Alcanivorax borkumensis and *Alteromonas Macleodii*	Acceleration	AISI-1020 steel coupons	Moura et al. (2018)
Desulfovibrio, Desulfobacter and *Firmicutes* genera	Acceleration	Steel	Li et al. (2017b)
Vibrio neocaledonicus sp.	Inhibition	Carbon steel	Moradi et al. (2015)
Bacillus sp.	Inhibition	Aluminum alloy	Örnek et al. (2002)
Bacillus mycoides	Inhibition	Aluminium	Juzeliunas et al. (2006)
Bacillus mycoides	Acceleration	Zinc	Juzeliunas et al. (2006)
Bacillus cereus	Inhibition	Q235 carbon steel	Hu et al. (2022)

study, monitoring, and prevention activities of MIC. Damage of corrosion can lead to the economic loss of independent companies and industries. Along with economic loss, damage to the environment is also associated with MIC. High remediation and cleaning costs result from MIC as release of toxic products cause damage to human health and environment (Thompson et al., 2022; Y. G. and Mulky, 2023). There are consequences of MIC in our surroundings in home kitchen plumbing, water sewage pipes, dental implants, copper pipes of fire sprinklers, etc. The increasing rate of corrosion due to MIC was observed in 2006 when microbes were thought to be responsible for the Alaskan pipeline leakage caused by a 12.7 mm hole (Dou et al., 2021b).

In nuclear power reactors numerous corrosion-associated instrument failures have been reported since the 1970s leading to a loss of billions of euros (Féron et al., 2009). Biocorrosion is continuously assessed and studied as MIC-associated corrosion is paired with huge economic loss reported in various environments and industries (Lekbach et al., 2021).

6.5 INHIBITION OF MIC

Traditional corrosion inhibition strategies are complicated and exhibit side effects. While corrosion inhibition measures are effective on most corrosion-related problems, they may exhibit environmental problems due to the release of toxic molecules (Lou et al., 2021b). Corrosion inhibition methods include the use of resistant alloys, coatings using organic materials, protections using anode or cathode, and inhibitors of corrosion (Lou et al., 2021a). The surface treatment methods include periodic cleaning of corrosion products (physical treatment procedure) and cathodic protection methods. Anti-corrosion coatings of organic, inorganic, and mixed type are employed. Anti-bacterial elements such as anti-bacterial metal addition and tuning of substrate surface to increase the corrosion resistance are also practiced (Liu et al., 2023). Corrosion inhibitor chemicals are used to reduce the rate of corrosion by interfering with cathodic or anodic or both (oxidation/reduction) reactions (Chugh et al., 2020). These methods possess high risk of contamination, are high in cost, and are difficult to perform. Inhibition of corrosion using microorganisms are novel and environmentally friendly (Lou et al., 2021a)

6.6 ROLE OF MICROBES IN CORROSION INHIBITION

Some microbes are involved in the mitigation or prevention of corrosion rate. Corrosion resistance is mediated by certain microbes present in biofilms by eliminating corrosive agents such as oxygen. Some produce antimicrobial agents inhibiting corrosion-causing bacterial growth and by generating a type of physical barrier that protects metal surfaces from corrosion (Li et al., 2013a). Certain beneficial microorganisms and their biofilms are involved in the protection of metals against corrosion. Zuo and colleagues reported the ability of bacterial biofilm to reduce pitting corrosion by interfering with oxygen reduction at metal surface. Biofilm produced by the bacteria *Bacillus subtilis* WB600 inhibited corrosion of aluminium 2024 in artificial seawater. The results were confirmed when corrosion was detected on metal surface as bacterial biofilms were removed by antibiotic treatment (Zuo et al., 2005). Aerobic and anaerobic biofilms grown on metals show inhibition of corrosion. The key mechanism by which microbial-mediated inhibition of corrosion works is by the removal of oxygen. Anti-microbial products produced by certain biofilms inhibit the growth of SRB on metals and thereby inhibit corrosion (Kip and Van Veen, 2015). Gramicidin S-producing *Bacillus brevis* biofilms were able to inhibit the growth of SRB on mild and stainless steel as reported earlier. The study inhibited the growth of SRBs *Desulfosporosinus orientis* and *Desulfovibrio vulgaris* and prevented the metal from corrosion by 2 to 10 fold (Mansfeld et al., 2004). Another strain of *B. brevis* was reported for their ability to protect mild steel from corrosion by the growth of *D. orientis* and *Leptothrix discophora* SP-6 and the inhibitory activity was contributed by the production of gramicidin S (Morikawa, 2006).

Another approach focuses on the competitive growth of beneficial bacteria over SRB. In oil industries, SRBs are eliminated by the growth of nitrate-reducing bacteria. Recently, exopolysaccharides isolated from certain biofilms have exhibited anti-corrosion properties such as from the biofilms of *Desulfovibrio alaskensis* (Stadler et al., 2010). Low carbon steel coated with EPS isolated from *Leuconostoc mesenteroides* inhibited the process of corrosion. The reaction of EPS with iron caused the formation of iron passive layer, which prevented corrosion of steel. The inhibition of corrosion was strain dependent in the study because some strains reacted to form rust on steel (Finkenstadt et al., 2011).

6.7 CONCLUSION

MIC is associated with equipment/instrument failures in industries and hospitals where metals pipelines or tanks are corroded. Together with the pipeline failures, safety hazards and economic loss are also associated with corrosion. Excessive corrosion may affect human health and environment due to the release of toxic chemicals from corroded pipes. Among several microorganisms studied, sulfur-reducing bacteria exert severe corrosion loss in industries leading to economic and environmental consequences. The dual roles of bacterial biofilms in metal corrosion protection and induction are discussed. Some strains of bacteria are considered as beneficial in corrosion protection. The utilization of biofilms for protecting metals from corrosion attack is one of the reliable and alternate methods of conventional inhibition strategies. This biofilm-mediated protection is compatible with the environment and long-lasting.

REFERENCES

Alasvand Zarasvand, K., and V. R. Rai. 2014. Microorganisms: Induction and inhibition of corrosion in metals. *Int. Biodeterior. Biodegrad.* 87: 66–74.

Anusha Y. G., and L. Mulky. 2023. Biofilms and beyond: A comprehensive review of the impact of Sulphate Reducing Bacteria on steel corrosion. *Biofouling.* 0: 1–19.

ASTM International (ASTM). 2022. Standard Terminology and Acronyms Relating to Corrosion. www.astm.org/g0193-22.html, Accessed on 20th December 2023.

Augustinovic, Z., ø. Birketveit, K. Clements, M. Freeman, S. Gopi, T. Ishoey, et al. 2012. Microbes-oilfield enemies or allies? *Oilf. Rev.* 24: 4–17.

Beech, I. B., and J. Sunner. 2004. Biocorrosion: Towards understanding interactions between biofilms and metals. *Curr. Opin. Biotechnol.* 15: 181–186.

Chugh, B., S. Thakur, and A. K. Singh. 2020. "Microbiologically Influenced Corrosion Inhibition in Oil and Gas Industry," in *Corrosion Inhibitors in the Oil and Gas Industry* (Wiley), 321–338.

Conforte, J. J., C. A. Sousa, A. C. R. da Silva, A. V. Ribeiro, C. Duque, and W. G. Assunção. 2023. Effect of enterococcus faecalis biofilm on corrosion kinetics in titanium Grade 4 alloys with different surface treatments. *Materials (Basel)* 16: 4532.

Dao, V. H., H. K. Ryu, and K. B. Yoon. 2021. Leak failure at the TP316L welds of a water pipe caused by microbiologically influenced corrosion. *Eng. Fail. Anal.* 122: 105244.

Diao, C., W. Ye, J. Yan, T. Hao, L. Huang, Y. Chen, et al. 2023. Application of microbial sulfate-reduction process for sulfate-laden wastewater treatment: A review. *J. Water Process Eng.* 52: 103537.

Dou, W., D. Xu, and T. Gu. 2021a. Biocorrosion caused by microbial biofilms is ubiquitous around us. *Microb. Biotechnol.* 14, 803–805.

Dou, W., Xu, D., and Gu, T. 2021b. Biocorrosion caused by microbial biofilms is ubiquitous around us. *Microb. Biotechnol.* 14(3): 803–805 doi: 10.1111/1751-7915.13690.

Du, Z., H. Li, and T. Gu. 2007. A state of the art review on microbial fuel cells: A promising technology for wastewater treatment and bioenergy. *Biotechnol. Adv.* 25:464–482.

Eckert, R. B. 2015. Emphasis on biofilms can improve mitigation of microbiologically influenced corrosion in oil and gas industry. *Corros. Eng. Sci. Technol.* 50: 163–168.

Enning, D., H. Venzlaff, J. Garrelfs, H. T. Dinh, V. Meyer, K. Mayrhofer, et al. 2012. Marine sulfate-reducing bacteria cause serious corrosion of iron under electroconductive biogenic mineral crust. *Environ. Microbiol.* 14: 1772–1787.

Féron, D., D. Crusset, and J. M. Gras. 2009. Corrosion issues in the french high-level nuclear waste program. *Corrosion.* 65: 213–223.

Finkenstadt, V. L., G. L. Côté, and J. L. Willett. 2011. Corrosion protection of low-carbon steel using exopolysaccharide coatings from *Leuconostoc mesenteroides*. *Biotechnol. Lett.* 33: 1093–1100.

Fu, Q., J. Xu, B. Wei, Q. Qin, L. Gao, Y. Bai, Y., et al. 2021. The effect of nitrate reducing bacteria on the corrosion behavior of X80 pipeline steel in the soil extract solution of Shenyang. *Int. J. Press. Vessel. Pip.* 190: 104313.

Groysman, A. 2010. *Corrosion for Everybody*. Dordrecht: Springer Netherlands.

Gu, J. D. 2003. Microbiological deterioration and degradation of synthetic polymeric materials: Recent research advances. *Int. Biodeterior. Biodegrad.* 52: 69–91.

Guo, Z., T. Liu, Y. F. Cheng, N. Guo, and Y. Yin. 2017. Adhesion of *Bacillus subtilis* and *Pseudoalteromonas lipolytica* to steel in a seawater environment and their effects on corrosion. *Coll. Surf. B Biointerfaces.* 157: 157–165.

Hamilton, W. A. 2003. Microbially influenced corrosion as a model system for the study of metal microbe interactions: A unifying electron transfer hypothesis. *Biofouling.* 19: 65–76.

Harsimran, S., K. Santosh, and K. Rakesh. 2021. Overview of corrosion and its control: A Critical review. *Proc. Eng. Sci.* 3: 13–24.

Hooper, D. G., J. Shane, J., D. C. Straus, K. H. Kilburn, V. Bolton, J. S. Sutton, et al. 2010. Isolation of sulfur reducing and oxidizing bacteria found in contaminated drywall. *Int. J. Mol. Sci.* 11: 647–655.

Hu, Y., C. Chen, S. Liu, Y. Zhou, W. Jia, and Y. Cao. 2022. Biofilm-induced corrosion inhibition of Q235 carbon steel by anaerobic *Bacillus cereus* inoculum in simulated cooling water. *Environ. Sci. Pollut. Res.* 30: 20833–20848.

Ibrahimi, B. El, J. V. Nardeli, and L. Guo. 2021. "An Overview of Corrosion," in *ACS Symposium Series* (American Chemical Society), 1403: 1–19.

Iino, T., K. Ito, S. Wakai, H. Tsurumaru, M. Ohkuma, and S. Harayama. 2015. Iron corrosion induced by nonhydrogenotrophic nitrate-reducing *Prolixibacter* sp. strain MIC1–1. *Appl. Environ. Microbiol.* 81: 1839–1846.

Jia, R., D. Yang, D. Xu, and T. Gu. 2017. Anaerobic corrosion of 304 stainless steel caused by the *Pseudomonas aeruginosa* biofilm. *Front. Microbiol.* 8: 2335.

Jia, R., J. L. Tan, P. Jin, D. J. Blackwood, D. Xu, and T. Gu. 2018. Effects of biogenic H2S on the microbiologically influenced corrosion of C1018 carbon steel by sulfate reducing *Desulfovibrio vulgaris* biofilm. *Corros. Sci.* 130: 1–11.

Jia, R., D. Wang, P. Jin, T. Unsal, D. Yang, J. Yang, et al. 2019. Effects of ferrous ion concentration on microbiologically influenced corrosion of carbon steel by sulfate reducing bacterium *Desulfovibrio vulgaris*. *Corros. Sci.* 153: 127–137.

Juzeliunas, E., R. Ramanauskas, A. Lugauskas, K. Leinartas, M. Samulevičiene, and A. Sudavičius. 2006. Influence of wild strain *Bacillus mycoides* on metals: From corrosion acceleration to environmentally friendly protection. *Electrochim. Acta* 51: 6085–6090.

Kanwal, M., R. A. Khushnood, F. Adnan, A. G. Wattoo, and A. Jalil. 2023. Assessment of the MICP potential and corrosion inhibition of steel bars by biofilm forming bacteria in corrosive environment. *Cem. Concr. Compos.* 137: 104937.

Keith, B., N. Aneta, and C. Ian. 2021. "Investigation into a Microbiologically Induced Corrosion (MIC) Failure of an Onshore Pipeline," in *Institution of Chemical Engineers Symposium Series*.

Khan, M. S., T. Liang, Y. Liu, Y. Shi, H. Zhang, H. Li, et al. 2022. Microbiologically influenced corrosion mechanism of ferrous alloys in marine environment. *Metals* (Basel). 12: 1458.

Kip, N., and J. A. Van Veen. 2015. The dual role of microbes in corrosion. *ISME J.* 9: 542–551.

Knisz, J., R. Eckert, L. M. Gieg, A. Koerdt, J. S. Lee, E. R. Silva, et al. 2023. Microbiologically influenced corrosion – more than just microorganisms. *FEMS Microbiol. Rev.* 47: fuad041.

Lan, X., J. Zhang, Z. Wang, R. Zhang, W. Sand, L. Zhang, et al. 2022. Corrosion of an AZ31B magnesium alloy by sulfate-reducing prokaryotes in a mudflat environment. *Microorganisms* 10: 1–12.

Lavanya, M. 2021. A brief insight into microbial corrosion and its mitigation with eco-friendly inhibitors. *J. Bio- Tribo-Corrosion.* 7: 125.

Lekbach, Y., T. Liu, Y. Li, M. Moradi, W. Dou, D. Xu, et al. 2021. Microbial corrosion of metals: The corrosion microbiome. *Adv. Microbial Physiol.* 78: 317–390.

Lenhart, T. R., K. E. Duncan, I. B. Beech, J. A. Sunner, W. Smith, V. Bonifay, et al. 2014. Identification and characterization of microbial biofilm communities associated with corroded oil pipeline surfaces. *Biofouling.* 30: 823–835.

Li, H., C. Yang, E. Zhou, C. Yang, H. Feng, Z. Jiang, et al. 2017a. Microbiologically influenced corrosion behavior of S32654 super austenitic stainless steel in the presence of marine *Pseudomonas aeruginosa* biofilm. *J. Mater. Sci. Technol.* 33: 1596–1603.

Li, K., M. Whitfield, and K. J. Van Vliet. 2013a. Beating the bugs: Roles of microbial biofilms in corrosion. *Corros. Rev.* 31: 73–84.

Li, K., M. Whitfield, and V. Vliet. 2013b. Beating the bugs: Roles of microbial biofilms in corrosion. *Corros. Rev.* 31: 73–84.

Li, X., J. Duan, H. Xiao, Y. Li, H. Liu, F. Guan, et al. 2017b. Analysis of bacterial community composition of corroded steel immersed in Sanya and Xiamen Seawaters in China via method of illumina MiSeq sequencing. *Front. Microbiol.* 8: 1737.

Li, Y., D. Xu, C. Chen, X. Li, R. Jia, D. Zhang, et al. 2018. Anaerobic microbiologically influenced corrosion mechanisms interpreted using bioenergetics and bioelectrochemistry: A review. *J. Mater. Sci. Technol.* 34: 1713–1718.

Little, B. J., and J.S. Lee. 2009. "Microbiologically Influenced Corrosion," in *Kirk-Othmer Encyclopedia of Chemical Technology* (Singapore: Wiley), 1–42.

Liu, H. W., D. K. Xu, B. J. Zheng, M. Asif, F. P. Xiong, G. A. Zhang, et al. 2018. A synergistic acceleration of corrosion of Q235 carbon steel between magnetization and extracellular polymeric substances. *Acta Metall. Sin.* (English Lett). 31:456–464.

Liu, P., H. Zhang, Y. Fan, and D. Xu. 2023. Microbially influenced corrosion of steel in marine environments: A review from mechanisms to prevention. *Microorganisms.* 11: 2299.

Liu, X., C. Qiu, M. Zhang, E. Zhou, D. Xu, Y. Fan, et al. 2023. synergistic antibacterial activity of benzalkonium bromide and Cu-bearing duplex stainless steel against *Pseudomonas aeruginosa. Microorganisms.* 11: 711.

Loto, C. A. 2017. Microbiological corrosion: Mechanism, control and impact—A review. *Int. J. Adv. Manuf. Technol.* 92: 4241–4252.

Lou, Y., W. Chang, T. Cui, H. Qian, L. Huang, L. Ma, et al. 2021a. Microbiologically influenced corrosion inhibition of carbon steel via biomineralization induced by *Shewanella putrefaciens. npj Mater. Degrad.* 5: 59.

Lou, Y., W. Chang, T. Cui, J. Wang, H. Qian, L. Ma, et al. 2021b. Microbiologically influenced corrosion inhibition mechanisms in corrosion protection: A review. *Bioelectrochemistry* 141: 107883.

Mansfeld, F. B., T. K. Wood, R. Zuo, D. Ornek, B. C. Syrett, R. M. Green, et al. 2004. Inhibiting mild steel corrosion from sulfate-reducing bacteria using antimicrobial-producing biofilms in Three-Mile-Island process water. *Appl. Microbiol. Biotechnol.* 64: 275–283.

Moradi, M., Z. Song, and X. Tao. 2015. Introducing a novel bacterium, *Vibrio neocaledonicus* sp., with the highest corrosion inhibition efficiency. *Electrochem. Commun.* 51: 64–68.

Morikawa, M. 2006. Beneficial biofilm formation by industrial bacteria *Bacillus subtilis* and related species. *J. Biosci. Bioeng.* 101: 1–8.

Moura, V., I. Ribeiro, P. Moriggi, A. Capão, C. Salles, S. Bitati, et al. 2018. The influence of surface microbial diversity and succession on microbiologically influenced corrosion of steel in a simulated marine environment. *Arch. Microbiol.* 200: 1447–1456.

Mumford, A. C., I. J. Adaktylou, and D. Emerson. 2016. Peeking under the iron curtain: Development of a microcosm for imaging the colonization of steel surfaces by *Mariprofundus* sp. strain DIS-1, an oxygen-tolerant Fe-oxidizing bacterium. *Appl. Environ. Microbiol.* 82: 6799–6807.

Ogawa, A., K. Takakura, N. Hirai, H. Kanematsu, D. Kuroda, T. Kougo, et al. 2020. Biofilm formation plays a crucial rule in the initial step of carbon steel corrosion in air and water environments. *Materials (Basel).* 13: 923.

Omran, B. A., and M. O. Abdel-Salam. 2020. "The Catastrophic Battle of Biofouling in Oil and Gas Facilities: Impacts, History, Involved Microorganisms, Biocides and Polymer Coatings to Combat Biofouling," In: A New Era for Microbial Corrosion Mitigation Using Nanotechnology. Advances in Material Research and Technology. Springer, Cham. 47–99.

Örnek, D., A. Jayaraman, B. Syrett, C. H. Hsu, F. Mansfeld, T. and Wood. 2002. Pitting corrosion inhibition of aluminum 2024 by *Bacillus* biofilms secreting polyaspartate or γ-polyglutamate. *Appl. Microbiol. Biotechnol.* 58: 651–657.

Pal, M. K., and M. Lavanya. 2022. Microbial influenced corrosion: Understanding bioadhesion and biofilm formation. *J. Bio- Tribo-Corrosion.* 8: 76.

Procópio, L. 2020. The era of 'omics' technologies in the study of microbiologically influenced corrosion. *Biotechnol. Lett.* 42: 341–356.

Pu, Y., W. Dou, T. Gu, S. Tang, X. Han, and S. Chen. 2020. Microbiologically influenced corrosion of Cu by nitrate reducing marine bacterium *Pseudomonas aeruginosa. J. Mater. Sci. Technol.* 47: 10–19.

Rajala, P., L. Carpén, M. Vepsäläinen, M. Raulio, E. Sohlberg, and M. Bomberg. 2015. Microbially induced corrosion of carbon steel in deep groundwater environment. *Front. Microbiol.* 6: 647.

Revie, R. W. 2011. *Uhlig's Corrosion Handbook*, Wiley (ed. R. W. Revie).

Ruan, X., L. Yang, Y. Wang, Y. Dong, D. Xu, and M. Zhang. 2023. Biofilm-induced corrosion inhibition of Q235 carbon steel by *Tenacibaculum mesophilum* D-6 and *Bacillus* sp. Y-6. *Metals* (Basel). 13: 649.

Saleem Khan, M., Z. Li, K. Yang, D. Xu, C. Yang, D. Liu, et al. 2019. Microbiologically influenced corrosion of titanium caused by aerobic marine bacterium *Pseudomonas aeruginosa*. *J. Mater. Sci. Technol.* 35: 216–222.

Stadler, R., L. Wei, W. Fürbeth, M. Grooters, and A. Kuklinski, A. 2010. Influence of bacterial exopolymers on cell adhesion of *Desulfovibrio vulgaris* on high alloyed steel: Corrosion inhibition by extracellular polymeric substances (EPS). *Mater. Corros.* 61: 1008–1016.

Thauer, R. K., E. Stackebrandt, and A. W. Hamilton. 2007. "Energy metabolism and phylogenetic diversity of sulphate-reducing bacteria," in *Sulphate-Reducing Bacteria: Environmental and Engineered Systems*. 2007:1-38

Thompson, A. A., J. L. Wood, E. A. Palombo, W. K. Green, and S. A. Wade, S. A. 2022. From laboratory tests to field trials: A review of cathodic protection and microbially influenced corrosion. *Biofouling* 38: 298–320.

Tripathi, A. K., P. Thakur, P. Saxena, S. Rauniyar, V. Gopalakrishnan, R. N. Singh, et al. 2021. Gene sets and mechanisms of sulfate-reducing bacteria biofilm formation and quorum sensing with impact on corrosion. *Front. Microbiol.* 12: 754140.

Victoria, S. N., A. Sharma, and R. Manivannan. 2021. Metal corrosion induced by microbial activity – Mechanism and control options. *J. Indian Chem. Soc.* 98: 100083.

Videla, H. A., and L. K. Herrera, L. K. 2009. Understanding microbial inhibition of corrosion. A comprehensive overview. *Int. Biodeterior. Biodegrad.* 63: 896–900.

Wang, J., X. Li, X., F. Guan, Z. Yang, X. Zhai, Y. Zhang, et al. 2023. The isolation of anaerobic and facultative anaerobic sulfate-reducing bacteria (SRB) and a comparison of related enzymes in their sulfate reduction pathways. *Microorganisms.* 11: 2019.

Xu, D., Y. Li, and T. Gu. 2016. Mechanistic modeling of biocorrosion caused by biofilms of sulfate reducing bacteria and acid producing bacteria. *Bioelectrochemistry.* 110: 52–58.

Xu, D., T. Gu, and D. R. Lovley. 2023. Microbially mediated metal corrosion. *Nat. Rev. Microbiol.* 21: 705–718.

Yu, S., Y. Lou, D. Zhang, E. Zhou, Z. Li, C. Du, et al. 2020. Microbiologically influenced corrosion of 304 stainless steel by nitrate reducing *Bacillus cereus* in simulated Beijing soil solution. *Bioelectrochemistry.* 133: 107477.

Zuo, R., E. Kus, F. Mansfeld, and T. K. Wood. 2005. The importance of live biofilms in corrosion protection. *Corros. Sci.* 47: 279–287.

7 Metagenomics Approaches to Understanding the Biofilms of Aquatic Ecosystems

Akshaya Viswanathan, Naji Naseef Pathoor, Gulshan Wadhwa, and Pitchaipillai Sankar Ganesh

7.1 INTRODUCTION

Biofilms are intricate communities of microorganisms, primarily bacteria that adhere to surfaces and form a protective film made of extracellular polymeric substances (EPS) (Zhao et al., 2023). Quorum sensing (QS) triggers the production of virulence factors that stimulate the synthesis of crucial biofilm matrix molecules, thereby promoting biofilm formation across various microbial species (Yan et al., 2019). In aquaculture, biofilms have the capacity to develop on various underwater structures including tank surfaces, pipes, nets, as well as the gills and skin of aquatic organisms (Mahapatra, 2015). Aquatic biofilms serve an important role in a variety of aquatic environments, significantly influencing ecosystem function and dynamics (Ji et al., 2024). Biofilms within aquatic ecosystems adhere to diverse surfaces such as sand, rocks, and leaves, contributing significantly to the ecosystem. These aquatic biofilms serve as a source of energy and organic matter for the food chain, actively participate in the recycling of organic substances, and contribute to the enhancement of water quality (Caister Academic Press, 2016).

Microbial complexity and interactions within aquatic ecosystems are critical components of the intricate web of life found in water environments (Grossart et al., 2020). Aquatic ecosystems, including freshwater and marine environments, host a wide variety of microorganisms that perform crucial functions in nutrient cycling, energy flow, and overall ecosystem function. Examining these microorganisms is crucial for understanding the complex connections that support life in aquatic environments and for devising approaches to oversee and safeguard these vital ecosystems. Progress in metagenomics has supplied potent instruments for delving

DOI: 10.1201/9781003487203-7

into the concealed realm of microbial existence within water ecosystems (Mithila Pandey, 2023).

Metagenomic approaches have transformed the field of microbiology by providing a powerful, culture-independent method for understanding microbial complexities and interactions (Streit et al., 2004). Until now, culture-independent methods have unveiled insights into the genomes, habitats, distribution, and abundance of a diverse array of non-culturable microorganisms. Additionally, these methodologies have facilitated a comprehensive understanding and exploration of microbial ecology, leading to the identification of biotechnological assets such as valuable drugs, enzymes, and biomolecules (Schmeisser et al., 2007). As a result, metagenomic approaches are preferred over culture-dependent methods of studying microbial interactions (Costa et al., 2020). Given the critical role of biofilm development and the ongoing advances in metagenomics within aquatic ecosystems, this chapter is dedicated to elucidating the multifaceted objectives of metagenomics. The exploration encompasses: i) the diverse sequencing technologies employed in metagenomics, ii) the application of meta-transcriptomics to comprehend gene expression patterns, iii) the ecological insights through metagenomic analyses, and iv) the applications of metagenomics in ecosystem management.

7.2 OVERVIEW OF METAGENOMICS IN THE AQUATIC ENVIRONMENT

Within the context of aquatic ecosystem, metagenomics provides an approach to studying microbial communities by directly analyzing the genetic material present in environmental samples. Unlike traditional microbiological methods that rely on isolating and cultivating individual species, metagenomics is a culture-independent tool that allows for the simultaneous investigation of every single microorganism found in a given habitat (Garza et al., 2015; George et al., 2011). This approach is particularly valuable in aquatic ecosystems, where the complexity and diversity of microbial life make it challenging to capture the full spectrum of organisms through traditional culturing methods.

Metagenomics serves as a culture-independent tool by extracting and sequencing the DNA or RNA directly from samples, such as water, sediments, or biofilms. This enables us to investigate the genetic diversity and functional potential of whole microbial populations without having to isolate individual species in laboratory cultures (Yang et al., 2022). The culture-independent nature of metagenomics overcomes the limitations of traditional cultivation-based techniques, providing a more accurate representation of the microbial diversity within aquatic environments (Ercolini, 2013).

Metagenomics offers several unique advantages in studying microbial communities in aquatic ecosystems. One key advantage is the ability to uncover rare or unculturable microorganisms that may play crucial roles in ecological processes (Shuikan et al., 2023). Additionally, metagenomics provides information about the functional capabilities of microbial communities, including their metabolic pathways, resistance mechanisms, and interactions (Azli et al., 2022). This aids in understanding the ecological dynamics, biodiversity, and potential contributions of microbial communities to nutrient cycling, pollutant degradation, and overall ecosystem health in aquatic environments.

7.3 OBJECTIVES OF METAGENOMICS IN AQUATIC ECOSYSTEMS

In aquatic ecosystem biofilm research, metagenomics is an effective tool for determining the intricate dynamics, genetic variation, and functional potential of microbial populations within biofilm structures. Metagenomics in biofilm research aims to gain an in-depth knowledge of these complex microbial consortiums (Behera et al., 2021). Here are the main objectives.

7.3.1 Describing Biofilm Microbial Diversity

Identify and describe the taxonomic diversity of microbial communities found in biofilms. Metagenomics opens an examination of the whole microbial population, which includes both identified and unique microorganisms, resulting in a more complete understanding of biofilm biodiversity (Imchen et al., 2022).

7.3.2 Functional Genetic Identification

Functional genetic identification in biofilms entails characterizing and comprehending the genes and genetic elements responsible for various functions within the biofilm community. This includes genes that play a role in adherence, biofilm development, QS, and other biofilm-forming and maintaining processes (Chistoserdovai, 2010).

7.3.3 Horizontal Gene Transfer (HGT)

Examine the level of incidence and significance of horizontal gene transfer (HGT) within biofilms. Understanding HGT dynamics within biofilm communities offers insights into the adaptability and evolution of biofilm-associated microorganisms (Abe et al., 2020).

7.3.4 Environmental Stressors

Investigate the genetic responses of biofilm-associated microorganisms to environmental stressors. Metagenomic analysis aids in the identification of genes associated with resistance to factors such as nutrient fluctuations, temperature changes, and pollutant being subjected in aquatic settings (Le et al., 2016).

7.3.5 Quorum Sensing System

Examine the genetic foundation for QS and interaction between cells in biofilms. Metagenomics provides glimpses into the signaling mechanisms and regulatory networks that govern microbial interactions in the biofilm community (Pérez-Cobas et al., 2020).

7.3.6 Biotechnological Applications

Metagenomics in biofilm research could lead to applications for investigating new enzymes, as it enables researchers to gather information on microbial diversity including the discovery of new biomolecules that can be used in a variety of industrial processes or environmental applications (Prayogo et al., 2020).

Indeed, each of the outlined metagenomics objectives helps develop an improved comprehension of the genetic composition and functional properties of biofilms in aquatic ecosystems. This knowledge has implications for environmental management, bioremediation strategies, and the development of novel methods for controlling and utilizing biofilms in a variety of applications.

7.4 METAGENOMIC DNA EXTRACTION

Metagenomic DNA extraction plays an important role in the field of aquaculture research, specifically when targeting biofilm samples. This process is crucial for exploring the complex genetic composition of microbial communities residing in aquatic environments (Ashok Kumar et al., 2019). In the context of aquaculture, where biofilms form ecosystems, understanding the genetic complexities is of utmost importance (Prayogo et al., 2020). This section seeks to explain the intricate methodologies used during metagenomic DNA extraction, focusing predominantly on the techniques for attaining high-quality DNA and the measures implemented to preserve the purity and integrity of the extracted genetic material.

The unique characteristics of biofilm in aquaculture environments require specialized techniques for metagenomic DNA extraction. Mechanical disruption, enzymatic digestion, or a combination of both are commonly used to break down biofilm matrices and release microbial DNA (Sharma et al., 2023). Bead-beating methods, ultrasonication, or pressure cycling technology may be incorporated to enhance cell lysis efficiency, ensuring the recovery of DNA from diverse microorganisms within the biofilm (Nithila et al., 2014; Yao et al., 2021). Optimizing these techniques is crucial for capturing the genetic diversity embedded in biofilm communities.

Maintaining the quality of metagenomic DNA is important for downstream applications in aquaculture biofilm research. Robust quality control measures are implemented to assess purity and integrity. Spectrophotometry determines DNA concentration and assesses the presence of contaminants (Rieder et al., 2023). Gel electrophoresis aids in evaluating the size distribution and integrity of DNA fragments (Sehar et al., 2016). Additional purification steps, such as using specialized columns or magnetic beads, are often integrated to remove impurities, ensuring the extracted DNA is of the highest quality.

In brief, obtaining metagenomic DNA from aquaculture biofilm requires specialized techniques designed for the specific attributes of biofilm matrices. Simultaneously, steps are taken to ensure the purity and integrity of the extracted DNA, laying the foundation for accurate and insightful analyses of microbial communities present within biofilm structures in aquaculture settings.

7.5 SEQUENCING TECHNOLOGIES IN METAGENOMICS

Sequencing technologies are crucial in advancing our understanding of the complex genetic profiles within diverse microbial communities, particularly in the field of metagenomics. These technologies serve as instrumental tools, allowing researchers to uncover the genetic complexities of microbial populations and investigate the diverse genetic information present in complex environmental samples (Zhang et al., 2021). Scientists can use various sequencing platforms to discover the composition, structure, and functional potential of these microbial communities, resulting in a better understanding of the role microorganisms play in various ecosystems.

Short-read sequencing platforms, such as Illumina and 454, are widely embraced in metagenomic studies. These technologies are well known for their capacity to generate a vast number of sequences with relatively concise lengths, typically ranging from tens to hundreds of base pairs (Escobar-Zepeda et al., 2015). Utilizing reversible dye terminator chemistry, Illumina sequencing stands out for its capacity to generate large quantities of short reads simultaneously, ensuring efficient high-throughput data acquisition. On the other hand, the 454 platform employs pyrosequencing, a technique that measures the light emitted during nucleotide incorporation (Satam et al., 2023).

The appeal of short-read technologies lies in their cost-effectiveness and exceptional data output. However, they may encounter challenges when attempting to resolve repetitive genomic regions and accurately reconstruct complex genomic structures (Amarasinghe et al., 2020). Despite these challenges, short-read sequencing remains a powerful and widely utilized tool in metagenomics, providing valuable information about the composition and diversity of microbial communities.

In contrast, long-read sequencing technologies, showed by Nanopore and PacBio, provide extended read lengths that range from thousands to tens of thousands of base pairs. Nanopore sequencing operates by threading DNA through a nanopore, detecting changes in electrical current as individual bases pass through. PacBio, utilizing single-molecule real-time (SMRT) sequencing, captures the real-time incorporation of nucleotides during the sequencing process (Logsdon et al., 2020; Oehler et al., 2023).

The hallmark of long-read technologies lies in their ability to resolve complex genomic regions, including repetitive elements, offering a more comprehensive understanding of the genomic architecture within microbial communities (Bianconi et al., 2023). However, it is essential to consider that long-read technologies may come with higher error rates.

By combining the capabilities of short-read and long-read sequencing technologies, scientists could create a detailed depiction of the genetic makeup of biofilms in aquaculture (Rieder et al., 2023). This comprehensive understanding is highly valuable for uncovering the functional roles, interactions, and adaptations of the diverse microorganisms within biofilm ecosystems. These observations not only bring progress in aquaculture research but also hold importance for strategies in environmental management, promoting the sustainability and well-being of aquatic ecosystems (Nam et al., 2023). Figure 7.1 gives an outline of the sequencing technology (Figure 7.1).

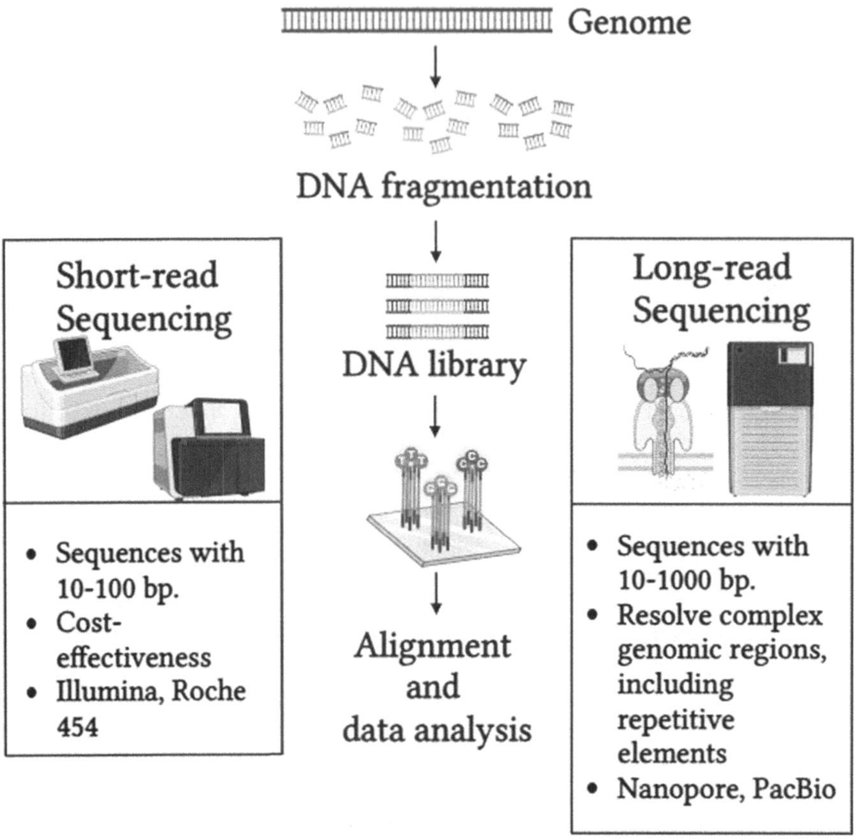

FIGURE 7.1 Represents the sequencing process and short vs. long-read comparison. This figure illustrates the generic sequencing process, showcasing the essential steps involved in DNA sequencing. A concise comparison between short-read sequencing and long-read sequencing technology is also provided. Sequencing platforms like Illumina and Roche 454 are used for short-read sequencing whereas Nanopore and PacBio are used for long-read sequencing.

7.6 APPLICATION OF BIOINFORMATICS TOOLS AND THEIR IMPORTANCE OF METAGENOMIC ANALYSIS

Bioinformatics is an interdisciplinary field connecting biology and computer science that plays a central role in explaining the abundant information present in metagenomic data originating from biofilm communities (Hiraoka et al., 2016). This includes a variety of computational tools and analytical methods aimed at gaining meaningful insights into the taxonomic layout, diversity, and functional properties of microbial communities found within biofilms.

- Taxonomic assignment involves identifying and categorizing the microbial taxa present in biofilm communities based on their genetic sequences. Bioinformatics tools, such as QIIME (Quantitative Insights into Microbial Ecology) and mothur, are commonly used for taxonomic assignment (Galloway-Peña et al., 2020). These tools compare genetic sequences against reference databases to assign taxonomy at various hierarchical levels, providing a picture of the microbial composition within biofilms. Understanding the taxonomic composition is fundamental for characterizing the diversity and community structure of biofilms, helping to identify key microbial players and their roles in ecosystem functioning (Velsko et al., 2018).
- Diversity analysis evaluates the richness and balance of microbial species in biofilm communities, providing an understanding of their ecological interactions and relationships. Alpha diversity measures (e.g., Shannon index, Simpson index) quantify species diversity within a single biofilm, while beta diversity measures compare diversity between different biofilm samples (Feranchuk et al., 2018; Galloway-Peña et al., 2020). Bioinformatics tools, such as diversity indices and ordination techniques like PCoA (Principal Coordinates Analysis), are utilized for conducting these analyses (Wen et al., 2023). Analyzing diversity helps us understand how different microbes are arranged and how they adapt to changes in the environment within biofilms.
- Functional annotation involves predicting the roles and functions of genes present in the metagenomic data from biofilm communities. Bioinformatics tools, including MG-RAST (Metagenomics Rapid Annotation using Subsystem Technology) and KEGG (Kyoto Encyclopedia of Genes and Genomes), facilitate functional annotation by associating genetic sequences with known functions and metabolic pathways (Niu et al., 2017). Understanding the functional potential of biofilm communities helps to decipher their contributions to nutrient cycling, biogeochemical processes, and other ecological functions within aquatic environments.

All of these contribute to a deeper understanding of the ecological roles and dynamics of microbial populations within biofilms in aquatic ecosystems.

7.7 META TRANSCRIPTOMICS AND IMPORTANCE OF GENE EXPRESSION OF AQUATIC ECOSYSTEM

Exploring gene expression in complex microbial communities, especially within aquatic biofilms, is significantly advanced through meta-transcriptomics, a crucial branch of metagenomics. Analyzing the transcriptomes of various microorganisms within biofilms enables us to unravel the active genetic pathways, regulatory networks, and adaptive responses that support the functionality of these biofilm structures in aquatic settings (Bashiardes et al., 2016). This thorough examination of gene expression not only enriches our understanding of microbial ecology but also has far-reaching implications for diverse fields such as environmental management, biotechnology, and the sustainable utilization of aquatic resource (Figure 7.2).

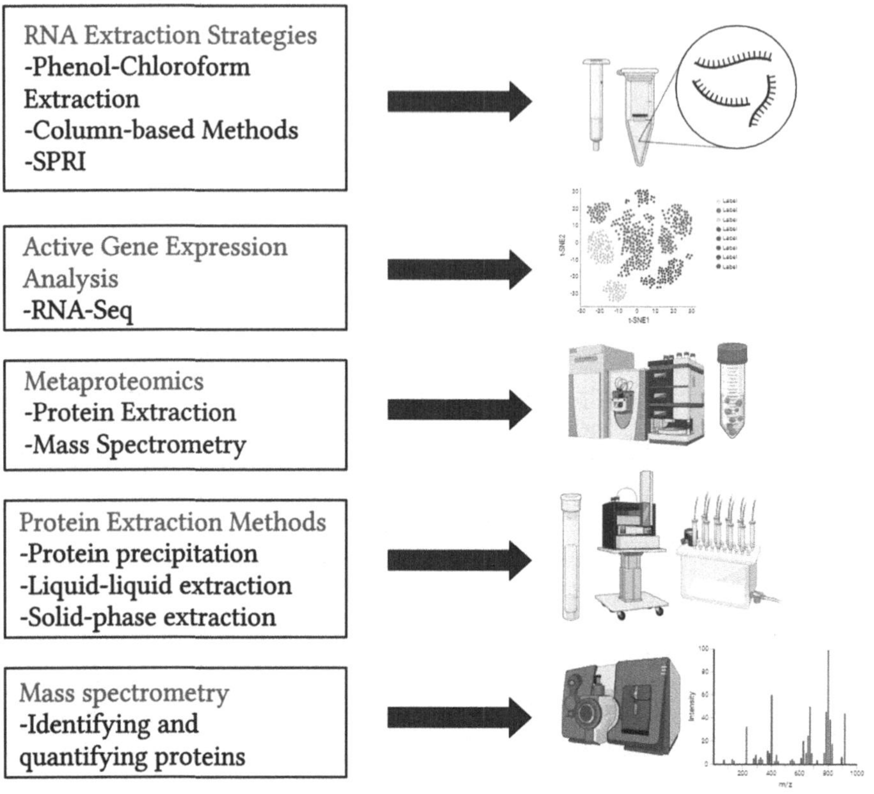

FIGURE 7.2 Various aspects of meta transcriptomics. This figure depicts the combined approach of meta-transcriptomics and metaproteomics in understanding gene expression within aquatic biofilms. It showcases the initial steps of RNA extraction and subsequent analysis through RNA-seq for transcript characterization. Additionally, metaproteomics employs mass spectrometry to identify and quantify proteins, offering a comprehensive view of biofilm microbial activities.

7.7.1 RNA Extraction Strategies from Biofilm Samples

In the context of meta-transcriptomic studies, understanding the dynamic gene expression within biofilms requires the use of effective RNA extraction strategies. Commonly adopted techniques include phenol-chloroform extraction, column-based methods, and solid-phase reversible immobilization (SPRI) (Aburayyan, 2023; Yao et al., 2021). These techniques are essential in ensuring the extraction of high-quality RNA from biofilm samples. This careful methodology enables a thorough understanding of the complex domain of gene expression within these microbial communities.

7.7.2 Analysis of Active Gene Expression in Aquatic Biofilms

Understanding dynamic gene expression in aquatic biofilms is critical for learning about the functional behavior of microbial communities. Meta-transcriptomic analysis employs sequencing technologies, such as RNA-seq, to both identify and characterize the transcripts actively expressed within biofilm environments (Lowe et al., 2017). This method enables researchers to explore the complex molecular mechanisms that contribute to the adaptability of the biofilm and its response to the surrounding aquatic environment.

7.7.3 Metaproteomics: Decoding Protein Landscapes

Metaproteomics serves as a complementary counterpart to meta-transcriptomics, concentrating on the analysis of proteins within biofilms. This method includes the extraction of proteins from biofilm samples, coupled with the application of mass spectrometry techniques to identify and quantify the proteins within the sample (Huang et al., 2014). By doing so, metaproteomics offers a complete view on the functional roles and activities of the microbial community present in the biofilm structure.

7.7.4 Extraction of Different Proteins from Aquatic Biofilms

Efficient extraction of proteins from biofilm samples is a key stage in metaproteomic studies. Essential methods for protein extraction include protein precipitation, liquid-liquid extraction, and solid-phase extraction (Almulhim et al., 2023). These techniques are designed to capture a broad spectrum of proteins, laying the foundation for comprehensive downstream analysis.

7.7.5 Mass Spectrometry Analysis

Mass spectrometry is a key analytical technique used in metaproteomics to identify and quantify proteins. It allows researchers to characterize the protein composition of biofilms, discern post-translational modifications, and learn about the microbial community's functional diversity (Pettersen et al., 2022; Shrestha et al., 2021).

In summary, meta-transcriptomics, coupled with metaproteomics, provides a comprehensive approach to understanding gene expression in aquatic biofilms. The integration of RNA and protein analyses offers a complete view of the functional roles and activities of microbial communities, contributing valuable knowledge to both aquatic ecology and biofilm-related research.

7.8 ECOLOGICAL INSIGHTS AND METAGENOMICS ANALYSIS

Metagenomics is dedicated to unravelling the ecological complexities through detailed analyses, offering valuable understanding of the biodiversity, interactions, and roles of microbial communities that shape ecosystems (Nwachukwu et al., 2022). By tapping into the genomic content of diverse organisms, this discipline significantly contributes to a complete understanding of ecological processes. It plays an important role in aiding environmental management and promotes sustainable practices by providing profound insights into the genetic information embedded in environmental samples.

By analyzing the genetic content of diverse microorganisms residing in biofilms, researchers gain an understanding into the relationships, symbiosis, and potential competition shaping these communities. This exploration extends beyond individual species, thereby providing a better understanding of the collective genomic contributions that drive the ecological balance within biofilm ecosystems (LeighAnne Olsen et al., 2012).

Additionally, metagenomic studies reveal networks of cooperation and competition among microbial species within biofilms. These networks are fundamental to understanding how different organisms collaborate or compete for resources, contributing to the stability or variability of biofilm communities (Schmeisser et al., 2017). Cooperation may involve shared metabolic pathways, biofilm formation, or mutualistic relationships, while competition may manifest in the utilization of similar resources or the production of inhibitory substances (Anju et al., 2022). Metagenomic analyses provide a comprehensive view of these intricate networks, aiding in the interpretation of ecological dynamics within biofilm environments.

This knowledge is necessary for advancing our understanding of ecological processes, promoting sustainable environmental management, and harnessing the potential of microbial communities in various applications.

7.9 APPLICATIONS OF METAGENOMICS IN AQUATIC ECOSYSTEM MANAGEMENT

Applications of metagenomics in aquatic ecosystem management extend to crucial aspects such as nutrient cycling, biogeochemical processes, and the advancement of conservation and sustainability practices (Vijayan et al., 2023). Understanding the genetic makeup of these communities helps identify key players involved in nutrient transformations, such as nitrogen and phosphorus cycling. This aids in predicting and managing nutrient dynamics, ensuring the balance of essential elements crucial for the health of aquatic environments (Ahmed et al., 2024). Metagenomic data offers a detailed look into the genes and metabolic pathways associated with nutrient cycling, contributing to more effective strategies for maintaining ecological balance.

Metagenomics is essential for providing conservation and sustainable management of aquatic ecosystems. By revealing the diversity, interactions, and functional roles of microbial communities, metagenomic data guides conservation efforts by identifying species and understanding ecosystem resilience (Brar et al., 2023). Moreover, metagenomic research provides understanding of sustainable management and helps to reduce the impact of human actions on aquatic environments. This includes developing strategies for pollution control, habitat restoration, and sustainable resource utilization (Bala et al., 2022). Metagenomics serves as a powerful tool for aligning conservation and sustainability goals with the intricate microbial dynamics within aquatic environments.

The applications of metagenomics in aquatic ecosystem management contribute to a better understanding of microbial contributions to nutrient cycling, biogeochemical

TABLE 7.1
Applications of Metagenomics in Aquatic Ecosystem Management

Aspect	Key Insights	Reference
Nutrient cycling	Identification of key microbial players in nitrogen and phosphorus cycling. Prediction and management of nutrient dynamics for a balanced aquatic environment. Understanding genes and metabolic pathways associated with nutrient cycling. Contribution to effective strategies for maintaining ecological balance.	Vijayan et al. (2023)
Biogeochemical processes	Detailed insights into genes and metabolic pathways influencing biogeochemical processes. Enhanced understanding of microbial roles in shaping aquatic ecosystem dynamics.	Vijayan et al. (2023), Ahmed et al. (2024)
Conservation and sustainability	Identification of species crucial for conservation efforts. Understanding interactions and functional roles of microbial communities for ecosystem resilience. Contribution to sustainable management by reducing the impact of human activities. Development of strategies for pollution control, habitat restoration, and sustainable resource utilization.	Brar et al. (2023), Bala et al. (2022)

processes, and the development of conservation strategies, ultimately promoting the sustainability of aquatic ecosystems. Table 7.1 summarizes the applications of metagenomics in aquatic ecosystem management (Table 7.1).

7.10 CONCLUSIONS

In conclusion, metagenomics is a powerful tool that has significantly advanced our understanding of biofilm dynamics in aquatic ecosystems. By revealing the genetic challenges of microbial populations, metagenomics has made it easier to understand the biodiversity, interactions, and roles of microorganisms in shaping biofilm structures. The exploration of microbial activities, gene expression, and protein landscapes within biofilms has provided valuable insights into the functional dynamics of these complex communities. Metagenomics has not only deepened our knowledge of ecological processes but also holds implications for environmental management, conservation, and sustainable practices.

Summarizing the contributions of metagenomics to biofilm research, this chapter has highlighted its role in deciphering microbial interactions, understanding gene expression, and exploring the ecological intricacies of biofilm ecosystems. The integration of meta transcriptomics and metaproteomics has allowed for a comprehensive understanding of both genetic and functional aspects, providing a complete view of biofilm dynamics.

Looking ahead, the future of metagenomics in biofilm research holds exciting opportunities. Continued advancements in sequencing technologies, bioinformatics, and data analysis techniques will enhance the depth and accuracy of metagenomic studies. Exploring the metagenome of biofilms in diverse aquatic environments, coupled with the application of novel methodologies, will broaden our knowledge of microbial communities. Additionally, leveraging metagenomics in conjunction with other omics approaches will enable a more comprehensive and integrated understanding of biofilm ecology.

As we embark on future directions in the field, collaboration across disciplines, the development of standardized protocols, and the incorporation of emerging technologies will further propel metagenomics into a forefront position in biofilm research. These will not only deepen our scientific understanding but also contribute to practical applications, offering solutions for sustainable ecosystem management and the preservation of aquatic environments. The journey of metagenomics in biofilm research continues to unfold, promising exciting discoveries and invaluable contributions to both scientific knowledge and environmental stewardship.

REFERENCES

Abe, K., Nomura, N., & Suzuki, S. 2020. Biofilms: Ht spots of horizontal gene transfer (HGT) in aquatic environments, with a focus on a new HGT mechanism. *FEMS Microbiology Ecology*, *96*(5). doi: 10.1093/femsec/fiaa031

Aburayyan, W. S. 2023. High-throughput RNA extraction method for P. aeruginosa and S. pyogenes biofilms. *Journal of Applied Pharmaceutical Science*. doi: 10.7324/JAPS.2023.28606

Ahmed, N., Zhang, B., Chachar, Z., Li, J., Xiao, G., Wang, Q., Hayat, F., Deng, L., Narejo, M.-N., Bozdar, B., & Tu, P. 2024. Micronutrients and their effects on Horticultural crop quality, productivity and sustainability. *Scientia Horticulturae*, *323*, 112512. doi: 10.1016/j.scienta.2023.112512

Almulhim, F., & Hong, P. 2023. Evaluation of protein extraction methods to improve meta-proteomics analysis of treated wastewater biofilms. *PROTEOMICS*, *23*(20). doi: 10.1002/pmic.202300191

Amarasinghe, S. L., Su, S., Dong, X., Zappia, L., Ritchie, M. E., & Gouil, Q. 2020. Opportunities and challenges in long-read sequencing data analysis. *Genome Biology*, *21*(1), 30. doi: 10.1186/s13059-020-1935-5

Anju, V. T., Busi, S., Imchen, M., Kumavath, R., Mohan, M. S., Salim, S. A., Subhaswaraj, P., & Dyavaiah, M. 2022. Polymicrobial infections and biofilms: Clinical significance and eradication strategies. *Antibiotics*, *11*(12), 1731. doi: 10.3390/antibiotics11121731

Aquatic Biofilms: Ecology, Water Quality and Wastewater Treatment. 2016. Caister Academic Press. doi: 10.21775/9781910190173

Ashok kumar, K., Vinaya kumar, N., Suganya, K., Nimisha, S., Ashok, A., Satheesha, Sujeet Kumar, Sahayarajan, J. J., Monendra Grover, Anil Rai, S. V. Alavandi, M. S. Shekhar, & Vijaya, K. K. 2019. *Metagenomics studies: an aquaculture perspective* (vijayan k k (ed.); vol. 30). New delhi: ICAR-CIBA.

Azli, B., Razak, M. N., Omar, A. R., Mohd Zain, N. A., Abdul Razak, F., & Nurulfiza, I. 2022. Metagenomics insights into the microbial diversity and microbiome network analysis on the heterogeneity of influent to effluent water. *Frontiers in Microbiology*, *13*. doi: 10.3389/fmicb.2022.779196

Bala, S., Garg, D., Thirumalesh, B. V., Sharma, M., Sridhar, K., Inbaraj, B. S., & Tripathi, M. 2022. Recent strategies for bioremediation of emerging pollutants: A review for a green and sustainable environment. *Toxics*, *10*(8), 484. doi: 10.3390/toxics10080484

Bashiardes, S., Zilberman-Schapira, G., & Elinav, E. 2016. Use of metatranscriptomics in microbiome research. *Bioinformatics and Biology Insights*, *10*, BBI.S34610. doi: 10.4137/BBI.S34610

Behera, B. K., Dehury, B., Rout, A. K., Patra, B., Mantri, N., Chakraborty, H. J., Sarkar, D. J., Kaushik, N. K., Bansal, V., Singh, I., Das, B. K., Rao, A. R., & Rai, A. 2021. Metagenomics study in aquatic resource management: Recent trends, applied methodologies and future needs. *Gene Reports*, *25*, 101372. doi: 10.1016/j.genrep.2021.101372

Bianconi, I., Aschbacher, R., & Pagani, E. 2023. Current uses and future perspectives of genomic technologies in clinical microbiology. *Antibiotics*, *12*(11), 1580. doi: 10.3390/antibiotics12111580

Brar, B., Kumar, R., Sharma, D., Sharma, A. K., Thakur, K., Mahajan, D., & Kumar, R. 2023. Metagenomic analysis reveals diverse microbial community and potential functional roles in Baner rivulet, India. *Journal of Genetic Engineering and Biotechnology*, *21*(1), 147. doi: 10.1186/s43141-023-00601-x

Chistoserdovai, L. 2010. Functional metagenomics: Recent advances and future challenges. *Biotechnology & Genetic Engineering Reviews*, *26*, 335–352.

Costa, O. Y. A., de Hollander, M., Pijl, A., Liu, B., & Kuramae, E. E. 2020. Cultivation-independent and cultivation-dependent metagenomes reveal genetic and enzymatic potential of microbial community involved in the degradation of a complex microbial polymer. *Microbiome*, *8*(1), 76. doi: 10.1186/s40168-020-00836-7

Ercolini, D. 2013. High-throughput sequencing and metagenomics: Moving forward in the culture-independent analysis of food microbial ecology. *Applied and Environmental Microbiology*, *79*(10), 3148–3155. doi: 10.1128/AEM.00256-13

Escobar-Zepeda, A., Vera-Ponce de León, A., & Sanchez-Flores, A. 2015. The road to metagenomics: From microbiology to DNA sequencing technologies and bioinformatics. *Frontiers in Genetics*, *6*. doi: 10.3389/fgene.2015.00348

Feranchuk, S., Belkova, N., Potapova, U., Kuzmin, D., & Belikov, S. 2018. Evaluating the use of diversity indices to distinguish between microbial communities with different traits. *Research in Microbiology*, *169*(4–5), 254–261. doi: 10.1016/j.resmic.2018.03.004

Galloway-Peña, J., & Hanson, B. 2020. Tools for analysis of the microbiome. *Digestive Diseases and Sciences*, *65*(3), 674–685. doi: 10.1007/s10620-020-06091-y

Garza, D. R., & Dutilh, B. E. 2015. From cultured to uncultured genome sequences: metagenomics and modeling microbial ecosystems. *Cellular and Molecular Life Sciences*, *72*(22), 4287–4308. doi: 10.1007/s00018-015-2004-1

George, I. F., Bouhajja, E., & Agathos, S. N. 2011. Metagenomics for bioremediation. In *Comprehensive Biotechnology* (pp. 47–57). Elsevier. doi: 10.1016/B978-0-08-088504-9.00364-0

Grossart, H., Massana, R., McMahon, K. D., & Walsh, D. A. 2020. Linking metagenomics to aquatic microbial ecology and biogeochemical cycles. *Limnology and Oceanography*, *65*(S1). doi: 10.1002/lno.11382

Hiraoka, S., Yang, C.-C., & Iwasaki, W. 2016. Metagenomics and bioinformatics in microbial ecology: Current status and beyond. *Microbes and Environments*, *31*(3), 204–212. doi: 10.1264/jsme2.ME16024

Huang, H.-J., Chen, W.-Y., & Wu, J.-H. 2014. Total protein extraction for metaproteomics analysis of methane producing biofilm: The effects of detergents. *International Journal of Molecular Sciences*, *15*(6), 10169–10184. doi: 10.3390/ijms150610169

Imchen, M., Anju, V., Busi, S., Mohan, M. S., Subhaswaraj, P., Dyavaiah, M., & Kumavath, R. 2022. Metagenomic insights into taxonomic, functional diversity and inhibitors of microbial biofilms. *Microbiological Research*, *265*, 127207. doi: 10.1016/j.micres.2022.127207

Ji, B., & Zhao, Y. 2024. Interactions between biofilms and PFASs in aquatic ecosystems: Literature exploration. *Science of The Total Environment*, *906*, 167469. doi: 10.1016/j.scitotenv.2023.167469

Le, P. T., Makhalanyane, T. P., Guerrero, L. D., Vikram, S., Van de Peer, Y., & Cowan, D. A. 2016. Comparative metagenomic analysis reveals mechanisms for stress response in hypoliths from extreme hyperarid deserts. *Genome Biology and Evolution*, *8*(9), 2737–2747. doi: 10.1093/gbe/evw189

Logsdon, G. A., Vollger, M. R., & Eichler, E. E. 2020. Long-read human genome sequencing and its applications. *Nature Reviews Genetics*, *21*(10), 597–614. doi: 10.1038/s41576-020-0236-x

Lowe, R., Shirley, N., Bleackley, M., Dolan, S., & Shafee, T. 2017. Transcriptomics technologies. *PLOS Computational Biology*, *13*(5), e1005457. doi: 10.1371/journal.pcbi.1005457

Mahapatra, A. 2015. Study of biofilm in bacteria from water pipelines. *Journal of Clinical and Diagnostic Research*. doi: 10.7860/JCDR/2015/12415.5715

Mithila Pandey. 2023. Impact of aquatic microbes in nutrient cycling and energy flow in ecosystems. *Journal of Marine Biology & Oceanography*, *12*(3).

Nam, N. N., Do, H. D. K., Loan Trinh, K. T., & Lee, N. Y. 2023. Metagenomics: An effective approach for exploring microbial diversity and functions. *Foods*, *12*(11), 2140. doi: 10.3390/foods12112140

Nithila, S. D. R., Anandkumar, B., Vanithakumari, S. C., George, R. P., Mudali, U. K., & Dayal, R. K. 2014. Studies to control biofilm formation by coupling ultrasonication of natural waters and anodization of titanium. *Ultrasonics Sonochemistry*, *21*(1), 189–199. doi: 10.1016/j.ultsonch.2013.06.010

Niu, S.-Y., Yang, J., McDermaid, A., Zhao, J., Kang, Y., & Ma, Q. 2017. Bioinformatics tools for quantitative and functional metagenome and metatranscriptome data analysis in microbes. *Briefings in Bioinformatics*. doi: 10.1093/bib/bbx051

Nwachukwu, B. C., & Babalola, O. O. 2022. Metagenomics: A tool for exploring key microbiome with the potentials for improving sustainable agriculture. *Frontiers in Sustainable Food Systems*, *6*. doi: 10.3389/fsufs.2022.886987

Oehler, J. B., Wright, H., Stark, Z., Mallett, A. J., & Schmitz, U. 2023. The application of long-read sequencing in clinical settings. *Human Genomics*, *17*(1), 73. doi: 10.1186/s40246-023-00522-3

Olsen, L.A., Choffnes, E. R., & Mack, A. 2012. *The Social Biology of Microbial Communities: Workshop Summary*. Washington: Institute of Medicine (US) Forum on Microbial Threats.

Pérez-Cobas, A. E., Gomez-Valero, L., & Buchrieser, C. 2020. Metagenomic approaches in microbial ecology: An update on whole-genome and marker gene sequencing analyses. *Microbial Genomics*, *6*(8). doi: 10.1099/mgen.0.000409

Pettersen, V. K., Antunes, L. C. M., Dufour, A., & Arrieta, M.-C. 2022. Inferring early-life host and microbiome functions by mass spectrometry-based metaproteomics and metabolomics. *Computational and Structural Biotechnology Journal*, *20*, 274–286. doi: 10.1016/j.csbj.2021.12.012

Prayogo, F. A., Budiharjo, A., Kusumaningrum, H. P., Wijanarka, W., Suprihadi, A., & Nurhayati, N. 2020. Metagenomic applications in exploration and development of novel enzymes from nature: A review. *Journal of Genetic Engineering and Biotechnology, 18*(1), 39. doi: 10.1186/s43141-020-00043-9

Rieder, J., Kapopoulou, A., Bank, C., & Adrian-Kalchhauser, I. 2023. Metagenomics and metabarcoding experimental choices and their impact on microbial community characterization in freshwater recirculating aquaculture systems. *Environmental Microbiome, 18*(1), 8. doi: 10.1186/s40793-023-00459-z

Satam, H., Joshi, K., Mangrolia, U., Waghoo, S., Zaidi, G., Rawool, S., Thakare, R. P., Banday, S., Mishra, A. K., Das, G., & Malonia, S. K. 2023. Next-generation sequencing technology: Current trends and advancements. *Biology, 12*(7), 997. doi: 10.3390/biology12070997

Schmeisser, C., Steele, H., & Streit, W. R. 2007. Metagenomics, biotechnology with non-culturable microbes. *Applied Microbiology and Biotechnology, 75*(5), 955–962. doi: 10.1007/s00253-007-0945-5

Schmeisser, C., Krohn-Molt, I., & Streit, W. R. 2017. Metagenome analyses of multispecies microbial biofilms: First steps toward understanding diverse microbial systems on surfaces. In *Functional Metagenomics: Tools and Applications* (pp. 201–215). Cham: Springer International Publishing. doi: 10.1007/978-3-319-61510-3_12

Sehar, S., & Naz, I. 2016. Role of the biofilms in wastewater treatment. In *Microbial Biofilms – Importance and Applications*. InTech. doi: 10.5772/63499

Sharma, S., Mohler, J., Mahajan, S. D., Schwartz, S. A., Bruggemann, L., & Aalinkeel, R. 2023. Microbial biofilm: A review on formation, infection, antibiotic resistance. *Control Measures, and Innovative Treatment. Microorganisms, 11*(6), 1614. doi: 10.3390/microorganisms11061614

Shrestha, H. K., Appidi, M. R., Villalobos Solis, M. I., Wang, J., Carper, D. L., Burdick, L., Pelletier, D. A., Doktycz, M. J., Hettich, R. L., & Abraham, P. E. 2021. Metaproteomics reveals insights into microbial structure, interactions, and dynamic regulation in defined communities as they respond to environmental disturbance. *BMC Microbiology, 21*(1), 308. doi: 10.1186/s12866-021-02370-4

Shuikan, M., Alshuwaykan, A.M., & Arif, A. 2023. The role of metagenomic approaches in the analysis of microbial community in extreme environment. In *Life in Extreme Environments – Diversity, Adaptability and Valuable Resources of Bioactive Molecules*. IntechOpen. doi: 10.5772/intechopen.108050

Streit, W. R., & Schmitz, R. A. 2004. Metagenomics – the key to the uncultured microbes. *Current Opinion in Microbiology, 7*(5), 492–498. doi: 10.1016/j.mib.2004.08.002

Velsko, I. M., Frantz, L. A. F., Herbig, A., Larson, G., & Warinner, C. 2018. Selection of appropriate metagenome taxonomic classifiers for ancient microbiome research. *MSystems, 3*(4). doi: 10.1128/mSystems.00080-18

Vijayan, J., Nathan, V. K., Ammini, P., & Ammanamveetil, A. M. H. 2023. Bacterial diversity in the aquatic system in India based on metagenome analysis—a critical review. *Environmental Science and Pollution Research, 30*(11), 28383–28406. doi: 10.1007/s11356-023-25195-2

Wen, T., Niu, G., Chen, T., Shen, Q., Yuan, J., & Liu, Y.-X. 2023. The best practice for microbiome analysis using R. *Protein & Cell, 14*(10), 713–725. doi: 10.1093/procel/pwad024

Yan, S., & Wu, G. 2019. Can biofilm be reversed through quorum sensing in pseudomonas aeruginosa? *Frontiers in Microbiology, 10*. doi: 10.3389/fmicb.2019.01582

Yang, S., Johnson, M. A., Hansen, M. A., Bush, E., Li, S., & Vinatzer, B. A. (2022). Metagenomic sequencing for detection and identification of the boxwood blight pathogen Calonectria pseudonaviculata. *Scientific Reports, 12*(1), 1399. doi: 10.1038/s41598-022-05381-x

Yao, Y., Rao, S., & Habimana, O. 2021. Active microbiome structure and functional analyses of freshwater benthic biofilm samples influenced by RNA extraction methods. *Frontiers in Microbiology*, *12*. doi: 10.3389/fmicb.2021.588025

Zhang, L., Chen, F., Zeng, Z., Xu, M., Sun, F., Yang, L., Bi, X., Lin, Y., Gao, Y., Hao, H., Yi, W., Li, M., & Xie, Y. 2021. Advances in metagenomics and its application in environmental microorganisms. *Frontiers in Microbiology*, *12*. doi: 10.3389/fmicb.2021.766364

Zhao, A., Sun, J., & Liu, Y. 2023. Understanding bacterial biofilms: From definition to treatment strategies. *Frontiers in Cellular and Infection Microbiology*, *13*. doi: 10.3389/fcimb.2023.1137947

8 Aquatic Biofilms and Their Role in Biogeochemical Cycling

Monika Mishra, Subhaswaraj Pattnaik, and Pradeep Kumar Naik

8.1 CONCEPT OF BIOFILMS

The life of bacteria has been categorized in two forms: free-floating form termed as planktonic state and the aggregate or sessile form termed as biofilm state. The biofilm or the sessile stage seems to be the chief architect of microbial life history in which the bacterial and archaeal cells have the tendency to form biofilms. Biofilms are basically the highly organized diverse congregations of microbial cells and are enclosed within a self-produced environment or matrix. The self-produced matrix by the biofilm-causing bacteria is predominantly made up of extracellular polymeric materials (ECMs) with the highest share of extracellular polysaccharides (EPS) materials, which are vital for the preservation of the structural integrity (Berlanga and Guerrero, 2016; Penesyan et al., 2021). The primary composition of the biofilm matrix includes the EPS followed by extracellular proteins, extracellular nucleic acids (with particular reference to eDNA), lipids, etc. The EPS contribute the majority of the total organic content of biofilm (Okuda et al., 2018). The composition of the EPS further varies physically as well as chemically based on the microorganisms as well as the microenvironment. The EPS is hydrophilic in nature because of the fusion of a large sum of water in its structure by hydrogen bonding. But EPS can be both hydrophilic and hydrophobic in nature (Kim et al., 2023).

Apart from the cellular materials, it is also made up of several non-cellular materials. The composition of these non-cellular materials is concomitantly dependent on the microenvironment in which the biofilm matrix has been developed. The non-cellular materials include clay, minerals, blood components, etc. The surface medium for the growth and establishment of the biofilm includes both abiotic surfaces as well as the biotic surface. Biotic surfaces include biomedical devices and living tissues; whereas the abiotic surfaces include the water pipelines and natural aquatic systems (Donlan, 2002). The matrix in bacterial biofilms provides a protective arsenal for survival under harsh environmental conditions including host

DOI: 10.1201/9781003487203-8

immune system and antimicrobial treatment. The host immunity-evading strategies employed by biofilm matrices corresponds to the modulation in metabolic activities and concomitantly exhibits resistance to antibiotics (Vestby et al., 2020). The biofilm system represents one of the most effective life-affirming strategies as they have the tendency to establish and survive in several extreme conditions. Among the different microenvironments, biofilms in aquatic ecosystem with special reference to marine biosystems significantly contribute to widespread microbial diversity and their highly complex systems have the ability to tackle the highly dynamic environmental conditions (Qian et al., 2022). The various stages of the development of biofilm are regulated by several pathophysiological and environmental parameters that collectively promote the survival mechanisms under harsh environmental conditions (Condinho et al., 2023).

8.2 ROLE OF QUORUM SENSING IN BIOFILMS

Similar to multicellular organisms, unicellular microorganisms also have the tendency to communicate with each other through the sensing of chemical signals. The cell density dependent production, sensing, and the response of specific signal molecules by microorganisms is labelled as quorum sensing (QS). The QS in pathogenic microorganisms not only regulates the metabolic activities of planktonic cells but also induces the formation and development of biofilms by producing several pathophysiological factors responsible for inferring bacterial virulence and chronic bacterial infections (Preda and Săndulescu, 2019; Law and Tan, 2022). As discussed earlier, the biofilm establishment and development process includes cascades of strategic phases with each phase metabolically controlled by several genetic, physiological, and environmental factors. For example, some physiological factors facilitate the initial adhesion of bacteria to different surfaces whereas several factors aptly regulate the successful establishment of microcolonies and their maintenance of structural integrity. Several other physiological factors are also responsible for biofilm development and successive scattering of biofilms (Ghosh et al., 2022). Since the QS regulatory network in microorganisms represents the community structure and its functional attributes, the QS phenomena have a strategic influence on the course of a biofilm community in terms of stability, maintenance, and pathophysiological behaviors (Lami et al., 2023). The elucidation of several virulence factors, activation of associated chronic infections, and relative expression to control collective traits in biofilm mode are triggered by QS regulatory behavior and pertinently modulate the course of antimicrobial therapy. Hence, it is important to control the several classes of microbial QS systems for the regulation of chronic biofilm infections (Rutherford and Bassler, 2012; Warrier et al., 2021).

8.3 BACTERIAL BIOFILMS: AN ENVIRONMENTAL PERSPECTIVE

As discussed earlier, the inherent capability of a microbial community to form biofilms provides a subsequent level of defense against physical and chemical stress and environmental persistence, and also promotes competent acquirement of nutrients (Martín-Rodríguez, 2023). Since biofilms constitute the interplay between the species of single microbial species and more frequently among the diverse complex microbial community in different environmental conditions, the physiological and physicochemical behavior of the biofilm-forming microbes is more or less governed by the environmental complexity (Balcázar et al., 2015). The environmental circumstances play a key role in the growth of bacterial biofilms. The factors that influence biofilm production include light penetration, hydrodynamics, pH, oxygen level, environmental temperature, availability of the nutrients, presence or absence of specific ions, osmolarity, water current, etc., and certain biological factors including autotrophs, heterotrophs, bacteria, algae, etc. The presence of the secondary messenger (i.e., c-di-GMP) is also controlled by several environmental factors that ultimately control the aspects involved in biofilm growth such as EPS, cell motility, surface proteins, and cell appendages (Koestler and Waters, 2013; Toyofuku et al., 2016). The primary adhesion of planktonic bacterial cells to the surface is greatly influenced by the environmental pH conditions. The biofilm slime that is produced by the microbial community during the biofilm formation is also dependent on the pH and subsequent change in the pCO_2 level of the environment as well as the physicochemical nature of the surface. The change in pH concomitantly alters the activity of different enzymes because enzymes are highly dependent on the variation in pH (Liu et al., 2020; Rath et al., 2022). The secretion of the extracellular polysaccharides during biofilm formation also depends on pH, but the pH varies from species to species. However, the standard pH for most bacteria is around 7–9, which is suitable for the maintenance of biofilm architecture (Ju et al., 2022).

Apart from pH, optimum temperature also affects the course of biofilm dynamics by modulating the nutrient acquisition. In addition, the temperature also affects the relative expression profiles of several biofilm-associated genes (Bisht et al., 2021). The optimum temperature for the formation of robust and recalcitrant biofilm matrices ranges between 25–42 °C (Roy et al., 2021). The intake of the nutrients and the rate of reaction of the different enzymes regulates the biochemical as well as the physiological cycle of the bacteria, which in turn is associated with the enhancement of the growth of the bacterial community leading to the rapid development of the biofilm matrix (Fathollahi and Coupe, 2021). Basically, nutrient-sufficient and nitrogen-deficient nutrients are optimal for biofilm growth and maintenance. Meanwhile, more nutrient uptake above the optimal state negatively affects the fate of biofilm dynamics (Shaikh et al., 2022). Hence, optimal temperature is required for the maintenance of bacterial growth and the inherent ability to form biofilm matrices, which aids in enabling adaptive traits for host colonization (Hernández-Cabanyero et al., 2020). The viscosity of the extracellular polysaccharides is also greatly affected by the temperature as the EPS forms a gel-like substance with increase in temperature and thereby provides intricate stability to the biofilm matrices (Garrett et al., 2008;).

The production of bacterial energy differs based on the availability of oxygen. For example, the biological activity of the bacteria can be decreased due to the lack of oxygen supply. Thus, oxygen can be a significant factor that can influence the development and composition of the biofilm matrix (Ahn and Burne, 2007). In addition to these environmental factors, several other factors such as humidity, oxidoreductive capacity, the action of water in the atmosphere, and hydrostatic pressure affect the biofilm dynamics (Stanaszek-Tomal et al., 2020; Silva et al., 2022).

8.4 MICROBIAL BIOFILMS IN AQUATIC ENVIRONMENT

Based on historical perspectives, biofilms can be considered as the most effective form of microbial life with aided advantages of withstanding extreme environmental circumstances. Apart from the terrestrial environment, biofilms have the tendency to establish in the most complex and dynamic aquatic ecosystems with considerably higher microbial biodiversity (Neagu et al., 2017). The rise of biofilms in the aquatic environment with special reference to marine environments has a direct influence on the world economy with catastrophic impacts on maritime transport, aquaculture systems, oil and gas industries, food industries, etc. (de Carvalho, 2018). In the successful establishment of biofilms in the marine environment, physicochemical parameters such as temperature, salinity, pH, nutrient availability, availability of light source, depth of water, seasonal variations, tidal movements, etc., are considered to be highly important. Besides the physicochemical parameters, several biogenic factors with particular reference to biofilm community composition and their diversity affect biofilm dynamics (Caruso, 2020). In aquatic environments, biofilm communities based on biomass composition and microbial biodiversity also significantly contribute to the maintenance of biogeochemical cycles in the aquatic atmosphere and multifaceted perspectives on environmental biodegradation (Newton et al., 2011; Debroy et al., 2022; Shilky et al., 2022; Sonawane et al., 2022).

The biofilm-producing bacteria in the aquatic environment are rarely planktonic, rather they are basically surface associated, which allows them to express pathogenicity by regulatory expression of virulence factors as they are highly tolerant to environmental stress (Johnson, 2008). For example, multidrug-resistant nosocomial pathogenic bacterium *Pseudomonas aeruginosa* has the tendency to produce alginates (provide stability to the ECM) by activation of QS regulatory mechanism only when attached to a surface (Gheorghita et al., 2022). The cholera-causing biofilm clusters are found in the Bay of Bengal. The seawater carries the cholera-causing bacteria and the disease is mostly found in the population of Bangladesh. *V. cholerae* survives for several years in the non-culturable state within bunches of biofilm in the costal belt, accounting for the widespread cholera in that geographical domain (Islam et al., 2007).

8.5 CONCEPT OF AQUATIC BIOFILMS

Since the aquatic ecosystem harbors a rich diversity of microorganisms with dynamic environment, the aquatic biofilms are considered to be most tolerant assemblages of heterogeneous microbial community (Rivera et al., 2021). The aquatic biofilms have a key role to play in the maintenance of aquatic ecosystem by affecting primary production, nutrient cycling, and water purification, which alter the microbial distribution (Sgier et al., 2018). Based on the diversity of aquatic ecosystems, the composition of the biofilm community, its ecological values, geographical distributions, ecological advantages, and negative implications also substantially differ.

8.5.1 Structure and Diversity of Aquatic biofilms

The structure and functional attributes of aquatic biofilms significantly differ in terms of the composition of microbial community with diverse range of bacteria, archaebacteria, fungi, algae, etc. (Besemer, 2015). Accordingly, freshwater- and marine ecosystem-based aquatic biofilms have their own characteristic structure and functional attributes.

8.5.1.1 Freshwater Biofilms

The occurrence of bacterial biofilms in freshwater habitats include diverse consortia of microorganisms distributed in different sectional compartments such as benthic, hyporheic, and aquifer sediments. These microorganisms facilitate the acquisition and distribution of inorganic as well as organic nutrients (Romaní, 2010). The biofilms that grow on the surface of leaf, wood, and rock are known as benthic surface biofilms. The organisms that grow on these surfaces are protozoa, bacteria, fungi, algae, and cyanobacteria. Epiphytic biofilm is considered as the main contributor of freshwater ecosystems with higher diversity of autotrophic (e.g. diatoms, cyanobacteria) and heterotrophic communities (e.g. bacteria, fungi) (Wijewardene et al., 2022). Since the freshwater biofilms maintain the cycling of organic materials with special reference to diverse enzymes associated with metabolic activities these biofilm communities are considered to remedial measures for the remediation of several environmental pollutants thereby maintaining the sustainability of aquatic ecosystem (Yin et al., 2022). Hence, these biofilms could be used as potential bioindicators of the health of the aquatic biosystems and could be helpful in water purification approaches (Steven et al., 2022). In this context, microplastics-based biofilms are being considered as strategic approaches for drinking water as well as wastewater treatment plants (Nguyen et al., 2023). Several ecological parameters including temperature, pH, light, availability of nutrients, water flow, surface properties, etc., are answerable for the structure and functional attributes of the biofilm matrices.

8.5.1.2 Marine Water Biofilms

As one of the most dynamic ecosystems in the world, marine ecosystems are the most diverse ecological niche for harboring a wide spectrum of microorganisms with the inherent tendency to form biofilms to tackle these dynamic environments (Qian et al., 2022). Recent advancements in big data analysis and metagenomic approaches have shown that the marine environment constitutes a highly diverse biofilm-forming microbial community that has still not yet been explored (Zhang et al., 2019). In the marine atmosphere, different species of micro- and macro-organisms have the tendency to quickly colonize on different surfaces associated with diverse ecological niches, which in turn results in the development of highly heterogenous biofilm communities (Antunes et al., 2019). The establishment of biofilm in the most dynamic environments affects the ecological biosystem and biogeochemical activities (Sushmitha et al., 2023). The marine biofilms are considered as the pioneer on the process of marine biofouling with critical implications on the environment and world economy with special reference to shipping industries by promoting corrosion.

Marine biofouling starts with the initial colonization of prokaryotic microorganisms followed by colonization of unicellular eukaryotes and the conscription of larvae and algal spores. In the process of establishment of marine biofilms to the state of marine biofouling, several aspects such as physicochemical properties of saline water and the surface possessions of the substratum are considered to be important parameters (Catão et al., 2019; Papadatou et al., 2021; Brown et al., 2023). In the marine ecosystem, the microbiologically influenced corrosion (MIC)-induced biofouling is an important event for establishment of microbial consortium and their implications on economy with an estimate of $0.5 trillion annually at the global stage (Tuck et al., 2022). The arrangement and architecture of marine biofilms not only rely on microbial species composition but also depend on physical factors like viscosity, hydrodynamic forces, flow rate of water, turbulence of water, EPS production, availability of the nutrients, and other biological factors (Stoodley et al., 2002).

8.5.2 Ecological and Environmental Perspectives on Aquatic Biofilms

Biofilms are mostly comprised of a single bacterial species but more often there are a group of multiple species that form the biofilm matrix such as bacteria, protozoa, algae, etc. (Teschler et al., 2015). The microorganisms in the aquatic environment have the tendency to form highly recalcitrant biofilms by colonizing several surfaces. The establishment of biofilms further modulates the aquatic environment by altering the energy supply chain and nutrient recycling leading to modulation in water quality. The biofilm-forming microbial species substantially alter their lifestyle based on alteration in aquatic environmental conditions, availability of nutrients, physiological status, etc. (Grossart et al., 2020). The aquatic biofilms not only develop in the benthic substrata but also in the free-moving microaggregates. Aquatic biofilms play a major role in the decomposition of organic matter, nutrient pedaling, and biogeochemical cycling (Romaní, 2010). According to Sabater et al. (2007), the presence of streambed biofilms increases water quality and ecological status of the surroundings. The aquatic biomes are frequently affected due to human activities including continuous discharge of industrial pollutants in water bodies, biomedical waste discharge, etc. As micro- and macroorganisms are regularly exposed by low concentrations of pollutants they change their behavior accordingly. This condition provides environmental stress to the organisms and they become resistant to several commonly used antibiotics (Osorio et al., 2014). Since antibiotics are also present in water bodies at low concentration they trigger stress conditions and ultimately accelerate the horizontal gene transfer (HGT) of antibiotic-resistant genes (ARGs) (Maiques et al., 2006).

8.5.3 Impact of Bacterial Biofilms on Aquatic Ecosystems

The presence of biofilm has some major benefits in the management of wastewater. The major benefits of biofilm-based wastewater treatment include operational flexibility, biomass residence time, reduced hydraulic retention time, ability to degrade recalcitrant compounds, and most importantly the lower sludge assembly due to the lower microbial development within the biofilm mass. (Chen and Chen, 2000). The reaction rate and population dynamics are also controlled by the biofilm system (Paula et al., 2020). There are specific biofilm reactors for the management of wastewater: fixed bed reactor and mobile bed reactors. The fixed bed comprises trickling filters, rotating biological contractors, underwater granular sieves while the mobile beds include fluidized beds, stormy mobile beds, and air lift reactors (Paula et al., 2020, Lusinier et al., 2021). In the fixed-bed system, the biofilm grows over the static surface. The continuous liquid flow provides nutrients and oxygen to the microorganisms. But in the moving-bed system, biofilm grows in the continuous moving media and is maintained by mechanical stirring and high air and water velocity (Rodgers and Zhan, 2003). The media or material in which the biofilm will grow is selected by their density, size, porosity, resistance to erosion, etc. (Ødegaard et al., 2000). By applying specific shear force, the thickness of the biofilm in the reactors can be maintained. Apart from these wastewater treatments, biofilms can also be used for the treatment of industrial wastes. There are several advantages of biofilm used in wastewater treatment including biomineralization, biodegradation, bioaccumulation, etc. (Singh et al., 2006). The biofilm matrix tends to absorb heavy metals and organic solvents. It also helps in the removal of alkanes, chlorophenols, pyrene, etc. (Jin and Englande, 1998). Bioaugmentation is a process of using the precise bacterial strains to increase the efficacy of the wastewater treatment with an intention to degrade the environmental pollutants under specific atmospheric conditions.

The biofilm matrices can be formed in the solid planes and can form a greasy and slippery coat. It plays a dynamic part in the transmission of human diseases. The formation of the biofilm matrix around the pathogenic bacteria protects itself from antimicrobial drugs or antibiotics. For example, the biofilm produced by *V. cholerae* in the gut of cholera patients is mostly protected from the high acidic condition in the gut, variations in the pH, etc. The aquatic ecosystem is a rich source of pathogenic microbes. Several human pathogenic diseases are caused by the aquatic environment, and the majority of the bacteria that are found in the aquatic bodies have the ability to produce biofilms (Fong et al., 2006). The formation of biofilms by the pathogenic bacteria in the aquatic environment plays a vibrant part in the transmission of diseases. The biofilm form bacteria by seeding a large number of cells and by dispersing the cells that can ultimately initiate or cause infections/diseases in the human body.

8.6 CONCEPT OF BIOGEOCHEMICAL CYCLING

8.6.1 Biogeochemical Cycling and Environmental Sustainability

The continuous movement and transformation of chemical elements between living organisms, atmosphere, and the various parts of the Earth's crust is termed as "biogeochemical cycle." In this cycle matter is always conserved and the elements to and from major pools. There is a continuous acceleration between the earth elements because of human activities. For example, the rate of reactive nitrogen has doubled in the biosphere since pre-industrial times (Canadell et al., 2021). In other words, the cycling of the essential chemical components from the biotic surfaces to the abiotic surfaces of the earth is essential for the maintenance sustainability of the life (Figure 8.1). Certain chemical, biological, and geological factors are involved in the biogeochemical cycle. The biological part includes the life and death of the plant and animal bodies; the geological part includes the earth, which helps in the decomposition of the dead plants and animal bodies; and lastly the chemical part includes the process by which the organic matters return back to the chemical elements in the earth (Wei et al., 2019). The chemicals that take part in the biogeochemical cycle include sulphur, water, nitrogen, phosphorus, carbon, and oxygen. The chemical cycle can be further sub-divided into two parts: gas cycle or sedimentary cycle. In the gaseous cycle, the chemical elements move in the atmospheric surface whereas in the sedimentary cycle the elements move from land, water, and sediment. The elements involved in the gas cycle include carbon, nitrogen, and oxygen. The chemical elements involved in the sedimentary cycle include phosphorus and sulphur. Globally the biogeochemical cycle impacts the environment. Climate is also influenced by the biogeochemical cycle; for example, the emission of carbon dioxide causes climate change, and cycles like phosphorus, nitrogen, and sulfur also influence the climatic changes. In marine biochemistry, the interactions are focused on the cycling of crucial components within seas and oceans (Achterberg, 2014). The primary element, carbon, is the key component in the biogeochemical cycle and affects climatic conditions and living organisms. During the photosynthetic process, the phytoplankton play a chief role in biogeochemical cycling through nitrogen and carbon fixation (Rousk and Bengtson, 2014). And interestingly, the zooplankton consume the phytoplankton.

The aquatic biogeochemical cycle occurs within water bodies and includes saltwater, brackish water, sea or ocean water. In this cycle, the chemical nutrients are imported or exported from the aquatic ecosystem. These exchanges occur above the atmosphere, below the ocean, and from the land surfaces. In the aquatic biogeochemical cycle certain chemicals are cycled such as calcium, selenium, carbon, phosphorus, oxygen, nitrogen, mercury, hydrogen, etc. In this cycle the marine microorganisms play a crucial role (Covey et al., 2021). The forces that are used for the biogeochemical cycle include the metabolic reactions that occur within the marine microorganisms, geological processes connecting the Earth's layer, and lastly the chemical reactions among the substances (Grossart et al., 2020).

8.6.2 Microorganisms and Their Role in the Biogeochemical Cycle

Archaebacteria play a key role in the biogeochemical cycle. They are considered as a diversified organism and are abundantly present in the biosphere. The archaea carry out the global nutrient cycle (i.e., sulfate-dependent anaerobic methane oxidation, methanogenesis, and aerobic ammonia oxidation). Although both the bacteria and archaea are responsible for this process they have certain metabolic and ecological differences and these differences need to be clarified. Archaea play an energetic role in biogeochemistry (Zhang et al., 2023). The archaeal metabolism also critically endures the production of the archaeomass, which ultimately determine the functional attributes of archaea in the biogeochemical cycles. Based on the different studies,

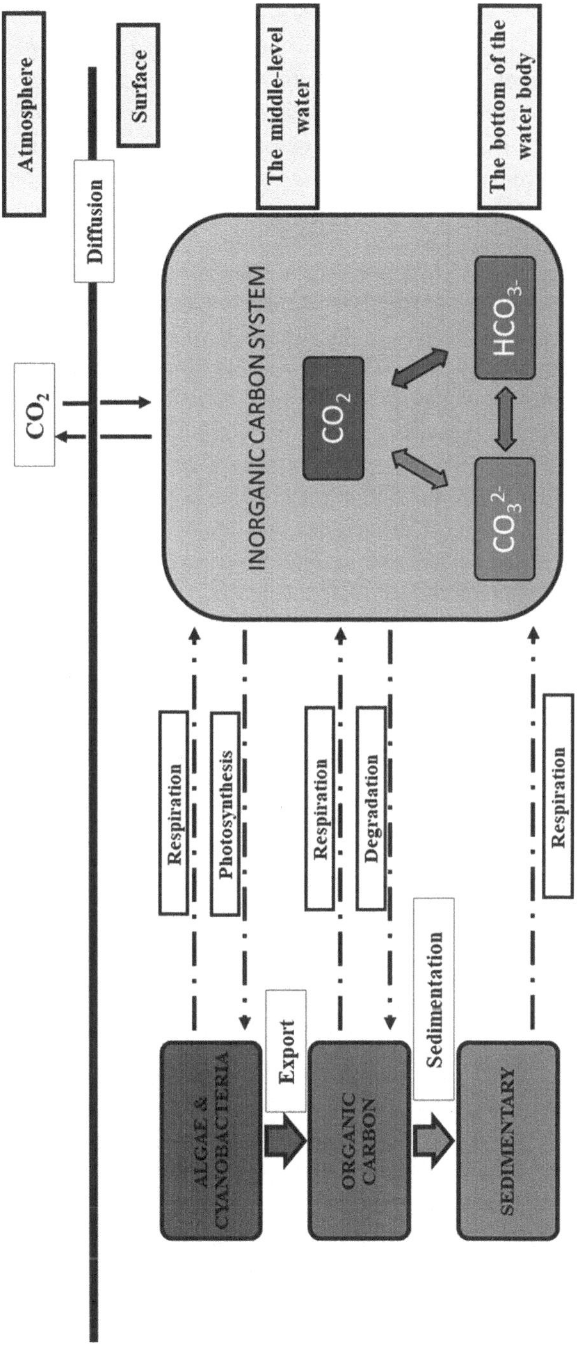

FIGURE 8.1 Schematic representation of marine biogeochemical cycle.

the archaea books for the more than 20% of all prokaryotes in aquatic ecosystem, 5% in the soil surfaces. Archaea dominate the production of methane but also play a significant role in the oxidation of hydrocarbon. Some of the archaeal organisms help in the production and mineralization of organic matter (Offre et al., 2013). The archaea helpful in carbon assimilation include *Crenarchaeota, Thaumarchaeota,* and *Euryarchaeota.* In the atmosphere methane is considered as an essential greenhouse gas and also a source of energy for human sustanibility. By conserving their metabolism, the methanogenic bacteria (which are anaerobes) produce the methane. The methanogenic archaea include *Thermoplasmata, Methanopyri, Methanomicrobia, Methanococci,* etc. (Laskar et al., 2018).

The anaerobic fungi seem to be present in deep environments or in the deep sea. The anaerobic fungi and the nitrogen cycle are correlated with each other as they treat nitrate as an electron acceptor in the deficiency of oxygen. These fungi are from the phylum neoallimastigomycota. These anaerobic fungi can survive in the depth of the ocean. These fungi have significant importance and are involved in the biogeochemical process and also possess a symbiotic relation between the prokaryotes and help in the overall biogeochemical cycling process (Drake and Ivarsson, 2018). Due to the deficiency of nutrients, the rate of metabolism in the deep sea is very slow and are completely dependent on hydrogen gas, but the abiotic source of the hydrogen gas is quite insufficient. The anaerobic fungi are termed as heterotrophs and produce the hydrogen gas that is required for their acceleration of the metabolic process. This confirms that the fungi play a vital role in the biogeochemical cycle in the deep biosphere (Jiang et al., 2023).

The marine bacteria play a dynamic role in the biogeochemical cycling of Fe in deep seas or oceans. The prokaryotic microbes contain the iron in offshore waters and these microbes uptake a large amount of iron from the oceanic water (Joye et al., 2022). According to the literature, in the North Pacific Ocean, they consume about 50% of the dissolved Fe. In the oligotrophic region, the photosynthetic bacteria are important as they play a major role in the recycling of iron (Boyd and Ellwood, 2010). Hence, microorganisms play a major role in the biogeochemical cycle (Table 8.1).

8.6.3 Bioremediation of Aquatic Ecosystem

The bioremediation process involves the restoration of the biological contaminant, re-integration of the polluted sites, and cleaning up of the contaminated areas by the direct or indirect use of the organic and inorganic chemicals. All of the abovementioned processes are employed in the presence of bacteria, fungi, algae, etc. So the process of bioremediation promotes the growth of the microorganisms in the polluted or the degraded sites in order to remediate the toxic pollutants from the environment (Skinder et al., 2020). The main principle of bioremediation is the use of biological sources for the mitigation of toxic pollutants in the given environmental sites. It uses the microorganisms to degrade the contaminants in the polluted environment (Figure 8.2). Microorganisms such as bacteria, fungi, algae, and yeasts break down the toxic pollutants into non-toxic elements. The bacteria use the toxic contaminants as their source of nutrients (Tyagi and Kumar, 2021).

TABLE 8.1 Microorganisms Involved In Different Biogeochemical Cycles

Sl. No.	Biogeochemical cycle	Microorganisms involved	Function	Reference
1.	Carbon cycle	*Bacterioides succinogenes, Clostridium butyrium, Syntrophomonas* sp.	It maintains the carbon balance which allows the aquatic ecosystem to remain friendly for the living organisms.	Cole et al., 2021
2.	Nitrogen cycle	*Azobacter, Rhizobium*	The nitrogen is converted into various forms because nitrogen is one of the essential nutrients in the aquatic ecosystem	Pajares and Ramos, 2019
3.	Oxygen cycle	*Cyanobacteria*	The concentration of the oxygen is maintained throughout the aquatic bodies as oxygen plays a key role for the sustainability of life	Naselli-Flores and Padisák, 2023
4.	Phosphorus cycle	*Actinomycetes, Pseudomonas, Bacillus, Aspergillus, Penicillium*	As phosphorus is the limiting nutrient in the aquatic environment, it helps in the determination of the productivity of the aquatic ecosystem	Jupp et al., 2021
5.	Sulphur cycle	*Proteus, Camphylobacter, Salmonella*	Sulphur acts as an essential element in the biogeochemical cycle as it is the key constituent of many proteins and it can be used as oxidants and reductants in the microbial respiration in the aquatic ecosystem	Zak et al., 2021

Bioaugmentation involves the direct addition of pre-grown microbial cultures with the ability to degrade pollutants and also accelerate the destruction of contaminants. In the absence of biodegrading microorganisms, specific microbes are added to the environment for the degradation of the pollutants. Thus, in this process microbes are added to complete the process of biodegradation (Behera et al., 2022).

In biostimulation, the microbes enrich the atmosphere by adding certain nutrients like nitrogen and phosphorus and boost the microbes to break down the chemical pollutants. One of the major factors in biostimulation is the perfect amount of nutrients for the development of the microbes. One of the major benefits of biostimulation is

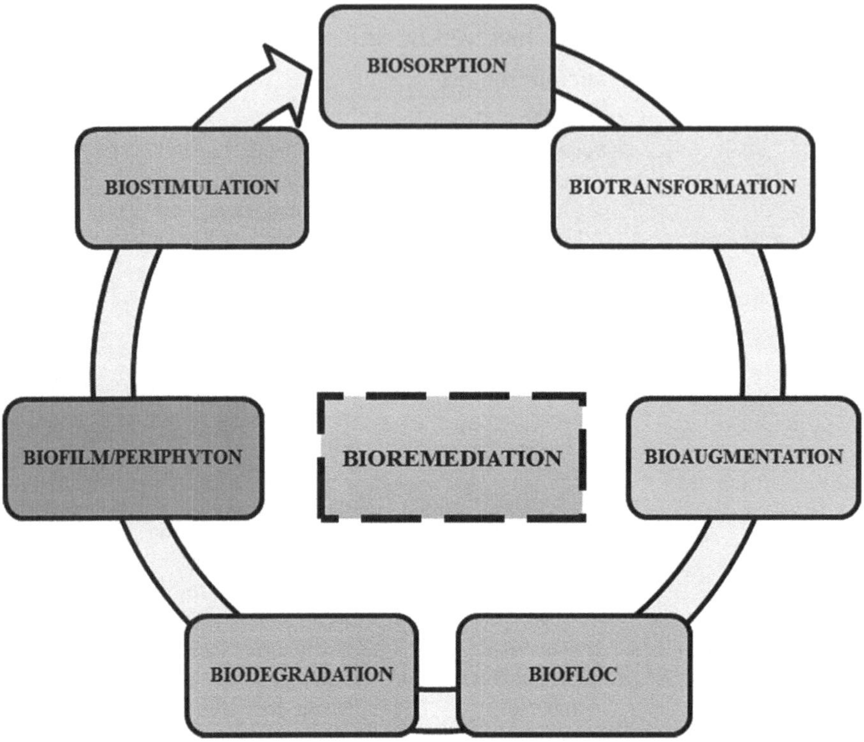

FIGURE 8.2 Schematic overview of bioremediation by aquatic toxicants.

biodegradation of the chemical pollutants due to the occurrence of the indigenous microbes (Adams et al., 2015).

In majority of polluted aquatic ecosystems, organic hydrocarbons are considered to be the most effective forms of pollutants along with inorganic pollutants. Since, bacterial biofilms tend to degrade the hydrocarbons by their hydrophobicity, these biofilms are considered to be effective sources for the bioremediation of the toxic hydrocarbons from the aquatic environment for the maintenance of the ecosystem. Certain bacteria are involved in the bioremediation of the hydrocarbons when released in the aquatic ecosystem. The basic condition of biodegradation is the contact between the bacteria and the contaminants (El-Sheekh et al., 2021).

8.6.4 Microplastics and Biofilms in Aquatic Ecosystems: Their Role in Biogeochemical Cycle

The presence of microplastics in the aquatic atmosphere is now a developing issue in modern society. The excessive use of plastic stuffs and plastic waste in aquatic ecosystems have hampered living organisms in various ways. Plastics less than the 5 mm are considered as microplastics. These microplastics are present abundantly in the sea, lakes, oceans, saltwater, and other aquatic bodies. These microplastics are sometimes ingested by aquatic organisms and are then transferred to the food chain (Barboza and Gimenez, 2015). Overexposure of microplastics to aquatic organisms can cause serious health issues and can also increase the mortality rate among them. Plastics like polyethylene, polystyrene, and polypropylene are usually less dense than the water. The less dense plastics are usually found in the sediments of the water bodies, which increases the life risk of the benthic organisms. Other than this, the microplastics provide a surface for the development of biofilm matrix (Alimba and Faggio, 2019). According to several reports, microbial biofilms rapidly grow on the surface of the microplastics as they are hydrophobic in nature. The presence of the biofilm layer changes the sinking nature of the microplastics present in the water bodies.

The microplastics are present in the aquatic atmosphere as suspended particles. These suspended particles absorb the nutrients and sometimes the organic matter from the marine ecosystem, which confirms the basic development and progress of the organisms (Debroy et al., 2022). These microplastics create a stable habitat for the growth of the microbes which enable them to withstand the pressure and stress caused by the surrounding environment. Subsequently, there is growth of the biofilm matrix in the external parts of the microplastics. The microbial communities attached in the microplastics have been found to be different from the surrounding environment (Wang et al., 2021). The geographical distinctions among the different samples were shown to be weakened due to the enrichment of the plastics in bacterial families such as Halobacteriaceae and Pseudoalteromonadaceae (Tu et al., 2020). Studies have shown that the species colonizing the microplastics had lower alpha diversity. Likewise, biofilm adhered to the microplastics were shown to have lower alpha diversity. The creation of the "plastisphere" encourages Horizontal Gene Transfer by the biofilm adhered to the plastic surfaces. The metabolic diversity of the different groups of organisms can be affected by the increase in the Horizontal Gene Transfer (Junaid et al., 2022). Several reports show that there is an effect of the microplastic-associated biofilms on the nitrogen and phosphorus cycle in freshwater bodies. The denitrification capacity is increased in the nitrogen cycle due to the potential of microplastic-associated biofilms (Yin et al., 2023). Depending on the environmental conditions, the nitrogen and phosphorus can be integrated or released to the surrounding environments by the microplastic-associated biofilms. Hence, it can be said that the presence of microplastics has increased pollutants in aquatic ecosystems and also provided a surface for the attachment of the biofilm matrix (Wang et al., 2022).

8.6.5 Impact of Quorum Sensing on Biogeochemical Cycling

The intercommunication between bacterial species is called quorum sensing (QS). QS contributes a major part in the maintenance of microbial life in dynamic aquatic environments. The bioluminescence properties in bacteria is due to the QS-signaling molecules. Reserachers have found that *Photobacterium phosphoreum* (Bioluminescence strain) exhibits light emission in the depth of the sea. But the bioluminescence properties may be density dependent or may be independent (Borges and Simões, 2019). Apart from the bioluminescence properties, the QS system also regulates the activities of the bacterial communities that are directly involved in the biogeochemical cycle in the aquatic environment. Certain bacterial species such as *Cyanobacter, Pseudomonas, Marinobacter, Vibrio, Roseobacter,* etc., have a QS signaling system and take part in the biogeochemical cycle (Urvoy et al., 2022). In the aquatic ecosystem, gene regulation by the QS signaling molecules influences the cycling of chemical components such as nitrogen, phosphorus, sulphur, carbon, etc. The QS signaling molecules secrete certain extracellular enzymes that are responsible for the breakdown of the organic material so that the microorganisms can competently utilize the breakdown molecules (Miller et al., 2001). The QS genes involved in the enzymatic actions improve the degradation of the molecules and help in the cycling of the organic compounds in the aquatic ecosystems. Chitin and polymesr like N-acetylglucosamine, which are abundantly present in the aquatic environment, are broken down by bacteria like *Pseudomonas* and *Vibrio* sp., which are directly associated with the carbon cycle (Raimundo et al., 2021). These microbes also play a role in the phosphorus cycle as they have the ability to produce enzymes such as phosphatase, which can cleave the phosphate groups and make them available for themselves and for other organisms (Stout et al., 2014). Genus like *Marinobacter* can degrade the hydrocarbons found in the oil spills and can participate in the carbon cycle. The nitrogen cycle is also influenced by the QS signaling molecules, as these molecules regulate processes such as nitrification and denitrification, and facilitate the availability of nitrogen in its biologically active forms (Visick and Ruby, 2006). The expression of the nitrogen fixation genes is controlled by the QS signaling molecules by bacteria like *Vibrio* sp. This system helps in the regulation of the expression of genes involved in nitrification and denitrification that disturb the equilibrium of the nitrogen species in the aquatic ecosystem such as *Pseudomonas* (Miller et al., 2001). The QS system also influences another biogeochemical process known as sulfur cycling (Israel et al., 2023). Gene expression by the sulfur oxidizing bacteria are controlled by the QS system in which the hydrogen sulfide is reduced to sulfate making it biologically active to all the aquatic organisms. A dominant sulfur compound known as dimethylsulphoniopropionate is broken down by the *Alphaproteobacteria* to dimethyl sulfide (Li et al., 2023). Hence, the QS signaling molecules play a key role in the biogeochemical cycling of the essential nutrients found in the aquatic environment.

8.7 CURRENT TRENDS AND FUTURE SCOPE

Today biofilm is present all over the environment. Microbial communities need shelter for biofilm formation. Due to the continuous discharge of chemical or pharmaceutical effluents in water bodies there is selective pressure generated in the residing bacterial communities and the bacterial communities exhibit resistance to various antibiotics (Flores-Vargas et al., 2021). There are advantages and disadvantages of microbial biofilm in water bodies. The major disadvantages of microbial biofilm include health issues in the living organisms, but the biofilm also plays a major role in biogeochemical cycling. Several studies have been recently done to determine the beneficial roles of biofilm in wastewater treatment. Biofilm acts as a natural filter in the treatment of wastewater (Verma et al., 2023). The biofilm matrix reduces the concentration of the biodegradable polymeric carbon present in the water. There is a connection between microplastics and bacteria. This interaction shows its effect on human health as well as its benefits in the biogeochemical cycles. Excess use of plastic products and their discharge in the aquatic environment affects the life of aquatic animals whereas the growth of the biofilm matrix on the microplastics helps in biogeochemical cycling (Chen et al., 2020).

Several studies have been done on the alterations of morphological and physiological conditions of biofilm-forming bacteria in aquatic systems but how these alterations affect the water bodies or how the biogeochemical cycle is affected are largely unknown. With the widespread incidence of chemical/organic/inorganic/pharmaceutical pollutants in the food chain, they travel from one organism to another through food but still not much is known about how many of these pollutants are being passed from one trophic level to another. The presence of microplastics greatly affects the biogeochemical cycle in aquatic ecosystems. The bioeochemical cycle is affected by factors like pH, temperature, climatic conditions, velocity of the water, etc. But molecular methodologies like proteomics, genomics, and metagenomics help in the understanding of the influence of microplastics on biogeochemical cycles.

8.8 CONCLUSION

Microbes have the inherent ability to maintain the sustainability of different ecosystems including aquatic ecosystems with valuable environmental services to mankind. In the aquatic ecosystem, these microbes have the tendency to invariably control the biogeochemical cycling of inorganic/organic matters and energy by significant contribution in the different phases of nitrogen fixation, nutrient acquisition, and mineral cycling. The communication among the aquatic microorganisms and their controlled metabolic activities are regulated by the QS regulatory network and their ability to form biofilms. The QS and biofilm mechanics invariably control the differential expression of several genes responsible for the maintenance of biogeochemical cycling. In this context, it is important to understand the insight mechanistic values associated with aquatic biofilms and their role in biogeochemical cycling under the influence of diverse environmental conditions.

ACKNOWLEDGMENTS

This work is funded by Department of Biotechnology (DBT-BUILDER), Govt. of India (Ref. Number: BT/INF/22/SP45357/2022). The author, Subhaswaraj Pattnaik, would also like to acknowledge DST-SERB for the award of National Post-Doctoral Fellowship (Ref. No. PDF/2021/001260).

REFERENCES

Achterberg, E.P. 2014. Grand challenges in marine geochemistry. *Front. Mar. Sci.*, 1, 007. https://doi.org/10.3389/fmars.2014.00007

Adams, G.O., Fufeyin, P.T., Okoro, S.E., and Ehinomen, I. 2015. Bioremediation, biostimulation and bioaugmention: A review. *Int. J. Environ. Bioremed. Biodegrad.*, 3(1), 28–39. https://doi.org/10.12691/ijebb-3-1-5

Ahn, S.J., and Burne, R.A. 2007. Effects of oxygen on biofilm formation and the AtlA autolysin of *Streptococcus mutans*. *J. Bacteriol.*, 189(17), 6293–6302. https://doi.org/10.1128/JB.00546-07

Alimba, C.G., and Faggio, C. 2019. Microplastics in the marine environment: Current trends in environmental pollution and mechanisms of toxicological profile. *Environ. Toxicol. Pharmacol.*, 68, 61–74. https://doi.org/10.1016/j.etap.2019.03.001

Antunes, J., Leão, P., and Vasconcelos, V. 2019. Marine biofilms: Diversity of communities and of chemical cues. *Environ. Microbiol. Rep.*, 11(3), 287–305. https://doi.org/10.1111/1758-2229.12694

Balcázar, J.L., Subirats, J., and Borrego, C.M. 2015. The role of biofilms as environmental reservoirs of antibiotic resistance. *Front. Microbiol.*, 6, 1216. https://doi.org/10.3389/fmicb.2015.01216

Barboza, L.G.A., and Gimenez, B.C.G. 2015. Microplastics in the marine environment: Current trends and future perspectives. *Mar. Pollut. Bull.*, 97(1–2), 5–12. https://doi.org/10.1016/j.marpolbul.2015.06.008

Behera, I.D., Nayak, M., Mishra, A., Meikap, B.C., and Sen, R. 2022. Strategic implementation of integrated bioaugmentation and biostimulation for efficient mitigation of petroleum hydrocarbon pollutants from terrestrial and aquatic environment. *Mar. Pollut. Bull.*, 177, 113492. https://doi.org/10.1016/j.marpolbul.2022.113492

Berlanga, M., and Guerrero, R. 2016. Living together in biofilms: The microbial cell factory and its biotechnological implications. *Microb. Cell. Fact.*, 15, 165. https://doi.org/10.1186/s12934-016-0569-5

Besemer, K. 2015. Biodiversity, community structure and function of biofilms in stream ecosystems. *Res. Microbiol.*, 166(10), 774–781. https://doi.org/10.1016/j.resmic.2015.05.006

Bisht, K., Moore, J.L., Caprioli, R.M., Skaar, E.P., and Wakeman, C.A. 2021. Impact of temperature-dependent phage expression on *Pseudomonas aeruginosa* biofilm formation. *npj Biofilms. Microbiomes.*, 7, 22. https://doi.org/10.1038/s41522-021-00194-8

Borges, A., and Simões, M. 2019. Quorum sensing inhibition by marine bacteria. *Mar. Drug.*, 17(7), 427. https://doi.org/10.3390/md17070427

Boyd, P.W., and Ellwood, M.J. 2010. The biogeochemical cycle of iron in the ocean. *Nat. Geosci.*, 3(10), 675–682. https://doi.org/10.1038/ngeo964

Brown, D.C., Okpala, G.N., Gieg, L.M., and Turner, R.J. 2023. Biofilms and their role in corrosion in marine environments. In: *Understanding Microbial Biofilms. Fundamentals to Applications*. Das, S., and Kungwani, N.A. (Eds.). Elsevier Inc. pp. 173–185. https://doi.org/10.1016/B978-0-323-99977-9.00036-3

Canadell, J.G., Monteiro, P.M., Costa, M.H., Da Cunha, L.C., Cox, P.M., et al. (2021). Global carbon and other biogeochemical cycles and feedbacks. *IPCC AR6 WGI, Final Government Distribution*, chapter 5, 2021. hal-03336145f

Caruso, G. 2020. Microbial colonization in marine environments: Overview of current knowledge and emerging research topics. *J. Mar. Sci. Eng.*, 8(2), 78. https://doi.org/10.3390/jmse8020078

Catão, E.C.P., Pollet, T., Misson, B., Garnier, C., Ghiglione, J.F., Barry-Martinet, R., et al. 2019. Shear stress as a major driver of marine biofilm communities in the NW Mediterranean sea. *Front. Microbiol.*, 10, 1768. https://doi.org/10.3389/fmicb.2019.01768

Chen, C.Y., and Chen, S.D. 2000. Biofilm characteristics in biological denitrification biofilm reactors. *Water. Sci. Technol.*, 41(4–5), 147–154. https://doi.org/10.2166/wst.2000.0438

Chen, X., Chen, X., Zhao, Y., Zhou, H., Xiong, X., and Wu, C. 2020. Effects of microplastic biofilms on nutrient cycling in simulated freshwater systems. *Sci. Total. Environ.*, 719, 137276. https://doi.org/10.1016/j.scitotenv.2020.137276

Cole, J.J., Hararuk, O., and Solomon, C.T. 2021. The carbon cycle: With a brief introduction to global biogeochemistry. In: Fundamentals of Ecosystem Science. Weathers, K.C., Strayer, D.L., and Likens, G.E. (Eds.). Elsevier Inc. pp. 131–160. https://doi.org/10.1016/B978-0-12-812762-9.00007-1

Condinho, M., Carvalho, B., Cruz, A., Pinto, S.N., Arraiano, C.M., and Pobre, V. 2023. The role of RNA regulators, quorum sensing and c-di-GMP in bacterial biofilm formation. *FEBS Open Bio.*, 13(6), 975–991. https://doi.org/10.1002/2211-5463.13389

Covey, K., Soper, F., Pangala, S., Bernardino, A., Pagliaro, Z., Basso, L., et al. 2021. Carbon and beyond: The biogeochemistry of climate in a rapidly changing amazon. *Front. For. Glob. Change.*, 4, 618401. https://doi.org/10.3389/ffgc.2021.618401

de Carvalho, C.C.C.C. 2018. Marine biofilms: A successful microbial strategy with economic implications. *Front. Mar. Sci.*, 5, 126. https://doi.org/10.3389/fmars.2018.00126

Debroy, A., George, N., and Mukherjee, G. 2022. Role of biofilms in the degradation of microplastics in aquatic environments. *J. Chem. Technol. Biotechnol.*, 97(12), 3271–3282. https://doi.org/10.1002/jctb.6978

Donlan, R.M. (2002). Biofilms: Microbial life on Surfaces. *Emerging Infect. Dis.*, 8(9), 881–889. https://doi.org/10.3201/eid0809.020063

Drake, H., and Ivarsson, M. 2018. The role of anaerobic fungi in fundamental biogeochemical cycles in the deep biosphere. *Fungal. Biol. Rev.*, 32(1), 20–25. https://doi.org/10.1016/j.fbr.2017.10.001

El-Sheekh, M., Abdel-Daim, M.M., Okba, M., Gharib, S., Soliman, A., and El-Kassas, H. (2021). Green technology for bioremediation of the eutrophication phenomenon in aquatic ecosystems: A review. *African. J. Aquatic. Sci.*, 46(3), 274–292. https://doi.org/10.2989/16085914.2020.1860892

Fathollahi, A., and Coupe, S.J. 2021. Effect of environmental and nutritional conditions on the formation of single and mixed-species biofilms and their efficiency in cadmium removal. *Chemosphere.*, 283, 131152. https://doi.org/10.1016/j.chemosphere.2021.131152

Flores-Vargas, G., Bergsveinson, J., Lawrence, J.R., and Korber, D.R. 2021. Environmental biofilms as reservoirs for antimicrobial resistance. *Front. Microbiol.*, 12, 766242. https://doi.org/10.3389/fmicb.2021.766242

Fong, J.C., Karplus, K., Schoolnik, G.K., and Yildiz, F.H. 2006. Identification and characterization of RbmA, a novel protein required for the development of rugose colony morphology and biofilm structure in *Vibrio cholerae*. *J. Bacteriol.*, 188(3), 1049–1059. https://doi.org/10.1128/JB.188.3.1049-1059.2006

Garrett, T.R., Bhakoo, M., and Zhang, Z. 2008. Bacterial adhesion and biofilms on surfaces. *Prog. Nat. Sci.*, 18(9), 1049–1056. https://doi.org/10.1016/j.pnsc.2008.04.001

Gheorghita, A.A., Li, Y.E., Kitova, E.N., Bui, D.T., Pfoh, R., Low, K.E., et al. 2022. Structure of the AlgKX modification and secretion complex required for alginate production and biofilm attachment in *Pseudomonas aeruginosa*. *Nat. Commun.*, 13, 7631. https://doi.org/10.1038/s41467-022-35131-6

Ghosh, D., Seth, M., Mondal, P., and Mukhopadhyay, S.K. 2022. Bacterial biofilms: Role of quorum sensing and quorum quenching. *J. Exp. Biol. Agric. Sci.*, 10(2), 278–293. https://doi.org/10.18006/2022.10(2).278.293

Grossart, H.P., Massana, R., McMahon, K.D., and Walsh, D.A. 2020. Linking metagenomics to aquatic microbial ecology and biogeochemical cycles. *Limnol. Oceanography.*, 65 (S1), S2–S20. https://doi.org/10.1002/lno.11382

Hernández-Cabanyero, C., Sanjuán, E., Fouz, B., Pajuelo, D., Vallejos-Vidal, E., Reyes-López, F.E., and Amaro, C. 2020. The effect of the environmental temperature on the adaptation to host in the zoonotic pathogen *Vibrio vulnificus*. *Front. Microbiol.*, 11, 489. https://doi.org/10.3389/fmicb.2020.00489

Islam, M.S., Jahid, M.I., Rahman, M.M., Rahman, M.Z., Islam, M.S., Kabir, M.S., et al. 2007. Biofilm acts as a microenvironment for plankton-associated *Vibrio cholerae* in the aquatic environment of Bangladesh. *Microbiol. Immunol.*, 51(4), 369–379. https://doi.org/10.1111/j.1348-0421.2007.tb03924.x

Israel, E., Ramganesh, S., Abia, A.L.K., and Chikere, C.B. 2023. Quorum sensing: Unravelling the intricacies of microbial communication for biofilm formation, biogeochemical cycling, and biotechnological applications. *J. Mar. Sci. Eng.*, 11(8), 1586. https://doi.org/10.3390/jmse11081586

Jiang, J.P., Leng, S., Liao, Y.F., Liu, X., Li, D.X., Chu, C., et al. 2023. The potential role of subseafloor fungi in driving the biogeochemical cycle of nitrogen under anaerobic conditions. *Sci. Total. Environ.*, 897, 165374. https://doi.org/10.1016/j.scitotenv.2023.165374

Jin, G., and Englande Jr, A.J. 1998. Carbon tetrachloride biodegradation in a fixed-biofilm reactor and its kinetic study. *Water. Sci. Technol.*, 38(8–9), 155–162. https://doi.org/10.1016/S0273-1223(98)00689-1

Johnson, L.R. 2008. Microcolony and biofilm formation as a survival strategy for bacteria. *J Theor. Biol.*, 251(1), 24–34. https://doi.org/10.1016/j.jtbi.2007.10.039

Joye, S.B., Bowles, M.W., and Ziervogel, K. 2022. Marine biogeochemical cycles. In: *The Marine Microbiome*. Stal, L.J., and Cretoiu, M.S. (Eds.). Springer, Cham. pp. 623–671. https://doi.org/10.1007/978-3-030-90383-1_15

Ju, Y., Shan, K., Liu, W., Xi, C., Zhang, Y., Wang, W., et al. 2022. Effect of different initial fermentation pH on exopolysaccharides produced by *Pseudoalteromonas agarivorans* Hao 2018 and identification of key genes involved in exopolysaccharide synthesis via transcriptome analysis. *Mar. Drugs.*, 20(2), 89. https://doi.org/10.3390/md20020089

Junaid, M., Siddiqui, J.A., Sadaf, M., Liu, S., and Wang, J. 2022. Enrichment and dissemination of bacterial pathogens by microplastics in the aquatic environment. *Sci. Total. Environ.*, 830, 154720. https://doi.org/10.1016/j.scitotenv.2022.154720

Jupp, A.R., Beijer, S., Narain, G.C., Schipper, W., and Slootweg, J.C. 2021. Phosphorus recovery and recycling–closing the loop. *Chem. Soc. Rev.*, 50(1), 87–101. https://doi.org/10.1039/D0CS01150A

Kim, B., Madukoma, C.S., Shrout, J.D., and Nerenberg, R. 2023. Effect of EPS production on the performance of membrane-based biofilm reactors. *Water Res.*, 240, 120101. https://doi.org/10.1016/j.watres.2023.120101

Koestler, B.J., and Waters, C.M. 2013. Exploring environmental control of cyclic di-GMP signaling in *Vibrio cholerae* by using the *ex vivo* lysate cyclic di-GMP assay (TELCA). *Appl. Environ. Microbiol.*, 79(17), 5233–5241. https://doi.org/10.1128/AEM.01596-13

Lami, R., Urios, L., Molmeret, M., and Grimaud, R. 2023. Quorum sensing in biofilms: A key mechanism to target in ecotoxicological studies. *Crit. Rev. Microbiol.*, 49(6), 786–804. https://doi.org/10.1080/1040841X.2022.2142089

Laskar, F., Das Purkayastha, S., Sen, A., Bhattacharya, M.K., and Misra, B.B. 2018. Diversity of methanogenic archaea in freshwater sediments of lacustrine ecosystems. *J. Basic. Microbiol.*, 58(2), 101–119. https://doi.org/10.1002/jobm.201700341

Law, S.K.K., and Tan, H.S. 2022. The role of quorum sensing, biofilm formation, and iron acquisition as key virulence mechanisms in *Acinetobacter baumannii* and the corresponding anti-virulence strategies. *Microbiol. Res.*, 260, 127032. https://doi.org/10.1016/j.micres.2022.127032

Li, C.Y., Mausz, M.A., Murphy, A., Zhang, N., Chen, X.L., Wang, S.Y., et al. 2023. Ubiquitous occurrence of a dimethylsulfoniopropionate ABC transporter in abundant marine bacteria. *ISME J.*, 17(4), 579–587. https://doi.org/10.1038/s41396-023-01375-3

Liu, Y., Zhang, J., and Ji, Y. 2020. Environmental factors modulate biofilm formation by *Staphylococcus aureus*. *Sci. Prog.*, 103(1), 36850419898659. https://doi.org/10.1177/0036850419898659

Lusinier, N., Seyssiecq, I., Sambusiti, C., Jacob, M., Lesage, N., and Roche, N. 2021. Application of moving bed biofilm reactor and fixed bed hybrid biological reactor for oilfield produced water treatment: Influence of total dissolved solids concentration. Energies. 14(21), 7297. https://doi.org/10.3390/en14217297

Maiques, E., Ubeda, C., Campoy, S., Salvador, N., Lasa, I., Novick, R.P., et al. 2006. Beta-lactam antibiotics induce the SOS response and horizontal transfer of virulence factors in *Staphylococcus aureus*. *J. Bacteriol.*, 188(7), 2726–2729. https://doi.org/10.1128/JB.188.7.2726-2729.2006

Martín-Rodríguez, A.J. 2023. Respiration-induced biofilm formation as a driver for bacterial niche colonization. *Trends. Microbiol.*, 31(2), 120–134. https://doi.org/10.1016/j.tim.2022.08.007

Miller, M.B., and Bassler, B.L. 2001. Quorum sensing in bacteria. *Annu. Rev. Microbiol.*, 55(1), 165–199. https://doi.org/10.1146/annurev.micro.55.1.165

Naselli-Flores, L., and Padisák, J. (2023). Ecosystem services provided by marine and freshwater phytoplankton. *Hydrobiol.*, 850(12–13), 2691–2706. https://doi.org/10.1007/s10750-022-04795-y

Neagu, L., Cirstea, D.M., Curutiu, C., Mitache, M.M., Lazăr, V., and Chifiriuc, M.C. 2017. Microbial biofilms from the aquatic ecosystems and water quality. In: *Water Purification*. Grumezescu, A.M. (Eds.). Elsevier Inc. pp. 621–642. https://doi.org/ https://doi.org/10.1016/B978-0-12-804300-4.00018-6

Newton, R.J., Jones, S.E., Eiler, A., McMahon, K.D., and Bertilsson, S. 2011. A guide to the natural history of freshwater lake bacteria. *Microbiol. Mol. Biol. Rev.*, 75(1), 14–49. https://doi.org/10.1128/MMBR.00028-10

Nguyen, H.T., Lee, Y.K., Kwon, J.H., and Hur, J. 2023. Microplastic biofilms in water treatment systems: Fate and risks of pathogenic bacteria, antibiotic-resistant bacteria, and antibiotic resistance genes. *Sci. Total. Environ.*, 892, 164523. https://doi.org/10.1016/j.scitotenv.2023.164523

Ødegaard, H., Gisvold, B., and Strickland, J. 2000. The influence of carrier size and shape in the moving bed biofilm process. *Water. Sci. Technol.*, 41(4–5), 383–391. https://doi.org/10.2166/wst.2000.0470

Offre, P., Spang, A., and Schleper, C. 2013. Archaea in biogeochemical cycles. *Annu. Rev. Microbiol.*, 67, 437–457. https://doi.org/10.1146/annurev-micro-092412-155614

Okuda, K., Nagahori, R., Yamada, S., Sugimoto, S., Sato, C., Sato, M., et al. 2018. The composition and structure of biofilms developed by *Propionibacterium acnes* isolated from cardiac pacemaker devices. *Front. Microbiol.*, 9, 182. https://doi.org/10.3389/fmicb.2018.00182

Osorio, V., Proia, L., Ricart, M., Pérez, S., Ginebreda, A., Cortina, J.L., et al. 2014. Hydrological variation modulates pharmaceutical levels and biofilm responses in a Mediterranean river. *Sci. Total. Environ.*, 472, 1052–1061. https://doi.org/10.1016/j.scitotenv.2013.11.069

Pagán, R., and García-Gonzalo, D. 2015. Influence of environmental factors on bacterial biofilm formation in the food industry: A review. *J. Postdoc. Journal.*, 3(6), 3–13.

Pajares, S., and Ramos, R. 2019. Processes and microorganisms involved in the marine nitrogen cycle: Knowledge and gaps. *Front. Mar. Sci.*, 6, 739. https://doi.org/10.3389/fmars.2019.00739

Papadatou, M., Robson, S.C., Dobretsov, S., Watts, J.E.M., Longyear, J., and Salta, M. 2021. Marine biofilms on different fouling control coating types reveal differences in microbial community composition and abundance. *Microbiol. Open.*, 10, e1231. https://doi.org/10.1002/mbo3.1231

Paula, A.J., Hwang, G., and Koo, H. 2020. Dynamics of bacterial population growth in biofilms resemble spatial and structural aspects of urbanization. *Nat. Commun.*, 11, 1354. https://doi.org/10.1038/s41467-020-15165-4

Penesyan, A., Paulsen, I.T., Kjelleberg, S., and Gillings, M.R. 2021. Three faces of biofilms: A microbial lifestyle, a nascent multicellular organism, and an incubator for diversity. *npj Biofilms Microbiomes.*, 7, 80. https://doi.org/10.1038/s41522-021-00251-2

Preda, V.G., and Săndulescu, O. 2019. Communication is the key: Biofilms, quorum sensing, formation and prevention. *Discoveries (Craiova).*, 7(3), e100. https://doi.org/10.15190/d.2019.13

Qian, P.Y., Cheng, A., Wang, R., Zhang, R. 2022. Marine biofilms: Diversity, interactions and biofouling. *Nat. Rev. Microbiol.*, 20, 671–684. https://doi.org/10.1038/s41579-022-00744-7

Raimundo, I., Silva, R., Meunier, L., Valente, S.M., Lago-Lestón, A., Keller-Costa, T., and Costa, R. 2021. Functional metagenomics reveals differential chitin degradation and utilization features across free-living and host-associated marine microbiomes. *Microbiome.*, 9, 1–18. https://doi.org/10.1186/s40168-020-00970-2

Rath, S., Palit, K., and Das, S. 2022. Variable pH and subsequent change in pCO_2 modulates the biofilm formation, synthesis of extracellular polymeric substances, and survivability of a marine bacterium *Bacillus stercoris* GST-03. *Environ. Res.*, 214(Pt 4), 114128. https://doi.org/10.1016/j.envres.2022.114128

Rivera, S.F., Vasselon, V., Mary, N., Monnier, O., Rimet, F., and Bouchez, A. 2021. Exploring the capacity of aquatic biofilms to act as environmental DNA samplers: Test on macroinvertebrate communities in rivers. *Sci. Total. Environ.*, 763, 144208. https://doi.org/10.1016/j.scitotenv.2020.144208

Rodgers, M., and Zhan, X.M. 2003. Moving-medium biofilm reactors. *Rev. Environ. Sci. Biotechnol.*, 2, 213–224. https://doi.org/10.1023/B:RESB.0000040467.78748.1e

Romaní, A.M. 2010. Freshwater biofilms. In: *Biofouling*. Dürr, S., and Thomason, J.C. (Eds.). Wiley-Blackwell, Oxford. pp. 137–153. https://doi.org/10.1002/9781444315462.ch10

Rousk, J., and Bengtson, J. 2014. Microbial regulation of global biogeochemical cycles. *Front. Microbiol.*, 5, 103. https://doi.org/10.3389/fmicb.2014.00103

Roy, P.K., Ha, A.J., Mizan, M.F.R., Hossain, M.I., Ashrafudoulla, M., Toushik, S.H., et al. 2021. Effects of environmental conditions (temperature, pH, and glucose) on biofilm formation of *Salmonella enterica* serotype Kentucky and virulence gene expression. *Poult. Sci.*, 100(7), 101209. https://doi.org/10.1016/j.psj.2021.101209

Rutherford, S.T., and Bassler, B.L. 2012. Bacterial quorum sensing: Its role in virulence and possibilities for its control. *Cold. Spring. Harb. Perspect. Med.*, 2(11), a012427. https://doi.org/10.1101/cshperspect.a012427

Sabater, S., Guasch, H., Ricart, M., Romaní, A., Vidal, G., Klünder, C., and Schmitt-Jansen M. 2007. Monitoring the effect of chemicals on biological communities. The biofilm as an interface. *Anal. Bioanal. Chem.*, 387(4), 1425–1434. https://doi.org/10.1007/s00216-006-1051-8

Sgier, L., Merbt, S.N., Tlili, A., Kroll, A., and Zupanic, A. 2018. Characterization of aquatic biofilms with flow cytometry. *J. Vis. Exp.* (136), 57655. https://doi.org/10.3791/57655

Shaikh, S., Rashid, N., Onwusogh, U., McKay, G., and Mackey, H.R. 2022. Effect of nutrients deficiency on biofilm formation and single cell protein production with a purple non-sulphur bacteria enriched culture. *Biofilm.*, 5, 100098. https://doi.org/10.1016/j.bioflm.2022.100098

Shilky., Patra, S., Harshvardhan, A., Kumar, A., Saikia, P. 2022. Role of microbes in controlling the geochemical composition of aquatic ecosystems. In: *Hydrogeochemistry of*

Aquatic Ecosystems. Madhav, S., Singh, V.B., Kumar, M., and Singh, S. (Eds.). Wiley. pp. 265–281. https://doi.org/10.1002/9781119870562.ch12

Silva, V., Pereira, J.E., Maltez, L., Poeta, P., and Igrejas, G. 2022. Influence of environmental factors on biofilm formation of *Staphylococci* isolated from wastewater and surface water. *Pathogens.*, 11(10), 1069. https://doi.org/10.3390/pathogens11101069

Singh, R., Paul, D., and Jain, R.K. 2006. Biofilms: Implications in bioremediation. *Trend. Microbiol.*, 14(9), 389–397. https://doi.org/10.1016/j.tim.2006.07.001

Skinder, B.M., Uqab, B., Ganai, B.A. 2020. Bioremediation: A sustainable and emerging tool for restoration of polluted aquatic ecosystem. In: *Fresh Water Pollution Dynamics and Remediation*. Qadri, H., Bhat, R., Mehmood, M., and Dar, G. (Eds.). Springer, Singapore. pp. 143–165. https://doi.org/10.1007/978-981-13-8277-2_9

Sonawane, J.M., Rai, A.K., Sharma, M., Tripathi, M., and Prasad, R. 2022. Microbial biofilms: Recent advances and progress in environmental bioremediation. *Sci. Total. Environ.*, 824, 153843. https://doi.org/10.1016/j.scitotenv.2022.153843

Stanaszek-Tomal, E. 2020. Environmental factors causing the development of microorganisms on the surfaces of national cultural monuments made of mineral building materials-Review. *Coatings.*, 10(12), 1203. https://doi.org/10.3390/coatings10121203

Steven, J.A.C., Thorn, R.M.S., Robinson, G.M., Turner, D., Lee, J.E., and Reynolds, D.M. 2022. The control of waterborne pathogenic bacteria in freshwater using a biologically active filter. *npj Clean Water.*, 5, 30. https://doi.org/10.1038/s41545-022-00169-y

Stoodley, P., Sauer, K., Davies, D.G., and Costerton, J.W. 2002. Biofilms as complex differentiated communities. *Annu. Rev. Microbiol.*, 56, 187–209. https://doi.org/10.1146/annurev.micro.56.012302.160705

Stout, L.M., Joshi, S.R., Kana, T.M., and Jaisi, D.P. 2014. Microbial activities and phosphorus cycling: An application of oxygen isotope ratios in phosphate. *Geochimica Cosmochimica. Acta.*, 138, 101–116. https://doi.org/10.1016/j.gca.2014.04.020

Sushmitha, T.J., Rajeev, M., and Pandian, S.K. 2023. Marine biofilms: Bacterial diversity and dynamics. In: *Understanding Microbial Biofilms: Fundamentals to Applications*. Das, S., and Kungwani, N.A. (Eds.). Elsevier Inc. pp. 3–21. https://doi.org/10.1016/B978-0-323-99977-9.00032-6

Teschler, J.K., Zamorano-Sánchez, D., Utada, A.S., Warner, C.J.A., Wong, G.C.L., Linington, R.G., et al. 2015. Living in the matrix: Assembly and control of *Vibrio cholerae* biofilms. *Nat. Rev. Microbiol.*, 13, 255–268. https://doi.org/10.1038/nrmicro3433

Toyofuku, M., Inaba, T., Kiyokawa, T., Obana, N., Yawata, Y., and Nomura, N. 2016. Environmental factors that shape biofilm formation. *Biosci. Biotechnol. Biochem.*, 80(1), 7–12. https://doi.org/10.1080/09168451.2015.1058701

Tu, C., Zhou, Q., Zhang, C., Liu, Y., and Luo, Y. 2020. Biofilms of microplastics. In: *Microplastics in Terrestrial Environments: Emerging Contaminants and Major Challenges*. He, D., and Luo, Y. (Eds.). Springer, Cham. pp. 299–317. https://doi.org/10.1007/698_2020_461

Tuck, B., Watkin, E., Somers, A., and Machuca, L.L. 2022. A critical review of marine biofilms on metallic materials. *npj Mater. Degrad.*, 6, 25. https://doi.org/10.1038/s41529-022-00234-4

Tyagi, B., and Kumar, N. 2021. Bioremediation: Principles and applications in environmental management. In: *Bioremediation for Environmental Sustainability. Toxicity, Mechanisms of Contaminants Degradation, Detoxification, and Challenges*. Saxena, G., Kumar, V., and Shah, M.P. (Eds.). Elsevier Inc. pp. 3–28. https://doi.org/10.1016/B978-0-12-820524-2.00001-8

Urvoy, M., Labry, C., L'helguen, S., and Lami, R. (2022). Quorum sensing regulates bacterial processes that play a major role in marine biogeochemical cycles. *Front. Mar. Sci.*, 9, 834337. https://doi.org/10.3389/fmars.2022.834337

Verma, S., Kuila, A., and Jacob, S. 2023. Role of biofilms in wastewater treatment. *Appl. Biochem. Biotechnol.*, 195(9), 5618–5642. https://doi.org/10.1007/s12010-022-04163-5

Vestby, L.K., Grønseth, T., Simm, R., and Nesse, L.L. 2020. Bacterial biofilm and its role in the pathogenesis of disease. *Antibiotics (Basel).*, 9(2), 59. https://doi.org/10.3390/antibiotics9020059

Visick, K.L., and Ruby, E.G. 2006. *Vibrio fischeri* and its host: It takes two to tango. *Curr. Opinion. Microbiol.*, 9(6), 632–638. https://doi.org/10.1016/j.mib.2006.10.001

Wang, J., Guo, X., and Xue, J. 2021. Biofilm-developed microplastics as vectors of pollutants in aquatic environments. *Environ. Sci. Technol.*, 55(19), 12780–12790. https://doi.org/10.1021/acs.est.1c04466

Wang, X., Xing, Y., Lv, M., Zhang, T., Ya, H., and Jiang, B. 2022. Recent advances on the effects of microplastics on elements cycling in the environment. *Sci. Total. Environ.*, 849, 157884. https://doi.org/10.1016/j.scitotenv.2022.157884

Warrier, A., Satyamoorthy, K., and Murali, T.S. 2021. Quorum-sensing regulation of virulence factors in bacterial biofilm. *Future. Microbiol.*, 16, 1003–1021. https://doi.org/10.2217/fmb-2020-0301

Wei, X., Zhu, Z., Wei, L., Wu, J., and Ge, T. 2019. Biogeochemical cycles of key elements in the paddy-rice rhizosphere: Microbial mechanisms and coupling processes. *Rhizosphere.*, 10, 100145. https://doi.org/10.1016/j.rhisph.2019.100145

Wijewardene, L., Wu, N., Fohrer, N., and Riis, T. 2022. Epiphytic biofilms in freshwater and interactions with macrophytes: Current understanding and future directions. *Aquatic. Bot.*, 176, 103467. https://doi.org/10.1016/j.aquabot.2021.103467

Yin, H., Wang, L., Zeng, G., Wang, L., and Li, Y. 2022. The roles of different fractions in freshwater biofilms in the photodegradation of methyl orange and Bisphenol A in aqueous solutions. *Int. J. Environ. Res. Public. Health.*, 19(20), 12995. https://doi.org/10.3390/ijerph192012995

Yin, M., Yan, B., Wang, H., Wu, Y., Wang, X., Wang, J., et al. 2023. Effects of microplastics on nitrogen and phosphorus cycles and microbial communities in sediments. *Environ. Pollut.*, 318, 120852. https://doi.org/10.1016/j.envpol.2022.120852

Zak, D., Hupfer, M., Cabezas, A., Jurasinski, G., Audet, J., and Kleeberg, A., et al. 2021. Sulphate in freshwater ecosystems: A review of sources, biogeochemical cycles, ecotoxicological effects and bioremediation. *Earth-Science Rev.*, 212, 103446. https://doi.org/10.1016/j.earscirev.2020.103446

Zhang, W., Ding, W., Li, Y.X., Tam, C., Bougouffa, S., Wang, R., et al. 2019. Marine biofilms constitute a bank of hidden microbial diversity and functional potential. *Nat. Commun.*, 10, 517. https://doi.org/10.1038/s41467-019-08463-z

Zhang, X., Zhang, C., Liu, Y., Zhang, R., and Li, M. 2023. Non-negligible roles of archaea in coastal carbon biogeochemical cycling. *Trend. Microbiol.*, 31(6), 586–600. https://doi.org/10.1016/j.tim.2022.11.008

9 Microbial Biofilm and Wastewater Treatment

Paramita Pakhira and Siddhardha Busi

9.1 INTRODUCTION

Population increases lead to increases in the production of wastewater (Van et al., 1993). Several different chemical, biological, and hybrid treatments are available, but currently bioremediation is the preferred wastewater treatment because it is a simple process, ecofriendly, and profitable (Mahto and Das 2022). Biological agent like bacteria, fungi, and plants are use for bioremediation (Shahot et al., 2014). Bacteria can be used in suspended or immobilized form and in living or dead forms (Verma, Kulia, and Jacob 2023). Bacteria in their suspended form are used for conventional activated sludge systems, but it has disadvantages like low pollutant removal capacity and high sludge production (Mahto and Das 2022). Bacterial biofilms have been developed with various strategies for the removal of toxic contaminant of wastewater (Saidula et a., 2021). Biofilms are complex structures that can form on any type of surface if it is in a contact with the water on regular basis (Verma, Kuila, and Jacob 2023). Several components are involved in the formation of biofilms, and extracellular polymeric substance (EPS) is an important component among them (Jayathilaka et al., 2017). EPS is composed of lipids, nucleic acids, polysaccharides, and proteins; these components of EPS determine strength, diffusibility, and elasticity of the biofilm as well as density (Limoli et al., 2015). Biofilm-based reactors are a promising strategy for overcoming the disadvantages of conventional activated sludge (CAS) systems. In a biofilm-based system, a high level of diversity is present, and amount of activated biomass is also high, which can decrease sludge production, cause low hydraulic retention time, and result in higher pollutant removal capacity (Zhao et al., 2019). Different types of biofilm-based reactors are used such as trickling filter (TF), membrane biofilm reactor (MBR), moving bed biofilm reactor (MBBR), and rotating biological contactor (RBC) for the effective removal of pollutants from wastewater (Abdelfattah, Hossain, and Cheng 2020).

9.2 PROCESS OF BIOFILM FORMATION

The biofilm matrix is made up of around 50% of extracellular polymeric matrix that consists of proteins and polysaccharides that are involved in irreversible adhesion and also maintain integrity of the biofilm (Mahto and Das 2022). In the

DOI: 10.1201/9781003487203-9

presence of charged and neutral polysaccharides a hydrated polymeric network is formed that helps to maintain mechanical stability, determine biofilm structure, and allow cell-to-cell communication (Jagaba et al., 2021). Water retention inside the biofilm occurs with the help of hydrophilic polysaccharides. Surrounding the biofilm a hydrated microenvironment forms by hydrophilic polysaccharides by which organisms can tolerate desiccation in water-deficient environment, leading to their tolerance of desiccation (Mahto and Das 2022). Nutrient accumulation occurs by charged or hydrophobic polysaccharides and provides the possibility of mineral formation and ion exchange and helps to form polysaccharide gel. Inside biofilm some enzymes are present that make the exogenous macromolecules available for digestion (Hussaini et al., 2021). Exopolysaccharides are sources of various nutrients like carbon, nitrogen, and phosphorus, which are necessary for the microbes living inside the biofilm. EPS is categorized into two groups: specific and non-specific polysaccharides. Specific polymers include glucose, mannose, glucuronic acid, and galacturonic acid and non-specific polysaccharides include homopolysaccharides and heteropolysaccharides (Veerachamy, Yarlagadda, and Manivasagam 2014).

Biofilm shape depends on various factors such as the primary structure, composition of solvent, media pH, temperature, and the amount of inorganic salts present (Toyofuku et al., 2016). Biofilms can react to the forces on the exopolysaccharide matrix by both reversible elatic responses and irreversible deformations (Muhammad et al., 2020). Biofilm density increases with increasing shear stress, calcium concentration, and biofilm age (Cydzik-kwiatkowska and Zielin 2016).

Biofilm formation consists of several steps. Macromolecules like proteins, nucleic acids, polysaccharides, and some micromolecules such as fatty acids, lipids, and polychlorinated biphenyls are absorbed onto the surface (Toyofuku et al., 2016). After adsorption of these molecules a conditioning film is formed that can play several distinct roles like changing the physiochemical properties of the surface, supplying the nutrient and vital trace elements for the biofilm, and triggering biofilm dispersion. Attachment of cells begins once the surface is ready (Muhammad et al., 2020).

After attachment, a slimy EPS is produced that helps the bacteria adhere. EPS is composed of polysaccharides and proteins. EPS assists biofilm formation by forming biofilm matrix, which is a naturally slimy substance (Espinosa-ortiz et al., 2023).

After the formation of mature biofilm propagation starts; propagation is much more complex than initial attachment (Cydzik-kwiatkowska and Zielin 2016).

Bacteria can swith from planktonic form to biofilm mode. Biofilm formation starts by reversible adherence process; after this surface attachment of the biofilm occurs and then colonization starts (Abdelfattah, Hossain, and Cheng 2020). Then the matrix of biofilm is formed, and irreversible adherence starts. Bacterial cells surrounded by the matrix contain proteins, extracellular polysaccharides, helps to form the three-dimensional scaffold structure of the biofilm (Peng et al., 2018). Secondary colonizers again adhere with the organisms after irreversible adhesion, and then coaggregation happens where cells attach to the surface of the biofilm (Morgan-sagastume 2018). After this, sequential detachment two other steps, erosion and disruption, occur.

There are various factors like shear stress of fluid, weak internal cohesion, depletion of nutrients, and oxygen are important for the process (Muhammad et al., 2020). At the stage where growth and the detachment occurred, the system reaches a pseudo steady state and the biofilm thickness is considered to be maximum (Verma, Kuila, and Jacob 2023).

So far there are three known types of biofilm structures: heterogeneous, mushroom like, and pseudo homogenous. These biofilm structures are also found in biofilm reactors. A dense, homogeneous biofilm that is exposed to the flowing solution is called heterogeneous biofilm; biofilms separated by a water channel are known as pseudo-homogeneous biofilms; whereas mushroom-like biofilms are surrounded by water (Andersson et al., 2008).

9.3 FACTORS AFFECTING BIOFILM FORMATION

Biofilm formation is not a simple process. Formation of biofilm can be affected by various factors including surface properties of the biocarrier, nutrient transport, character of the interface fluid, diffusion, cell properties, and growth rate of microorganisms. Rough hydrophobic surfaces that are covered by conditioning films are better for microorganism attachment (Muhammad et al., 2020). Electrostatic repulsion between the bacteria and the organic molecule of the carrier makes attachment difficult. Bacterial adhesion increases with increased velocity of the interface fluid. Bacterial surface shows electronegativity at higher pH of the liquid phase due to the ionization of the amino acids but normally bacterial surfaces have electropositivity. The presence of zeta potential influences microbial interaction and biofilm formation (Ercan and Demirci 2015). At the isoelectric point zeta potential becomes zero and disappearance of the surface salvation structure also occurs; at this condition microorganisms are present at their most stable state and can adhere to the carrier by decreasing surface free energy. Negative effects of biofilm on microbial communities are also observed, such as production of some bacteriotoxins, which can lower the pH by the production of acids (Ercan and Demirci 2015). Biofilm strength is more in the presence of mechanical stressors (Muhammad et al., 2020). Some organisms have maximum production of EPS at their optimal growth conditions. When bacterial cells face harmful or optimal environments, the production of EPS increases (Toyofuku et al., 2016). EPS can protect microorganisms from toxic metals and pollutants by binding of free cells with the ions present in the wastewater. Bacterial quorum sensing (QS) plays a major role in improving biofilm formation as the production of QS signaling molecules have a positive relation with EPS concentration (Verma, Kuila, and Jacob 2023). Intracellular c-di-GMP is important for biofilm formation process as well as phenotype regulation and shedding. Flagella, pilus, adhesion, and extracellular DNA such as extracellular matrix component formation can be regulated by c-di-GMP, which plays a major role in adhesion and aggregation of bacteria. Increased concentration of c-di-GMP promotes biofilm formation and decrease in the concentration can cause biofilm dispersal (Peng et al., 2017). In wastewater treatment it was observed that AHL molecules (mainly N-octanoyl-L-homoserine lacton) accelerate the initial biofilm formation process (Boltz et al., 2017).

9.4 BIOFILM REACTORS

Microbes naturally form biofilms on a solid surface; similarly they can form a thick layer of biofilm in bioreactors without the addition of chemicals. By formation of biofilm inside the reactor high cell concentration can be achieved in a reactor (Shahot et al., 2014). Biofilm reactors are known as "passive immobilized cell reactors" (Abdelfattah, Hossain, and Cheng 2020). Bacterial cells are attached with biofilm reactors with the help of different carriers. There are two types of these carriers: fixed or free moving biofilm carriers (Karadag et al., 2015). The biofilm matrix is made up of water and different soluble and particulate components, soluble microbial products, some inert material, and EPS. After a certain time attached bacterial biofilm detaches and exits the effluent stream (Cydzik-kwiatkowska and Zielin 2016).

Biofilm reactors can be characterized based on different phases present like gas, liquid, and solid phases. Also according to the linked biofilm as it is linked to a stationary or mobile transporter inside the bioreactor (Ercan and Demirci 2013).

9.4.1 Two-Phase Methodology

In this process biofilm-coated transporters are used and water flows through the bioreactor where electron acceptor and donors are also present (Butler and Boltz 2013). Figure 9.2 shows some examples. As both the electron donor and electron acceptors are present inside the biofilm reactor, the flow of water occurs through the biofilm reactor (e.g., denitrification filter). A proton exchange membrane is present that can separate anode-loaded biofilm from cathode, where the electron acceptor and electron donor are present on opposite sides (e.g., biofilm-based microbial fuel cell) (Verma, Kuila, and Jacob 2023).

9.4.2 Three-Phase System

In this system air, bulk amount of water, and fixed biofilm-laden carrier are the three phases. At first upward and downward movement of the two phases occurs and then in the third phase biofilm is present (Butler and Boltz 2013). Examples are defined in Figure 9.1. All three phases are fixed or semi-fixed. Gas bubbles flow through the water phase and then are circulated through the biofilm reactor. An example of a three-phase system is BAF (biologically active filter). In the three phase system, bioreactor is filled with water after that air is introduced as gas bubbles (Verma, Kuila, and Jacob 2023). Example is MBBR (aerobic moving bed biofilm reactor). In another type of system, a microporous hollow fiber membrane is present, which contains a gas arm and biofilm (Martin and Nerenberg 2012). An example is MBfR (membrane biofilm reactor) (Figure 9.1).

9.5 VARIOUS COMPARTMENTS OF BIOFILM REACTORS

Biofilm reactors have five main compartments (Figure 9.2): influent wastewater distribution system, structure of the containment, biofilm carrier, effluent water collection system, and aeration or mixing system.

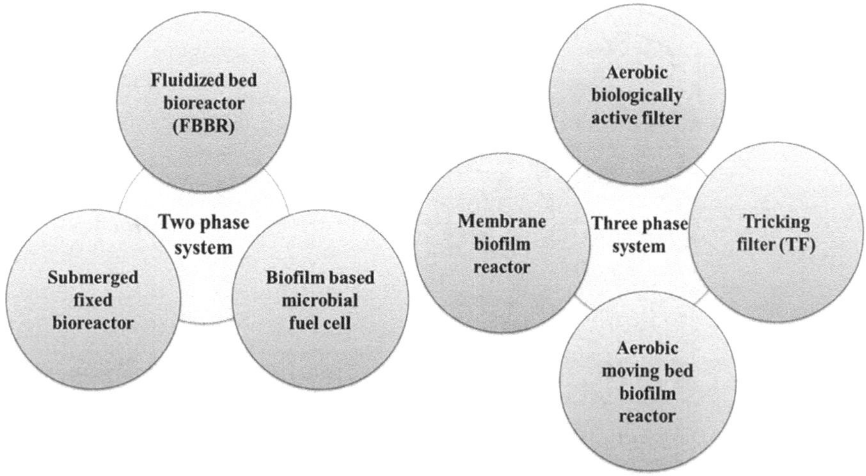

FIGURE 9.1 Example of various two-phase and three-phase biofilm reactor systems.

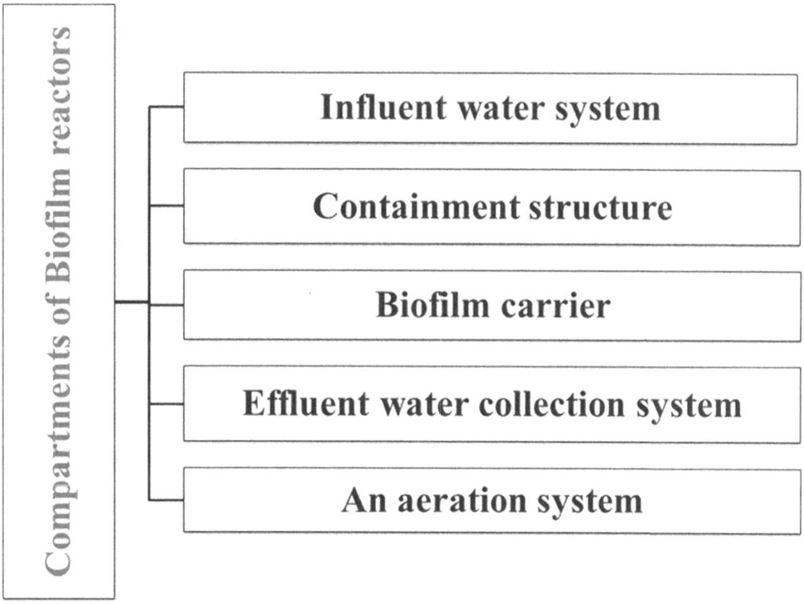

FIGURE 9.2 Five main components of biofilm reactor.

There are five components (Figure 9.3) that can influence the inside conditions of biofilm: (i) Substratum or the surface of biofilm carrier; (ii) the nature of the biofilm, which includes particulate and liquid fractions; (iii) the mass-transfer boundary layer; (iv) amount of bulk liquid; and (v) gas phase when it is present.

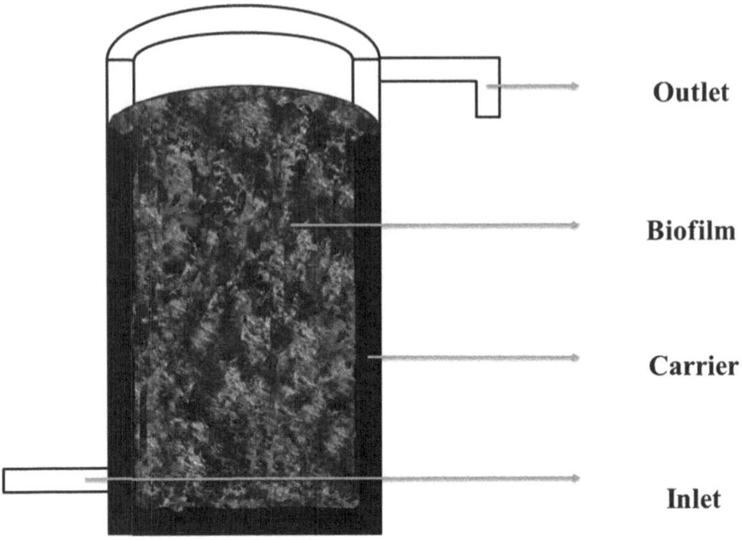

FIGURE 9.3 Components of biofilm reactor. Biofilm formation by the attached microorganisms, different types of biofilm carrier, presence of gas phase.

9.6 BIOLOGICALLY ACTIVE FILTERS

In BAFs filtration media contain natural mineral, structured or any types of plastic media present which can support the biofilm growth. Accumulation of the solid particles occurs at the time of backwashing (Kesaano and Sims 2014). Both the preliminary and primary treatments are needed for influent processing in BAFs. Previously BAFs were known as biological aerated filters; they are also known as aerated biofilters, which can be used for secondary treatment. BAFs can be characterized based on media configurations and flow regime (Jayathilake et al., 2017). During the 1980s downflow BAFs were used for both secondary and tertiary treatments including the biocarbones process, and if packed-bed tertiary. Backwashing of BAFs occurs by using an alternating counter-current flow. Upflow BAFs have media heavier than water (Butler and Boltz 2013). In these systems clay- and mineral-based media can be used. Floating media like polystyrene, polypropylene, or polyethylene are used for both the secondary and tertiary treatments in BAFs by the process of veolia biostyrs (Verma, Kuila, and Jacob 2023). In submerged aerated filters (SAFs) retention of solids is not possible. For promoting biological denitrification, methanol or any other carbon source is used (Ercan and Demirci 2013). Three different types of media can be used: expanded clay media, expanded shale media, and angular grains media. Coarse-bubble aeration diffusers present at the bottom of the media can increase the air contact, contact with water, and biomass (Al et al., 2020). For the backwashing process counter-current air scour and backwash flow of water are important. For preventing media attrition, on the top of reactor a metal grid is fixed. The media-containing accumulated solids can be drawn down to the bottom part of the filter (Figure 9.4). Wash box discharge is used to control the upflow of liquid inside the wash box.

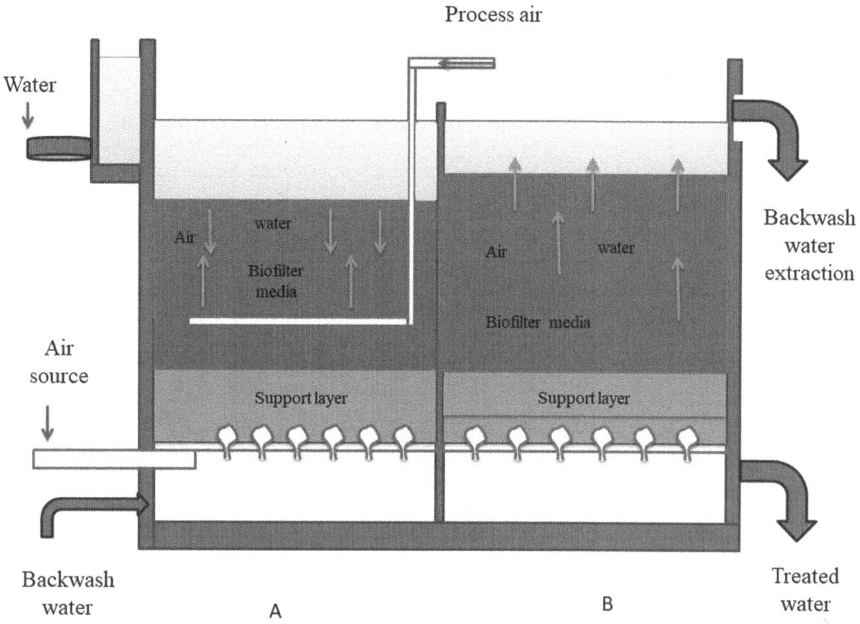

FIGURE 9.4 Biological active filter. (A) downflow of BAF and (B) upflow of BAF where water is lighter than media.

9.7 EXPANDED BED BIOFILM REACTORS AND FLUIDIZED BED BIOFILM REACTORS (EBBRS AND FBBRS)

In both EBBRs and FBBRs small particles of media are suspended in vertically flowing wastewater; this is the process for fluidized media and for bed expansion. After the drag force of the fast-flowing wastewater overcomes gravity, individual particles are suspended. Fluidized beds are mainly used for tertiary denitrification of municipal wastewater (Swain et al., 2018).

FBBRs are used to remove oxidized contaminants likes nitrate and perchlorate in the case of groundwater and industrial wastewater treatment. Contact surface of microorganisms and wastewater is increased by increasing amount of suspended media (Ercan and Demirci 2013). As there is notable comparative motion among the biofilm with the flowing wastewater treatment efficiency can be increased by improving bulk transfer. Small particles are present on the top and the large particles are present at the bottom due to stability of forces present in particle fluidization and for the bed expansion so that media particles can be sorted to a comparatively tight size range (Wa et al., 2021). The bed is fully expanded or fluidized depending on grade of bed expansion. Less bed expansion is more useful because it requires low flow velocity, and less energy is required, which can increase the concentration of the biomass and can reduce reactor footprint. The aerobic process system consists of a fluidized column that can retain a fixed and vertical hydraulic flow by maintaining a constant bed expansion and also biofilm-coated particles are present (Figure 9.5).

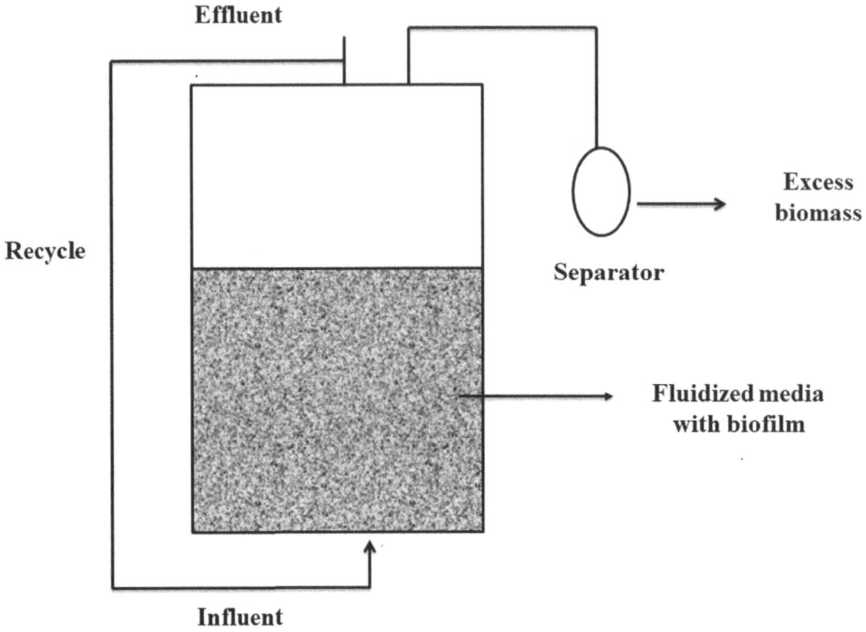

FIGURE 9.5 Parts of fluidized bed reactor with mechanism of action.

At the time of recycle mixing of influent wastewater and effluent aeration can be achieved (Abdelfattah, Hossain, and Cheng 2020). If the aeration occurs inside the fluidized bed, a noticeable amount of gas interrupts the fluidized phase by creating turbulence and collision between the interparticles is observed.

9.8 ROTATING BIOLOGICAL CONTACTORS (RBCS)

The RBC method can be used for effluent having average or low water quality standards. The RBC is of a cylindrical shape, and contains synthetic media attached to a horizontal shaft. A liquid–solid separation unit helps in the detachment of fragments of biofilm suspended in the RBC effluent system. The RBC process includes various operating stages (Huang et al., 2019). In each series one or more shafts are present. Rotation of media-supporting shafts occurs with the help of a mechanical drives. In RBCs both type of media (i.e., low- and high-density media) can be used. For the first stage mainly low-density media is used for BOD removal that can reduce media clogging and also reduce the extensive biofilm growth and high-density media mainly important for nitrification (Lu et al., 2021). Inside a mechanical shaft drive various components are present including an electric motor, a speed reducer, and a belt. For avoiding uneven weight distribution, development of a uniformly formed biofilm is needed. The RBC method is fully protected for avoiding UV light-mediated media corrosion and growth of algae as well as to stop extreme cooling (Ercan and Demirci 2013).

9.9 TRICKLING FILTERS (TFS)

The trickling filters are biofilm reactors containing three different phases and fixed carriers. First, wastewater goes into the bioreactor from a distribution system, then trickles down to where the biofilm is present, and air can move up or down in the last or third phase then disperse through the flowing liquid and then to the biofilm.

Important features of TFs include:

i) An influent system that helps in water distribution
ii) Structure of the containment is an important part of TF
iii) Media such as a rock or plastic type
iv) Underdrain and ventilation system

There are two main types of TF distribution systems: fixed-nozzle and rotary distributor. Fixed nozzle is not commonly used due to its low efficiency. A well-designed rotary distribution system needs proper media wetting for the wastewater treatment by using biofilm carrier. Advantages of TF media is that it can give more surface area with the advantage of low cost and high level of durability, with proper porosity for avoiding clogging, which can support ventilation (Verma, Kuila, and Jacob 2023). Various types of TF media include rock, random, vertical flow (VF) media, and cross-flow (XF) media. Both the VF and XF media are composed of plastic sheets that are smooth or corrugated in texture (Figure 9.6). One more synthetic type of media is a vertically hanging plastic strip, which is not commonly used. In modular synthetic media high amounts of surface area and void space provide greater hydraulic loading,

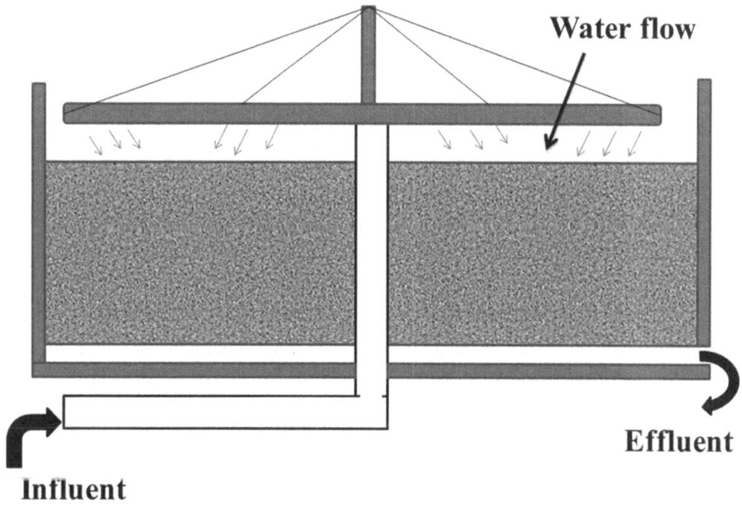

FIGURE 9.6 Trickling filter system used in wastewater treatment' water flow system used for equal distribution of wastewater to the media.

increasing the amount of oxygen transfer and also controlling biofilm thickness compare to rock media (Chaali, Naghdi, and Brar 2018). Rock media with high unit weight and low void space provides limited oxygen transfer and low bioreactor performance. Operational problems are less in TFs.

9.10 MEMBRANE BIOFILM REACTORS (MBFRS)

MBfRs are also well known as membrane aerated biofilm reactors. In biofilm reactors both co-diffusional and counter-diffusional biofilms are formed (Martin and Nerenberg 2012). In co-diffusional biofilms the active metabolic region is the exterior portion, but in counter-diffusional biofilms diffusion of one substrate occurs from base to biofilm and substrate diffusion occurs in the bulk liquid (Nerenberg 2016).

MBfR is one example of counter-diffusional biofilm process. In this system oxygen is supplied from the biofilm base so the organic carbon concentration is high, which can promote a higher level of nitrification. If the anoxic condition is maintained in the bulk liquid, nitrification and denitrification can occur simultaneously. The roughness of counter-diffusional biofilm can increase the liquid diffusional layer thickness (Espinosa-ortiz et al., 2023).

Both oxygen-based and hydrogen-based MBfR can be used for wastewater treatment. Air- or oxygen-based MBfRs have several advantages. They are used for the process of carbon oxidation, nitrification, anaerobic ammonia oxidation, concurrent COD, and total nitrogen removal (Nerenberg 2016). In the case of oxygen-based MBfRs gas transfer rate is high when gas supply pressure is high or pure oxygen is present, resulting in less energy consumption. By eliminating bubble formation stripping of greenhouse gases can be prevented. Foam formation is also lower in the presence of surfactants (Martin and Nerenberg 2012).

Hydrogen-based MBfRs are used for the treatment of inorganic oxidized contaminates and organic compounds. Advantages such as less toxic to humans, lower production of biomass, and extensive use are observed in this system. Autotrophic denitrification inside MBfRs can be used for wastewater treatment. Under anaerobic condition organic compounds can be degraded by using hydrogen-based MBfR (Verma, Kuila, and Jacob 2023).

In a conventional bubble aeration system oxygen transfer efficiency is 15–40% but in MBfR efficiency it is 100%. For carbon oxidation special measures are implemented to prevent biofouling due to high microbial growth (Muhammad et al., 2020). For nitrogen removal nitrifying bacteria develops biofilm at the membrane surface where the concentration of the oxygen is high. Growth of heterotrophic bacteria occurs at the outer side of the growing biofilm, which protects nitrifying bacteria from shear and chemical stress. Hybrid MBfRs are used to avoid fouling that occurs due to excess biofilm formation. Oxygen of air-based MBfR can support anaerobic ammonia oxidation (anammox) biofilms. Disadvantages of anammox is their slow growing rate, greater sensitivity, and less protection from detachment (Martin and Nerenberg 2012).

9.11 MOVING BED BIOFILM REACTORS (MBBRS)

MBBRs contains either a two- or three-phase system, where the two-phase system is an anoxic system, and the three-phase system is an aerobic system. In both of these systems a free-moving plastic biofilm carrier is present, and for the homogenous distribution of carrier throughout tank aeration or mechanical mixing is needed (Barwal and Chaudhary 2014). This submerged fully mixed biofilm reactor contains a separation unit that can separate liquid and solids. It has been observed that MBBRs are able to process wastewater (Safwat 2019). The MBBR has some similarity with the activated sludge method in case of carbon oxidation, nitrification process, and denitrification, but compared to activated sludge system MBBRs use smaller tank volume (Madan, Madan, and Hussain 2022). In the activated sludge process the biomass-holding capacity of the clarifier is dependent whereas the MBBR does not depend on clarifier (Leyva-daz et al., 2020). Differential operational cycle for the controlling of biofilm thickness is not needed in MBBR as the flowing process is continuous (Di et al., 2019). Process flow sheet can be changed in MBBR as operational complexity and hydraulic head loss is much less (Mahto and Das 2022).

In MBBR a single reactor or more than one reactor can be present in a series. The length X width of a MBBR is around 0.5:1 and 1.5:1. MBBRs have a plastic biofilm carrier that contains around 67% of the total liquid volume (Butler and Boltz 2013). The biofilm carriers are prepared from recycled high-density polyethylene (HDPE). Both the native and biofilm-coated carriers can float in quiescent water (Verma, Kuila, and Jacob 2023). The primary development of biofilm occurs inside the surface of the plastic biofilm carrier. There are two patterns present for the constant distribution of plastic biofilm carriers: a diffuser grid layout and drop pipe arrangement system (Butler and Boltz 2013). When MBBR is not in service and drained, it is not able to resist the weight of the biofilm carrier as coarse-bubble diffusers are constructed that can provide structural support (Deena et al., 2022). For denitrification mechanical mixers are used that can agitate bulk volume of liquid and can also uniformly distribute plastic biofilm carriers throughout the bioreactor tank (Leyva-daz et al., 2020). These mechanical mixers are mainly units of rail-mounted submersibles also known as wet motors. For avoiding damage to biofilm carrier and impeller wear, a stainless steel backward curve propeller is present (Barwal and Chaudhary 2014).

9.12 CARRIER FOR BIOFILM REACTORS

There are different sizes and shapes of biofilm carriers and they are also made of different materials that have been used for improving microorganism and nutrient mixing to enhance oxygen transfer (Barwal and Chaudhary 2014). Several carriers have been successfully used for municipal, paper mill, and dairy wastewater processing. Biofilm removal from flat surface carrier is comparatively easy (Huang et al., 2019). Biofilm carriers with crimped surface can increase stability and attachment of biofilm, which are essential for biofilm growth (Al et al., 2020). Biocarriers are made of polypropylene (PP), polyethylene (PE), or HDPE and come in different sizes and shapes. Depending on conditions biocarriers can be used for 10 to 30 years (Morgan-sagastume

2018). After various experiments it was observed that biocarrier made of PP can withstand temperatures over 100 C (Ødegaard, Gisvold, and Strickland 2000).

Carriers have been integrated into various reactors such as suspended carrier biofilm reactors, biological fluidized bed, aerated biofilters, fixed media reactors, rotating biological contactors reactors, integrated fixed film activated sludge processes, and trickling filters (Jagaba et al., 2021).

The main drawback of using biofilm carrier is clogging of carrier pores, which can lead to different problems like backwashing, shortcircuit, and excessive head loss, but the development of MBBR in the 1980s solved this problem by circulating the biofilm bed. MBBRs can perform the removal of nitrogen and organic components.

MBBRs fulfill all the requirements for an efficient bioreactor such as high specific surface area for the growth of biofilm, non-cloggable, and operations can be performed continuously (Mahto and Das 2022). MBBRs are also used in fish farms for wastewater treatment and for the production of potable water. Effectiveness of the treatment is determined by various factors such as media volume, shape of the carrier, and surface area. Based on media structure thickness, the depth of biofilm can be changed (Leyva-daz et al., 2020). As the charge of bacterial surface is negative, positively charged carriers are most preferable. It has been observed that for the proper growth of bacteria 50% filling ratio is optimum. Carrier surface texture also plays an important role in performance (Madan, Madan, and Hussain 2022).

Several carriers can be used in MBBRs (Table 9.1) such as sponge biocarrier, activated carbon particle, polystyrene carrier, and polymeric carrier such as

TABLE 9.1
Different Biofilm Carriers and Their Advantages in Wastewater Treatment

Biofilm carrier	Advantages	Reference
Diatomaceous earth (DE) as biomass carrier	High porosity, can remove turbidity more than 96.7%	(Barwal and Chaudhary 2014)
Polypropylene biofilm carrier	Good hydrodynamic properties	(Raj et al. 2022)
Polymeric nano-fibrous carrier	Active space area is large, resistance against toxic incidence is more	(Morgan-sagastume 2018)
Moving bed biofilm carrier	Can remove bisphenol, higher COD removal efficiency, better nitrification efficiency	(Raj et al., 2022)
Polyurethane sponge	Cost is low, specific surface area is high, absorbent capacity of hydrophobic compounds are high	(Ercan and Demirci 2013)
Polyethylene	Can generate biogas by anaerobic digestion, removal of petroleum hydrocarbon	(Ødegaard, Gisvold, and Strickland 2000)
Polyvinyl alcohol gel beads	Simultaneous nitrification and denitrification, more effective for cell immobilization	(Andersson et al., 2008)
Polyvinyl chloride	Can able to treat ammonia-rich wastewater	(Morgan-sagastume 2018)

polyhydroxybutyrate. These carriers are divided into two groups: porous and non-porous carriers. Non-porous carriers like polypropylene, polyethylene which are used in wastewater treatment have several disadvantages due to their smooth surface where biofilm can easily detach from the carriers (Safwat 2019). Porous carriers can overcome these drawbacks, but can trap biomass inside the pores (Barwal and Chaudhary 2014).

9.13 ADVANTAGES AND DISADVANTAGES OF BIOFILM REACTORS

Biofilm reactors are used in municipalities for the elimination of carbon-based organic substances and nitrogen-containing compounds from wastewater (Leyva-daz et al., 2020).

9.13.1 Advantages

Compared to other novel bioreactors (i.e., cell recycled and hollow fiber reactors) biofilm reactors are more advantageous. By using biofilm reactor carbon oxidation, combination of nitrification with carbon oxidation, tertiary nitrification, and denitrification can be performed (Boltz et al., 2017). Treatment objectives of biofilm reactors for the treatment of industrial and municipal wastewaters are similar to other industries. They can release the treated wastewater directly to sewers present in municipal system where further processing can be accomplished (Zhao et al., 2019). Biofilm reactors are useful in industrial applications as they are reliable, robust, easy to use, have operation stability, high biomass density, have resiliency to toxic loading, have less membrane fouling, and low capital cost (Chaali, Naghdi, and Brar 2018). High production as well as lower chance of washing out of the cells at the time of continuous fermentation occur with biofilm reactors. The presence of biofilm matrix provides resistance to microorganisms to survive in extreme conditions such as pH or temperature or in the presence of antibiotics and toxic substances (Espinosa-ortiz et al., 2023).

9.13.2 Disadvantages

Biofilm reactors are unable to remove phosphorus. Biofilm growth is unavoidable in water even though controlled biofilm growth is very essential in wastewater treatment (Mahto and Das 2022). Uncontrolled growth of biofilm can cause problems related to material performance as well as health-related and esthetical-related problems.

Biofilm-related problems are, formation of biofilms on the surface of metal, Biofilms formation on the surface of concrete, Formation of biofilms on the filtration membranes present in drinking water treatment, presence of biofilms on the filtration membranes used for wastewater treatment.

The common solution to treating unwanted accumulation of biomass is by producing bubbling air near the surface of the membrane, which can create a high level of shear force. Some autoinducers or QS molecules are involved in biofilm formation (Karadag et al., 2015). There are three strategies for interfering with these molecules: blocking of autoinducer production, inactivating autoinducers, and interfering with the signal receptors (Peng et al., 2017). Autoinducer inactivation in a MBRR is used to decrease deposition of EPS on the membrane and increase the fouling of filtration membranes. Scaling up of the support material is also challenging (Espinosa-ortiz et al., 2023).

9.14 CONCLUSION

Biofilm-based processes have advantages when the treatment is used for large streams. More than biofilm concentration, biofilm surface area is important for increasing productivity of biofilm reactors. Industrial wastewater contains a high amount of contaminants where the activated sludge process is not economically wise due to the requirement of high aeration energy. However, biofilm-based treatment provides an alternative to other treatment strategies. Some improvements in reactor design and solid supports are essential for the equal distribution of biofilm through the reactors. It has been observed in many cases that suspended solids can negatively affect performance of the biofilm so in the future research is needed to find alternatives to overcome these problems for developing biofilm-based wastewater treatment with low cost and high efficiency.

REFERENCES

Abdelfattah, Abdallah, Md Iqbal Hossain, and Liang Cheng. "High-strength wastewater treatment using microbial biofilm reactor: A critical review." *World Journal of Microbiology and Biotechnology* 36 (2020): 1–10.

Al-Amshawee, M.Y. Yunus, and A.A. Azoddein. "A novel microbial biofilm carrier for wastewater remediation." In *IOP Conference Series: Materials Science and Engineering*, vol. 736, no. 7, p. 072006. IOP Publishing, 2020.

Al-Amshawee, Sajjad, Mohd Yusri Bin Mohd Yunus, Dai-Viet N. Vo, and Ngoc Han Tran. "Biocarriers for biofilm immobilization in wastewater treatments: A review." *Environmental Chemistry Letters* 18 (2020): 1925–1945.

Andersson, Sofia, Gunaratna Kuttuva Rajarao, Carl Johan Land, and Gunnel Dalhammar. "Biofilm formation and interactions of bacterial strains found in wastewater treatment systems." *FEMS Microbiology Letters* 283, no. 1 (2008): 83–90.

Barwal, Anjali, and Rubina Chaudhary. "To study the performance of biocarriers in moving bed biofilm reactor (MBBR) technology and kinetics of biofilm for retrofitting the existing aerobic treatment systems: A review." *Reviews in Environmental Science and Bio/Technology* 13 (2014): 285–299.

Boltz, Joshua P., Barth F. Smets, Bruce E. Rittmann, Mark CM Van Loosdrecht, Eberhard Morgenroth, and Glen T. Daigger. "From biofilm ecology to reactors: A focused review." *Water Science and Technology* 75, no. 8 (2017): 1753–1760.

Butler, C.S., and J.P. Boltz. 3.6 biofilm processes and control in water and wastewater treatment. In *Comprehensive Water Quality and Purification 2013 Jan 1* (pp. 90–107). Elsevier.

Chaali, Mona, Mitra Naghdi, Satinder K. Brar, and Antonio Avalos-Ramirez. "A review on the advances in nitrifying biofilm reactors and their removal rates in wastewater treatment." *Journal of Chemical Technology & Biotechnology* 93, no. 11 (2018): 3113–3124.

Cydzik-Kwiatkowska, Agnieszka, and Magdalena Zielińska. "Bacterial communities in full-scale wastewater treatment systems." *World Journal of Microbiology and Biotechnology* 32 (2016): 1–8.

Deena, Santhana Raj, Gopalakrishnan Kumar, A.S. Vickram, Reeta Rani Singhania, Cheng–Di Dong, Karunakaran Rohini, K. Anbarasu, S. Thanigaivel, and Vinoth Kumar Ponnusamy. "Efficiency of various biofilm carriers and microbial interactions with substrate in moving bed-biofilm reactor for environmental wastewater treatment." *Bioresource Technology* 359 (2022): 127421.

di Biase, Alessandro, Maciej S. Kowalski, Tanner R. Devlin, and Jan A. Oleszkiewicz. "Moving bed biofilm reactor technology in municipal wastewater treatment: A review." *Journal of Environmental Management* 247 (2019): 849–866.

Ercan, Duygu, and Ali Demirci. "Current and future trends for biofilm reactors for fermentation processes." *Critical Reviews in Biotechnology* 35, no. 1 (2015): 1–14.

Espinosa-Ortiz, Erika J., Robin Gerlach, Brent M. Peyton, Luke Roberson, and Daniel H. Yeh. "Biofilm reactors for the treatment of used water in space: potential, challenges, and future perspectives." *Biofilm* 6 (2023): 100140.

Huang, Hui, Chong Peng, Pengcheng Peng, Yuan Lin, Xuxiang Zhang, and Hongqiang Ren. "Towards the biofilm characterization and regulation in biological wastewater treatment." *Applied Microbiology and Biotechnology* 103 (2019): 1115–1129.

Jagaba, Ahmad Hussaini, Shamsul Rahman Mohamed Kutty, Azmatullah Noor, Abdullahi Haruna Birniwa, Augustine Chioma Affam, Ibrahim Mohammed Lawal, Mubarak Usman Kankia, and Abdullahi Usman Kilaco. "A systematic literature review of biocarriers: Central elements for biofilm formation, organic and nutrients removal in sequencing batch biofilm reactor." *Journal of Water Process Engineering* 42 (2021): 102178.

Jayathilake, P.G., S. Jana, S. Rushton, D. Swailes, B. Bridgens, T. Curtis, and J. Chen. Extracellular polymeric substance production and aggregated bacteria colonization influence the competition of microbes in biofilms. *Frontiers in Microbiology* 8 (2017) Sep 27: 1865.

Karadag, Dogan, Oguz Emre Köroğlu, Bestami Ozkaya, and Mehmet Cakmakci. "A review on anaerobic biofilm reactors for the treatment of dairy industry wastewater." *Process Biochemistry* 50, no. 2 (2015): 262–271.

Kesaano, Maureen, and Ronald C. Sims. "Algal biofilm based technology for wastewater treatment." *Algal Research* 5 (2014): 231–240.

Leyva-Díaz, Juan Carlos, A. Monteoliva-García, J. Martín-Pascual, M.M. Munio, J.J. García-Mesa, and J.M. Poyatos. "Moving bed biofilm reactor as an alternative wastewater treatment process for nutrient removal and recovery in the circular economy model." *Bioresource Technology* 299 (2020): 122631.

Limoli, DH, C.J. Jones, and D.J. Wozniak. Bacterial extracellular polysaccharides in biofilm formation and function. *Microbial Biofilms* (2015) Oct 7: 223–47.

Lu, Duowei, Hao Bai, Fangong Kong, Steven N. Liss, and Baoqiang Liao. "Recent advances in membrane aerated biofilm reactors." *Critical Reviews in Environmental Science and Technology* 51, no. 7 (2021): 649–703.

Madan, Sangeeta, Richa Madan, and Athar Hussain. "Advancement in biological wastewater treatment using hybrid moving bed biofilm reactor (MBBR): A review." *Applied Water Science* 12, no. 6 (2022): 141.

Mahto, Kumari Uma, and Surajit Das. "Bacterial biofilm and extracellular polymeric substances in the moving bed biofilm reactor for wastewater treatment: A review." *Bioresource Technology* 345 (2022): 126476.

Martin, K.J., and R. Nerenberg. The membrane biofilm reactor (MBfR) for water and wastewater treatment: Principles, applications, and recent developments. *Bioresource Technology* 122 (2012) Oct 1: 83–94.

Morgan-Sagastume, Fernando. "Biofilm development, activity and the modification of carrier material surface properties in moving-bed biofilm reactors (MBBRs) for wastewater treatment." *Critical Reviews in Environmental Science and Technology* 48, no. 5 (2018): 439–470.

Muhammad, Musa Hassan, Aisha Lawan Idris, Xiao Fan, Yachong Guo, Yiyan Yu, Xu Jin, Junzhi Qiu, Xiong Guan, and Tianpei Huang. "Beyond risk: Bacterial biofilms and their regulating approaches." *Frontiers in Microbiology* 11 (2020): 928.

Nerenberg, Robert. "The membrane-biofilm reactor (MBfR) as a counter-diffusional biofilm process." *Current Opinion in Biotechnology* 38 (2016): 131–136.

Ødegaard, H., B. Gisvold, and J. Strickland The influence of carrier size and shape in the moving bed biofilm process. *Water Science and Technology* 41, no. 4–5 (2000) Feb 1: 383–91.

Peng, Yuke, Jie Li, Junling Lu, Lin Xiao, and Liuyan Yang. "Characteristics of microbial community involved in early biofilms formation under the influence of wastewater treatment plant effluent." *Journal of Environmental Sciences* 66 (2018): 113–124.

Safwat, Safwat M. "Moving bed biofilm reactors for wastewater treatment: A review of basic concepts." *International Journal of Research* 6, no. 10 (2019): 85–90.

Saidulu, Duduku, Abhradeep Majumder, and Ashok Kumar Gupta. "A systematic review of moving bed biofilm reactor, membrane bioreactor, and moving bed membrane bioreactor for wastewater treatment: Comparison of research trends, removal mechanisms, and performance." *Journal of Environmental Chemical Engineering* 9, no. 5 (2021): 106112.

Shahot, Khaled, Azni Idris, Rozita Omar, and Hamdan M. Yusoff. "Review on biofilm processes for wastewater treatment." *Life Sci J* 11, no. 11 (2014): 1–13.

Swain, Anup Kumar, Abanti Sahoo, Hara Mohan Jena, and Hemalata Patra. "Industrial wastewater treatment by aerobic inverse fluidized bed biofilm reactors (AIFBBRs): A review." *Journal of Water Process Engineering* 23 (2018): 61–74.

Toyofuku, Masanori, Tomohiro Inaba, Tatsunori Kiyokawa, Nozomu Obana, Yutaka Yawata, and Nobuhiko Nomura. "Environmental factors that shape biofilm formation." *Bioscience, Biotechnology, and Biochemistry* 80, no. 1 (2016): 7–12.

Van Loosdrecht, Mark C.M., and Sef J. Heijnen. "Biofilm bioreactors for waste-water treatment." *Trends in Biotechnology* 11, no. 4 (1993): 117–121.

Veerachamy, Suganthan, Tejasri Yarlagadda, Geetha Manivasagam, and Prasad KDV Yarlagadda. "Bacterial adherence and biofilm formation on medical implants: a review." *Proceedings of the Institution of Mechanical Engineers, Part H: Journal of Engineering in Medicine* 228, no. 10 (2014): 1083–1099.

Verma, Samakshi, Arindam Kuila, and Samuel Jacob. "Role of biofilms in waste water treatment." *Applied Biochemistry and Biotechnology* 195, no. 9 (2023): 5618–5642.

Wu, Lan, Wei Wei, Juan Xu, Xueming Chen, Yiwen Liu, Lai Peng, Dongbo Wang, and Bing-Jie Ni. "Denitrifying biofilm processes for wastewater treatment: developments and perspectives." *Environmental Science: Water Research & Technology* 7, no. 1 (2021): 40–67.

Zhao, Yingxin, Duo Liu, Wenli Huang, Ying Yang, Min Ji, Long Duc Nghiem, Quang Thang Trinh, and Ngoc Han Tran. "Insights into biofilm carriers for biological wastewater treatment processes: Current state-of-the-art, challenges, and opportunities." *Bioresource Technology* 288 (2019): 121619.

10 Aquatic Biofilms and Plastisphere

Simi Asma Salim and Siddhardha Busi

10.1 INTRODUCTION

The planet Earth is mostly occupied by ocean. A considerable part is occupied by ponds, rivers, streams, glaciers, etc., which altogether make the aquatic environment. The irrational use of plastics, their improper disposal, and related environmental impacts are topics of continuous discussion. Accumulation of plastics in aquatic environments include direct and indirect sources. As aquatic ecosystems are interconnected with terrestrial ecosystems, the activities on land obviously contribute to aquatic pollution. Major contributors of marine plastic pollutions include inputs from freshwater, domestic and residential activities, tourism, and economic actions such as harbour activities. It is estimated that more than 70% of plastics in marine environments come from land-based activities (Thushari and Senevirathna, 2020). According to the size of plastic pollutants, various researchers have classified these pollutants as macroplastics, microplastics, nanoplastics, etc. Shwarz and his group conducted a review study on the sources, transport, and accumulation of various plastic pollutants in aquatic ecosystems. They found packaging and consumer products as the most commonly encountered categories in rivers, whereas, in the oceans, plastics related to fishery items dominate. In all observed aquatic environments, they reported polyethylene and polypropylene as the most commonly encountered polymer types (Schwarz et al., 2019).

Plastics of small size (micro- and nanoplastics) have huge and direct impacts on aquatic life. Microplastics are plastics of size less than 5mm. They can either be of primary source, which have been originally produced in small size (including powder forms) for various applications, or secondary source, which are fragments of large plastics formed due to mechanical action, weathering, etc. (Issac and Kandasubramanian, 2021). The recently coined term plastisphere community refers to various microbial communities colonized on plastic fragments or debris, which later on form biofilms (Amaral-Zettler, 2020; Du et al., 2022). Algae, fungi, cyanobacteria, protozoans, bacteria, and viruses colonize and attach to the hydrophobic plastic surface. The plastisphere thus acts as a microbial community assemblage and includes autotrophs, heterotrophs, predators, and pathogens. Plastics can serve as a 'raft' and unique substrate for microorganisms. Organic matter deposition and surface properties of the plastics are exploited by microbes to overcome environmental

 DOI: 10.1201/9781003487203-10

hurdles and stress. In addition, various nutrients including iron, phosphorous, etc., also get deposited on the plastic surface over time, which also makes it a favourable substrate for microbes (Li et al., 2021; Dey et al., 2022) (Figure 10.1).

10.2 FORMATION AND ESTABLISHMENT OF PLASTISPHERE

The plastisphere microbiome is composed of both eukaryotic and prokaryotic members. A detailed explanation of the community composition is described in the section 10.3.. Attachment, secretion of exoploymeric substances, and microbial proliferation on plastic surface will establish plastisphere biofilm. Pioneer colonizers include primary producers including diatoms and cyanobacteria. Proteobacteria, especially gamma and alphaproteobacteria, are the pioneer bacterial members. Colonization and attachment of these pioneer members decreases the hydrophobicity of plastics. Furthermore, these organisms produce exopolysaccharides increasing adhesion properties and helping in the entrapment of secondary colonizers. Enrichment of bacteroidetes, especially of *Flavobacteriaceae*, occurs later on. Predominance of members of families *Rhodobacteraceae*, *Microtrichaceae*, *Bacillaceae,* and *Moraxellaceae* also occurs. These are all secondary colonizers. They form pili, produce proteins for adhesion, and other strategies including synthesis of extracellular polymeric substances for the irreversible attachment onto the plastic surface. Over time, recruitment or loss of species on the biofilm will occur, improving adhesion to the plastic surface and eventually resulting in the development of a mature biofilm due to competition or combined action of various microbes (Du et al., 2022).

10.3 MICROBIAL COMPOSITION OF PLASTISPHERE

Initially, phototrophic diatoms and cyanobacteria colonize the plastic surface (Zhai et al., 2023). To an extent, geographic location and exposure time determines the diatom composition of plastisphere. Early colonizers of marine samples include *Cylindrotheca, Navicula,* and *Nitzschia* spp. (Laroche et al., 2023). *Navicula* and *Nitzschia* species dominate in lentic samples as well (Di Pippo et al., 2020). These species are common aquatic biofilm formers (Zettler et al., 2013). Cyanobacteria is also one of the dominant members of plastisphere biofilm, especially during earlier stages. Chroococcalean and oscillatorialean cyanobacteria were the dominant phototropic members of Italian lentic systems (Di Pippo et al., 2020). Earlier studies reported the presence of the oscillatorialean cyanobacteria *Phormidium* in marine plastic samples, but not in the seawater (Zettler et al., 2013).

Although the surrounding environment and polymer characteristics influence the plastisphere community, a group of common core microbiome will be present. Pioneer bacterial communities of the plastisphere include members of Alphaproteobacteria and Gammaproteobacteria. Later on diversity occurs on maturation of plastisphere, which incorporates the group Bacteriodetes (Zhai et al., 2023). Evaluation of plastisphere of Italian lentic systems showed Alphaproteobacteria and Gammaproteobacteria as the dominant ones in both biofilm and planktonic communities. Among Alphaproteobacteria, the family Sphingomonadaceae predominated. On

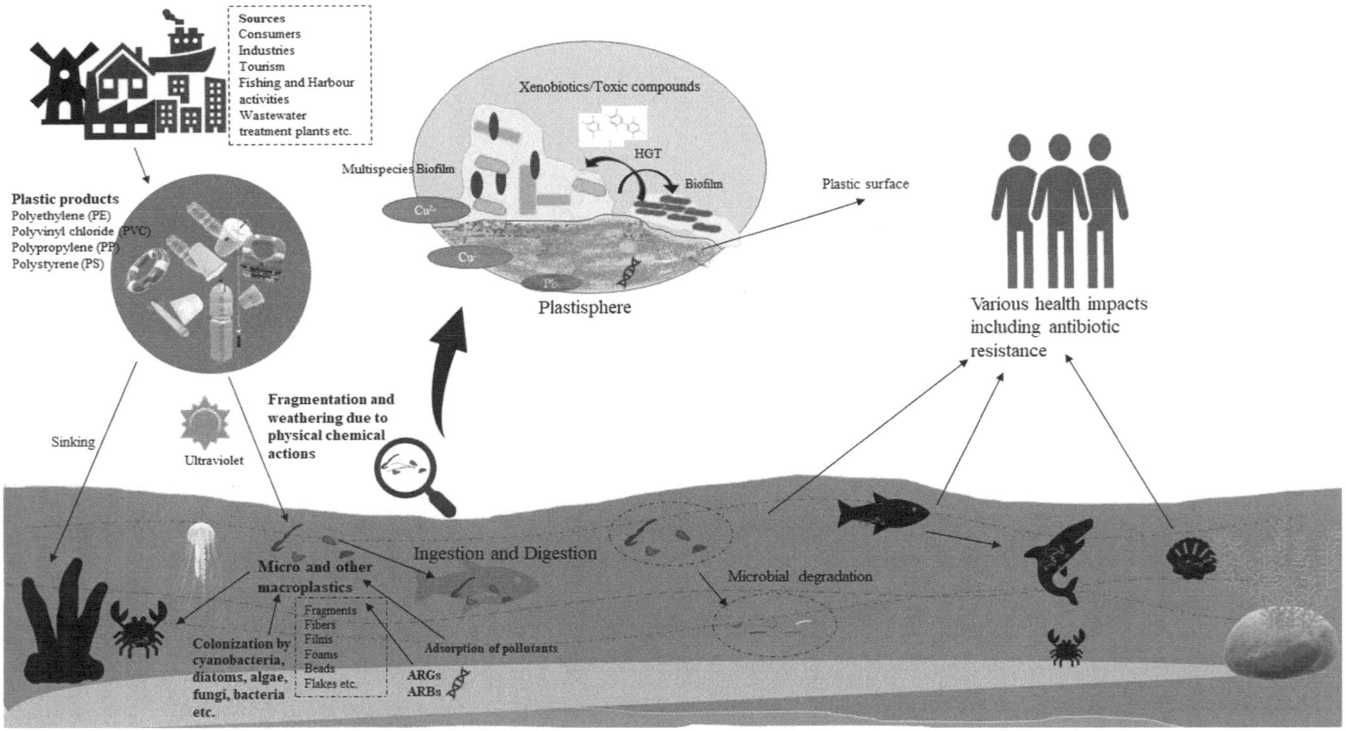

FIGURE 10.1 Various sources of plastics in aquatic systems and plastisphere development. Microbial colonization, biofilm formation, and adsorption of various pollutants and pathogens to the plastic surface occur. Later on, ingestion of the microplastics by aquatic organisms occurs and eventually to humans.

the other hand, in Gammaproteobacteria the family Burkholderiaceae predominated. The dominant genus were *Sphingomonas, Sphingorabdus, Limnohabitan,* etc. (Di Pippo et al., 2020). Riverine microplastics of larger size showed an abundance of the *genus Acinetobacter, Pseudomonas, Thiothrix,* and *Alkanindiges,* all of which belong to Gammaproteobacteria, and the members of genus *Roseateles, Massilia, Hydrogenophaga,* and *Acidovorax* belonging the Betaproteobacteria. Bacteroidetes and Actinobacteria predominated on small microplastic samples (Mughini-Gras et al., 2021). Plastic-type specific microbial abundance from samples incubated in freshwater were reported. Similar to the aforementioned bacterial composition and pattern, in the polypropylene (PP) sheets (macroplastic) Gamma and Betaproteobacteria predominated. In the polyethylene (PE) sheets (macroplastic) and microplastic-associated biofilms, Alpha and Betaproteobacteria predominated (Wang et al., 2021).

Exploration of microbial diversity in deep sea plastisphere was done by Kelly et al., 2022. Like any other aquatic systems, the core taxa was predominated by Gammaproteobacteria, Alphaproteobacteria, and Bacteroidia, which accounted for 78% of all classes. Some of the most abundant genus includes *Sediminibacterium, Extensimonas, Bradyrhizobium, Vibrionimonas, Pseudoalteromonas, Ralstonia,* and *Novosphingobium* (Kelly et al., 2022). Irrespective of location and polymer type, consistent domination of Alphaproteobacteria, Gammaproteobacteria, and Bacteroidia were reported from the marine plastisphere (Coons et al., 2021). These classes of bacteria also dominated in the samples from riverine systems. Selective enrichment of plastic-degrading genera were also reported by various studies. The bacteria *Ideonella* showed enrichment on microplastics incubated in urban rivers (Xu et al., 2023).

In the case of non-natural water environment as in the case of treated wastewater systems, microplastic biofilm communities showed higher abundance of the bacterial families *Flavobacteriaceae* followed by *Comamonadaceae* and *Rhodocyclaceae* (Galafassi et al., 2021).

Fungal communities of riverine plastisphere were evaluated by Xue et al. (2021). Results showed Ascomycota and Basidiomycota as the most abundant phyla. Similar previous studies on different polymer types also showed these phyla as the dominating fungal members. Mucuromycota were also among the dominating members. Co-occurrence network analysis revealed the shared keystone taxa of PE and PP sheets—*Malasseziales* and *Ascomycota_norank*. They also reported the presence of potential plastic degrader *Cladosporium cladosporioides* (Wang et al., 2021).

10.4 FACTORS INFLUENCING BIOFILM FORMATION IN PLASTISPHERE

Various factors influence biofilm formation on plastic surfaces, and include both environmental factors and material (plastic polymer) characteristics. Environmental factors comprise physicochemical aspects (like temperature, dissolved oxygen), nutrient availability, presence of toxins, metals, etc. Whereas characteristics of plastics that affect the microbial colonization and biofilm formation include polymer type, size, and surface property (Zhai et al., 2023). In a study in the coastal Northern Adriatic, polyvinyl chloride (PVC) surface exhibited more biofilm formation compared to other plastics such as polypropylene (PP), low-density polyethylene (LDPE) etc., after one week of incubation regardless of the ambient and dim light conditions (Pinto et al., 2019). Higher quantity of biofilm on PVC in freshwater environment was also reported (Miao et al., 2021). Seawater microcosm experimental studies on polyethylene, polypropylene, and polystyrene MPs showed influence of time and polymer type on biofilm formation. Although biofilm formation and biomass increased with time despite of plastic type, after the initial 10-day incubation, bacterial colonization and biofilm formation was higher in polystyrene MPs. The study also confirmed earlier hypotheses that hydrophobicity of plastic affects bacterial diversity but not abundance (Tu et al., 2021). Shape and size of the plastic have a slight influence on the abundance and diversity of plastisphere communities (Cheng et al., 2021). A distinct study by Wen et al. (2020) reported that colour of the plastic can also affect development of biofilm on plastic surface. LDPE microplastics of different color (blue, yellow, and transparent) showed significant distinction of plastisphere community. Interestingly, blue microplastics showed unique species, plausibly due to plastic color/additive-driven selection of microorganisms. In addition, functional characteristic analysis showed an enrichment of metabolic pathways; for instance, transport and metabolism of amino acids, carbohydrates, especially in blue microplastics (Wen et al., 2020). Size matters as well in the community structure and biofilm formation on the plastic surface. Larger particles (100–500 μm) were found to have a different microbial community and low diversity compared with smaller (10–100 μm) particles (Mughini-Gras et al., 2021).

Seasonal and geographical variations in plastisphere communities were studied by various researchers. MP microbial composition of different geographical locations, Yellow Sea and South China Sea, showed clear differences. Diversity of the microbial community showed a positive correlation with the exposure time. Significant influence of season was also reported (Xu et al., 2019). Evaluation of marine microplastic microbial communities from four different geographic locations (i.e., Cape Verde, Chile, Japan, and South Africa) also exposed differences in microbial composition (Coons et al., 2021).

Organic content, salinity, and dissolved oxygen influencing the communities of riverine microplastic plastisphere were evaluated. To no surprise, the most significant parameter in shaping the biofilm was organic content. Salinity stood as the second determining factor followed by concentration of dissolved oxygen. Furthermore, plastisphere communities of riverine water with high organic content showed enrichment of various pathogens including foodborne pathogen *B. cereus* (Nguyen et al., 2022). Copper content in seawater has been found to have an influence on colonizing microbial community and biofilm formation. Higher Cu concentrations were found to elevate bioaccumulation of Cu to the microplastic surface, which enriched the microbial colonizers and thereby shaped the biofilm (Djaoudi et al., 2022).

10.5 PLASTIC DEGRADATION BY PLASTISPHERE COMMUNITIES

Plastic has become an uncontrollable material in the modern world. Although alternatives for plastics have been identified, a complete solution to avoid plastics completely has not been found. For this reason, emphasis is put on recycling and biodegradation. Plastisphere communities can be explored as potential plastic degraders because various studies have shown selective enrichment of plastic-degrading microbes. The plastic degradation ability of these organisms can be identified and characterized by various methods such as analyzing biomass growth through enrichment culture techniques and using plastic as a sole source of carbon. Surface growth pattern can be observed by scanning electron microscopy (SEM), changes in surface physical properties such as roughness can be identified by SEM and atomic force microscopy (AFM). Isotope-labeled carbon assays give precise information on degradation products, but it is a high-cost approach. High-temperature gel permeation chromatography, differential scanning calorimetric analysis, thermogravimetric analysis, etc., can be used to determine the physicochemical changes in the plastic to provide information on degradation potential of different microbes (Zeghal et al., 2021).

The southwest Atlantic Ocean deep sea plastisphere core microbiome showed the presence of families Oleiphilaceae and Hyphomonadaceae, which include plastic-degrading members (Agostini et al., 2021). Early colonizers on weathered polyethylene plastics samples incubated in coastal seawater showed the enrichment of obligate hydrocarbon-degrading bacteria (OHCB) – *Oleiophilus* capable of degrading n-alkanes in the C11–C20 range. The genera *Aestuariibacter* and *Roseobacter*, which are proven for their hydrocarbon degradation abilities, were also found to be enriched on the samples (Erni-Cassola et al., 2020). Riverine microplastic samples showed selective enrichment of the genus *Ideonella,* the well-known plastic degrader (Xu et al., 2023). It can produce the PET-degradation enzymes and use the PET as source of carbon (Walter et al., 2022). Marine LDPE and PET samples showed an enrichment of potential hydrocarbon-degrading genera *Alcanivorax, Marinobacter*, and *Arenibacter*. The bacterium *Alcanivorax borkumensis* was found to have efficient biofilm formation on LDPE and their potential to degrade it was also stated (Delacuvellerie et al., 2019). Other than bacteria, various fungi can also degrade plastic polymers. About 3% of the eukaryotic plastisphere community is composed of fungi. Isolation of plastic-degrading fungus from PET waste of Korean seacoasts was performed by a group of researchers. Polycaprolactone (PCL) degradation analysis revealed the presence of many potential biodegraders. Isolates such as *Phaeophleospora eucalypticola*, *Alternaria* spp., *Aureobasidium* spp., and *Cladosporium* spp. were identified as having strong PCL degradation abilities (Kim et al., 2022).

The plastic degradation efficiency of the selectively enriched microorganisms can be exploited for mitigation of plastic pollution through eco-friendly approaches. Isolation of the plastisphere microbes followed by formulation of a potential consortium of plastic degraders will undoubtedly help in the biodegradation of plastic. A plastisphere-associated marine bacterial consortium was developed by Joshi et al. (2022). The consortium was formulated by choosing the top individual degraders.

They showed significant LDPE degradation on about four-month incubation. Exploration of the terrestrial-marine ecotone (i.e., from a salt marsh plastisphere community) showed the presence of excellent plastic degraders such as *Jonesia* and *Streptomyces* (Pang et al., 2023). Likewise, isolation and utilization of the plastisphere communities from this type of environment may provide unique and efficient plastic-biodegrading microbial flora.

10.6 ACCUMULATION OF POLLUTANTS BY PLASTISPHERE COMMUNITIES

Biofilms on plastisphere augments adsorption of various pollutants including various toxic contaminants and other heavy metals and thereby reinforces the vector potential of macro/microplastics (Bhagwat et al., 2021; Liu et al., 2022). The role of plastisphere biofilms in absorbing trace metals in freshwater systems was demonstrated by Liu et al. (2022). They found significant influence of biofilm on the adsorption kinetics of metals to the plastic. Regarding estuarine environment, studies showed the increased accumulation of heavy metals on plastisphere with time. An *in-situ* experiment revealed the accumulation of heavy metals in the order Cr > Mn > Zn > Cu > Pb > Cd, representing the role of biofilm in facilitating metal mobilization (Forero-López et al., 2022). Higher absorption of perfluorooctane sulfonate (PFOS) and lead (Pb) was demonstrated by aged biofilm covered microplastics than virgin microplastics. Enhanced adsorption of contaminant can be related to the transformed surface characteristics (Bhagwat et al., 2021). Positive correlation of accumulated metal with formed biofilm was also observed in other estuarine studies (Richard et al., 2019). Studies on microplastics in coastal seawater provided information on enhanced copper absorption efficacy by biofilm (Zhou et al., 2022). The microalgae of the plastisphere play a key role in accumulation potential of toxic chemicals owing to their ability to produce extracellular polymeric substances (Tarchi et al., 2023). Adsorption of lead was potentiated by biofilms. Moreover, biofilm also increased combined toxicity of Pb(II) and microplastics (Qi et al., 2021). Verdú et al. demonstrated the role of biofilm in adsorption of triclosan, a chlorinated biphenyl ether used as antimicrobial agents in consumer products. Triclosan was found to possess carcinogenic, endocrine toxicity (Fang et al., 2010; Verdú et al., 2023). Even though not in aquatic systems, the role of plastisphere microbes in enhanced production of methylmercury in agricultural soil was explained. Runoff of this will certainly affect aquatic environments (Hao et al., 2023). Ten-year-old HDPE plastic harbored high concentrations of metal(loid)s, polycyclic aromatic hydrocarbons (PAHs), and per- and polyfluoroalkyl substances (PFAS) (Bhagwat et al., 2021). Although other factors also contribute to the higher occurrence of contaminants, we can assume in light of other aforesaid studies that biofouling or biofilm formation on the plastic surface plays significant role.

Accumulation of heavy metals, antibiotic resistance genes (ARGs), and other toxic pollutants can have a significant impact on aquatic flora and fauna and thereby human health. Biofilm formation on plastics generates chemosensory responses and thus increases chance of ingestion by various higher and lower organisms. Thus, bioaccumulation of various pollutants in aquatic environments occurs, eventaully reaching humans (Bhagwat et al., 2021).

10.7 PLASTISPHERE AS A VECTOR OF POTENTIAL PATHOGENS AND ANTIBIOTIC RESISTOMES

It is well known that plastic pollution can contribute to the spread of potential pathogens and antibiotic resistomes, chiefly through aquatic environments. Because of their hydrophobic nature, plastics can accumulate various organic and inorganic pollutants including heavy metals, pesticides, antibiotics, and other persistent organic pollutants that have already been proven to play a role in driving the process of selection/co-selection of resistance to antibiotics (Yang et al., 2020). Metagenomic analysis of a riverine plastisphere biofilm community showed an enrichment of pathogenic microbes, especially *Pseudomonas aeruginosa, Acinetobacter, Aeromonas,* and ARG subtypes conferring resistance to macrolides/lincosamides, rifamycin, sulfonamides, and disinfecting agents and glycopeptides. On the other hand, the planktonic community of plastisphere had a distinct community of potential pathogens including *Escherichia, Salmonella, Klebsiella, Streptococcus,* and ARGs of aminoglycosides, tetracycline, aminocoumarin, fluoroquinolones, nitroimidazole, oxazolidinones, and fosfomycins (Zadjelovic et al., 2023). A temporal dynamics study of ARGs, mobile genetic elements (MGEs), and microbial diversity of plastisphere in urban water revealed the presence of 82 ARGs, 12 MGEs, and 63 pathogenic bacteria. MGEs are positively co-related with horizontal gene transfer of ARGs. Class 1 integron-integrase genes (*intI1*), which have been reported to play a crucial role in transmission of ARG cassettes among Gram-negative bacteria, were also reported to be among the five persistent MGEs. Furthermore, *trb-C* genes playing a key role in the type IV secretion system-mediated conjugation can favor the transconjugation of ARG plasmids into pathogens. Metallo-β-lactamase confers resistance to last-resort antibiotics (carbapenems) through ARG (*cphA)* (Yang et al., 2020).

Selective enrichment of pathogens by microplastic biofilm was reported by Wu and co-workers. They showed enrichment of specific ARG subtypes in microplastic biofilm compared to natural substrate biofilms. The enriched human pathogens were *Pseudomonas monteilii* and *Pseudomonas mendocina* (opportunistic pathogens). Of note, a plant pathogen, *Pseudomonas syringae,* was also reported to be abundant in the microplastic biofilm (Wu et al., 2019).

Treated wastewater is destined to open waters. Only a few studies are available on the plastisphere in treated wastewater and its contribution to the transmission of ARGs and pathogens. Although chances for transmission of ARGs and pathogens as planktonic form through water are higher than for biofilm forms, attention has to be given to the microplastic-attached biofilm forms due to their stress-resistant abilities and horizontal gene transfer (HGT) efficiency. Studies report the enrichment of potential pathogens, *Chryseobacterium* and ARG *sul2*, which confer resistance against sulphonamides on MPs of treated wastewater (Galafassi et al., 2021). Meta-analysis of ARGs in microplastics of aquatic environment revealed higher abundance of ARGs in MPs than that of inorganic substrate and natural water environment (but less than natural organic substrate). Moreover, freshwater MP samples showed higher degree of ARG enrichments than saline and sewage MP samples (Yu et al., 2022).

It is well known that antibiotic concentration in the surrounding environment positively correlates with the ARGs in plastisphere. The correlation of ARGs is even stronger than other substrates like glass slides and limestone rocks. Notable enrichment of *sul1* and *dfrA* genes was observed from the plastic samples on river site having higher anthropogenic impact (Martínez-Campos et al., 2023). Similar evidence has also been found from other studies. Antibiotic-resistant priority pathogens were shown to be more enriched in MPs than in sand particles. The most encountered ARBs from river site before wastewater treatment plant discharge were *Aeromonas* and *Pseudomonas*, whereas *Enterobacteriaceae* predominated in the river samples after the wastewater treatment plant discharge. *Escherichia coli*, *Klebsiella pneumoniae*, *Citrobacter* spp., *Enterobacter* spp., and *Shigella* sp. were identified as the Ciprofloxacin- and/or cefotaxime-resistant *Enterobacteriaceae* members (Silva, I et al., 2023).

As we discussed earlier, plastic characteristics influence its community composition and further affects ARG composition. Among the studied PE, PP, and PET samples in estuarine environment, PP demonstrated a higher relative abundance of ARGs. Furthermore a correlation of MGEs with absolute and relative abundance of ARGs were also demonstrated, suggesting the role of MGEs in ARGs migration within biofilm (Guo et al., 2022).

10.8 ROLE OF NOVEL TOOLS IN THE UNVEILING OF PLASTISPHERE

High-throughput sequencing technologies helped the emergence of novel techniques including metabarcoding, metagenomics, and other omics technologies, studies on plastisphere structural composition, functional studies including biodegradation properties, understanding of the role in the transmission of ARBs and ARGs, etc., and have provided scientific insights.

The highly conserved 16S rRNA is typically analyzed to understand prokaryotic composition. Because of its larger size and difficulty in sequencing the entire gene using high-throughput sequencing technologies, amplicon-specific sequencing has been introduced. Here, using particular primer sets, one or more variable (V) regions of the 16S rRNA are amplified and sequenced (Di Bella et al., 2013).

DNA metabarcoding helps in the taxonomic characterization within a sample. Studies of plastisphere in a subtropical estuary (Sérvulo et al., 2023), floating plastics of Western South Atlantic, etc. Lacerda et al. (2022) used DNA metabarcoding to reveal plastisphere composition. The colonization dynamics of prokaryotes in plastisphere was revealed by 16s rRNA gene metabarcoding (Silva, V et al., 2023). Metabarcoding studies based on third-generation sequencing, which permits sequencing of long regions of the same barcode, is expected to give better taxa classification and thereby more accurate species classification (Du et al., 2022). An example is the DNA metabarcoding with Nanopore MinION (Oxford Nanopore Technology) technologies for analyzing species associated with plastics in the

Mediterranean Sea (Davidov et al., 2020). Although 18s rRNA gene is typically used for identifying eukaryotes, for fungal identification and classification internal transcribed spacer (ITS) technique using specific primers is also utilized, because of the limitation of 18s rRNA gene in taxonomic resolution for fungal groups (Du et al., 2022).

Metagenomic, metaproteomic, and metatranscriptomic analysis of plastisphere will provide additional functional information related to antibiotic resistomes, plastic degradation, biogeochemical cycling, etc. Wu et al. used 16s rRNA gene sequencing and metagenomic and metatranscriptomic sequencing methods to evaluate the actively expressed antibiotic resistomes in river plastisphere (Wu et al., 2022). Metagenomic sequencing helps in exploring the viral communities of plastisphere (Li et al., 2022A). Moreover, the potential environmental risk of plastisphere viruses has also been elucidated. Hydrocarbon degradation, polymer degradation, and virulence properties of marine plastisphere was studied using metagenomics and metaproteomics (Delacuvellerie et al., 2022). A shortgun metagenomics-based study also showed the potential of plastic degradation, especially production of biosurfactant rhamnolipids, which can help in the biodegradation of plastics. The study also revealed the possibilities of exploring plastisphere for bioactive secondary metabolites (Saleem et al., 2023). The biodegrdation efficiency of plastisphere isolates were confirmed by metabolomics analysis coupled with proteogenomics (Wright et al., 2021). Table 10.1 summarizes different studies of aquatic plastisphere based on metagenomics, metaproteomics, and metatranscriptomic approaches (Table 10.1).

10.9 CONCLUSION

Understanding the plastisphere is important because of the tremendous amount of plastics reaching aquatic and other ecosystems. The plastisphere is composed of autotrophs, heterotrophs, predators, and pathogens. Initially pioneer colonizers attach followed by the establishment of secondary colonizers, and further competition and selection of microbial communities makes a mature plastisphere. Various environmental and polymer characteristics influence the community structure of the plastisphere. As the microbial communities in the plastisphere occur as biofilm sessile forms, they can better resist environmental stresses and hurdles than the planktonic communities. Apart from its role in transmitting potential pathogens, the plastisphere is reported to accumulate various environmental toxins and pollutants including heavy metals and antibiotic resistomes,- which enhance the vector potential of macro and microplastics. Emergence high-throughput sequencing, matabarcoding, and various omics approaches are useful in understanding the plastisphere community. Isolation of the plastic-biodegrading organisms from the plastisphere may help in the ecofriendly mitigation of plastic pollution.

TABLE 10.1
Studies Based on Omics Approaches for the Exploration of Plastisphere Communities in Different Aquatic Systems

Approach	Description of study	Aquatic environment	Experimental strategy	Reference
Metagenomics	Study on microplastic plastisphere composition and their environmental risk assessment by evaluating ARGs and virulence factors	River	Field experiment (Samples collected)	Li et al., 2022 B
Metagenomics	Evaluation of viral composition and environmental risk assessment	River	Field experiment (samples collected)	Li et al., 2022 A
Metagenomics	Evaluation of the incidence of ARGs, bacterial communities, and mobile genetic elements (MGEs) on MPs	River	Field experiments (Samples collected)	Xu et al., 2022
Metagenomics	Evaluation of tetracycline degradation and ARG occurrence	River	Laboratory incubation experiment	Wang et al., 2022
Integrated metagenomics and metatranscriptomics	Exploration of the occurrence, abundance, and transcriptional level of ARGs, plasmid associated ARGs, microbial composition and ARG bacterial hosts in the plastisphere.	River	Laboratory bioreactor experiment	Wu et al., 2022
Metagenomics	Dynamics of the plastisphere communities in microplastics over time	Wastewater	Microcosm	Xia et al.,2022
Metagenomics	Distribution profile of plastic-degrading genes	Estuary	Collected plastic samples from estuaries in China	Zhou et al.,2023
Metagenomics	Study on microbial community response to plastic and bioplastics	Marine	Microcosm	Pinnell and Turner, 2019
Metagenomics and Metaproteomics	Study on the community structure and functioning (plastic degradation) and pathogenicity of plastisphere microbes	Marine	Field samples collected from Mediterranean sea	Delacuvellerie et al., 2022
Metagenomics	Evolution of plastic biofilm assemblage over time	Marine	Combined laboratory and Field experiments	Bos et al.,2022

REFERENCES

Agostini, L., Moreira, J.C.F., Bendia, A.G., Kmit, M.C.P., Waters, L.G., Santana, M.F.M., Sumida, P.Y.G., Turra, A. and Pellizari, V.H.2021. Deep-sea plastisphere: long-term colonization by plastic-associated bacterial and archaeal communities in the Southwest Atlantic Ocean. *Sci. Total Environ.* 793:148335. https://doi.org/10.1016/j.scitotenv.2021.148335

Amaral-Zettler, L.A., Zettler, E.R. and Mincer, T.J. 2020. Ecology of the plastisphere. *Nat. Rev. Microbiol.* 18:139–151. https://doi.org/10.1038/s41579-019-0308-0

Bhagwat, G., Carbery, M., Anh Tran, T.K., Grainge, I., O'Connor, W. and Palanisami, T. 2021. Fingerprinting plastic-associated inorganic and organic matter on plastic aged in the marine environment for a decade. *Environ. Sci. Technol.* 55:7407–7417. https://doi.org/10.1021/acs.est.1c00262

Bhagwat, G., Tran, T.K.A., Lamb, D., Senathirajah, K., Grainge, I., O'Connor, W., Juhasz, A. and P Alanisami, T. 2021. Biofilms enhance the adsorption of toxic contaminants on plastic microfibers under environmentally relevant conditions. *Environ. Sci. Technol.* 55:8877–8887. https://doi.org/10.1021/acs.est.1c02012

Bos, R.P., Kaul, D., Zettler, E.R., Hoffman, J.M., Dupont, C.L., Amaral-Zettler, L.A. and Mincer, T.J. 2022. Tracking genomic characteristics across Oceanic provinces: Contrasting early and mature plastic biofilm communities. *BioRxiv*. 2022-01. https://doi.org/10.1101/2022.01.23.477364

Cheng, J., Jacquin, J., Conan, P., Pujo-Pay, M., Barbe, V., George, M., Fabre, P., Bruzaud, S., Ter Halle, A., Meistertzheim, A.L. and Ghiglione, J.F. 2021. Relative influence of plastic debris size and shape, chemical composition and phytoplankton-bacteria interactions in driving seawater plastisphere abundance, diversity and activity. *Front.Microbiol.* 11:610231.

Coons, A.K., Busch, K., Lenz, M., Hentschel, U. and Borchert, E. 2021. Biogeography rather than substrate type determines bacterial colonization dynamics of marine plastics. *PeerJ.* 9:e12135. https://doi.org/10.7717/peerj.12135

Davidov, K., Iankelevich-Kounio, E., Yakovenko, I., Koucherov, Y., Rubin-Blum, M. and Oren, M. 2020. Identification of plastic-associated species in the Mediterranean Sea using DNA metabarcoding with Nanopore MinION. *Sci. Rep.* 10:17533. https://doi.org/10.1038/s41598-020-74180-z

Delacuvellerie, A., Cyriaque, V., Gobert, S., Benali, S. and Wattiez, R. 2019. The plastisphere in marine ecosystem hosts potential specific microbial degraders including Alcanivorax borkumensis as a key player for the low-density polyethylene degradation *J. Hazard. Mater.* 380:120899. https://doi.org/10.1016/j.jhazmat.2019.120899

Delacuvellerie, A., Geron, A., Gobert, S. and Wattiez, R. 2022. New insights into the functioning and structure of the PE and PP plastispheres from the Mediterranean Sea. *Environ. Pollut.* 295:118678. https://doi.org/10.1016/j.envpol.2021.118678

Dey, S., Rout, A.K., Behera, B.K. and Ghosh, K. 2022. Plastisphere community assemblage of aquatic environment: Plastic-microbe interaction, role in degradation and characterization technologies. *Environ. Microbiome* 17:1–21. https://doi.org/10.1186/s40793-022-00430-4

Di Bella, J.M., Bao, Y., Gloor, G.B., Burton, J.P. and Reid, G. 2013. High throughput sequencing methods and analysis for microbiome research. *J. Microbiol. Methods* 95:401–414. https://doi.org/10.1016/j.mimet.2013.08.011

Di Pippo, F., Venezia, C., Sighicelli, M., Pietrelli, L., Di Vito, S., Nuglio, S. and Rossetti, S. 2020. Microplastic-associated biofilms in lentic Italian ecosystems. *Water Res.* 187:116429. https://doi.org/10.1016/j.watres.2020.116429

Djaoudi, K., Onrubia, J.A.T., Boukra, A., Guesnay, L., Portas, A., Barry-Martinet, R., Angeletti, B., Mounier, S., Lenoble, V. and Briand, J.F. 2022. Seawater copper content controls biofilm bioaccumulation and microbial community on microplastics.*Sci. Total Environ.* 814:152278. https://doi.org/10.1016/j.scitotenv.2021.152278

Du, Y., Liu, X., Dong, X. and Yin, Z., 2022. A review on marine plastisphere: Biodiversity, formation, and role in degradation.*Comput Struct Biotechnol J.* 20:975–988. https://doi.org/10.1016/j.csbj.2022.02.008

Erni-Cassola, G., Wright, R.J., Gibson, M.I. and Christie-Oleza, J.A. 2020. Early colonization of weathered polyethylene by distinct bacteria in marine coastal seawater. *Microb. Ecol.* 79:517–526. https://doi.org/10.1007/s00248-019-01424-5

Fang, J.L., Stingley, R.L., Beland, F.A., Harrouk, W., Lumpkins, D.L. and Howard, P. 2010. Occurrence, efficacy, metabolism, and toxicity of triclosan. *J Environ Sci Health C Environ Carcinog Ecotoxicol Rev.* 28:147–171. https://doi.org/10.1080/10590501.2010.504978

Forero-López, A.D., Brugnoni, L.I., Abasto, B., Rimondino, G.N., Lassalle, V.L., Ardusso, M.G., Nazzarro, M.S., Martinez, A.M., Spetter, C.V. and Biancalana, F. 2022. Plastisphere on microplastics: In situ assays in an estuarine environment. *J. Hazard. Mater.* 440:129737. https://doi.org/10.1016/j.jhazmat.2022.129737

Galafassi, S., Sabatino, R., Sathicq, M.B., Eckert, E.M., Fontaneto, D., Dalla Fontana, G., Mossotti, R., Corno, G., Volta, P. and Di Cesare, A. 2021. Contribution of microplastic particles to the spread of resistances and pathogenic bacteria in treated wastewaters. *Water Res.* 201:117368. https://doi.org/10.1016/j.watres.2021.117368

Guo, X.P., Chen, Y.R., Sun, X.L., Li, C.L., Hou, L.J., Liu, M. and Yang, Y. 2022. Plastic properties affect the composition of prokaryotic and eukaryotic communities and further regulate the ARGs in their surface biofilms. *Sci. Total Environ.* 839:156362. https://doi.org/10.1016/j.scitotenv.2022.156362

Hao, Y.Y., Liu, H.W., Zhao, J., Feng, J., Hao, X., Huang, Q., Gu, B. and Liu, Y.R. 2023. Plastispheres as hotspots of microbially-driven methylmercury production in paddy soils. *J. Hazard. Mater.* 457:131699. https://doi.org/10.1016/j.jhazmat.2023.131699

Issac, M.N. and Kandasubramanian, B. 2021. Effect of microplastics in water and aquatic systems. *Environ. Sci. Pollut. Res.* 28:19544–19562. https://doi.org/10.1007/s11356-021-13184-2

Joshi, G., Goswami, P., Verma, P., Prakash, G., Simon, P., Vinithkumar, N.V. and Dharani, G. 2022. Unraveling the plastic degradation potentials of the plastisphere-associated marine bacterial consortium as a key player for the low-density polyethylene degradation. *J. Hazard. Mater.* 425:128005. https://doi.org/10.1016/j.jhazmat.2021.128005

Kelly, M.R., Whitworth, P., Jamieson, A. and Burgess, J.G. 2022. Bacterial colonisation of plastic in the Rockall Trough, North-East Atlantic: An improved understanding of the deep-sea plastisphere. *Environ. Pollut.* 305:119314. https://doi.org/10.1016/j.envpol.2022.119314

Kim, S.H., Lee, J.W., Kim, J.S., Lee, W., Park, M.S. and Lim, Y.W. 2022. Plastic-inhabiting fungi in marine environments and PCL degradation activity. *Antonie Van Leeuwenhoek* 115:1379–1392. https://doi.org/10.1007/s10482-022-01782-0

Lacerda, A.L., Taylor, J.D., dS Rodrigues, L., Kessler, F., Secchi, E. and Proietti, M.C. 2022. Floating plastics and their associated biota in the Western South Atlantic. *Sci. Total Environ.* 805:150186. https://doi.org/10.1016/j.scitotenv.2021.150186

Laroche, O., Pantos, O., Kingsbury, J.M., Zaiko, A., Wallbank, J., Lear, G., Thompson-Laing, J., Audrezet, F., Maday, S., Doake, F. and Abbel, R. 2023. A spatio-temporal analysis of marine diatom communities associated with pristine and aged plastics. *Biofouling* 39:427–443. https://doi.org/10.1080/08927014.2023.2226069

Li, C., Wang, L., Ji, S., Chang, M., Wang, L., Gan, Y. and Liu, J. 2021. The ecology of the plastisphere: Microbial composition, function, assembly, and network in the freshwater and seawater ecosystems. *Water Res.* 202:117428. https://doi.org/10.1016/j.watres.2021.117428

Li, R., Zhu, L., Cui, L. and Zhu, Y.G. 2022A. Viral diversity and potential environmental risk in microplastic at watershed scale: Evidence from metagenomic analysis of plastisphere. *Environ. Int.* 161:107146. https://doi.org/10.1016/j.envint.2022.107146

Li, R., Zhu, L., Wang, Y. and Zhu, Y.G. 2022B. Metagenomic insights into environmental risk of field microplastics in an urban river. *Water Res.* 223:119018. https://doi.org/10.1016/j.watres.2022.119018

Liu, Z., Adyel, T.M., Wang, Z., Wu, J., Liu, J., Miao, L. and Hou, J. 2022. Effects of biofilms on trace metal adsorption on plastics in freshwater systems. *Int. J. Environ. Res. Public Health* 19:13752. https://doi.org/10.3390/ijerph192113752

Martínez-Campos, S., González-Pleiter, M., Rico, A., Schell, T., Vighi, M., Fernández-Piñas, F., Rosal, R. and Leganés, F. 2023. Time-course biofilm formation and presence of antibiotic resistance genes on everyday plastic items deployed in river waters. *J. Hazard. Mater.* 443:130271. https://doi.org/10.1016/j.jhazmat.2022.130271

Miao, L., Gao, Y., Adyel, T.M., Huo, Z., Liu, Z., Wu, J. and Hou, J. 2021. Effects of biofilm colonization on the sinking of microplastics in three freshwater environments. *J. Hazard. Mater.* 413:125370. https://doi.org/10.1016/j.jhazmat.2021.125370

Mughini-Gras, L., van der Plaats, R.Q., van der Wielen, P.W., Bauerlein, P.S. and de Roda Husman, A.M. 2021. Riverine microplastic and microbial community compositions: a field study in the Netherlands. *Water Res.* 192:116852. https://doi.org/10.1016/j.watres.2021.116852

Nguyen, H.T., Choi, W., Kim, E.J. and Cho, K. 2022. Microbial community niches on microplastics and prioritized environmental factors under various urban riverine conditions. *Sci. Total Environ.* 849:157781. https://doi.org/10.1016/j.scitotenv.2022.157781

Pang, G., Li, X., Ding, M., Jiang, S., Chen, P., Zhao, Z., Gao, R., Song, B., Xu, X., Shen, Q. and Cai, F.M., 2023. The distinct plastisphere microbiome in the terrestrial-marine ecotone is a reservoir for putative degraders of petroleum-based polymers. *J. Hazard. Mater.* 453:131399. https://doi.org/10.1016/j.jhazmat.2023.131399

Pinnell, L.J. and Turner, J.W. 2019. Shotgun metagenomics reveals the benthic microbial community response to plastic and bioplastic in a coastal marine environment. *Front Microbiol.* 10:1252. https://doi.org/10.3389/fmicb.2019.01252

Pinto, M., Langer, T.M., Hüffer, T., Hofmann, T. and Herndl, G.J. 2019. The composition of bacterial communities associated with plastic biofilms differs between different polymers and stages of biofilm succession. *PloS one* 14:e0217165. https://doi.org/10.1371/journal.pone.0217165

Qi, K., Lu, N., Zhang, S., Wang, W., Wang, Z. and Guan, J. 2021. Uptake of Pb (II) onto microplastic-associated biofilms in freshwater: adsorption and combined toxicity in comparison to natural solid substrates. *J Hazard Mater.* 411:125115. https://doi.org/10.1016/j.jhazmat.2021.125115

Richard, H., Carpenter, E.J., Komada, T., Palmer, P.T. and Rochman, C.M. 2019. Biofilm facilitates metal accumulation onto microplastics in estuarine waters. *Sci. Total Environ.* 683:600–608. https://doi.org/10.1016/j.scitotenv.2019.04.331

Saleem, M., Yahya, S., Razzak, S.A., Khawaja, S. and Ali, A. 2023. Shotgun metagenomics and computational profiling of the plastisphere microbiome: unveiling the potential of enzymatic production and plastic degradation. *Arch. Microbiol.* 205:359. https://doi.org/10.1007/s00203-023-03701-x

Schwarz, A.E., Ligthart, T.N., Boukris, E. and Van Harmelen, T. 2019. Sources, transport, and accumulation of different types of plastic litter in aquatic environments: A review study. *Mar. Pollut. Bull.* 143:92–100. https://doi.org/10.1016/j.marpolbul.2019.04.029

Sérvulo, T., Taylor, J.D., Proietti, M.C., dS Rodrigues, L., Puertas, I.P., Barutot, R.A. and Lacerda, A.L. 2023. Plastisphere composition in a subtropical estuary: Influence of season, incubation time and polymer type on plastic biofouling.*Environ. Pollut.* 332:121873. https://doi.org/10.1016/j.envpol.2023.121873

Silva, I., Rodrigues, E.T., Tacão, M. and Henriques, I. 2023. Microplastics accumulate priority antibiotic-resistant pathogens: Evidence from the riverine plastisphere. *Environ. Pollut.* 332:121995. https://doi.org/10.1016/j.envpol.2023.121995

Silva, V., Pérez, V. and Gillanders, B.M., 2023. Short-term plastisphere colonization dynamics across six plastic types. *Environ. Microbiol.* 25:2732–2745. https://doi.org/10.1111/1462-2920.16445

Tarchi, M., Zaaboub, N., Alsubih, M., Brik, B., Martins, M.V.A., Aleya, L. and Trabelsi, L. 2023. Microalgae colonization and trace element accumulation on the plastisphere of marine plastic debris in Monastir Bay (Eastern Tunisia). *Environ. Sci. Pollut. Res.* 30:32427–32451. https://doi.org/10.1007/s11356-022-23930-9

Thushari, G.G.N. and Senevirathna, J.D.M. 2020. Plastic pollution in the marine environment. *Heliyon* 6:e04709. https://doi.org/10.1016/j.heliyon.2020.e04709

Tu, C., Liu, Y., Li, L., Li, Y., Vogts, A., Luo, Y. and Waniek, J.J. 2021. Structural and functional characteristics of microplastic associated biofilms in response to temporal dynamics and polymer types. *Bull Environ Contam Toxicol.* 107:633–639.

Verdú, I., Amariei, G., Rueda-Varela, C., González-Pleiter, M., Leganés, F., Rosal, R. and Fernández-Piñas, F. 2023. Biofilm formation strongly influences the vector transport of triclosan-loaded polyethylene microplastics. *Sci. Total Environ.* 859:160231. https://doi.org/10.1016/j.scitotenv.2022.160231

Walter, A., Sopracolle, L., Mutschlechner, M., Spruck, M. and Griesbeck, C. 2022. Biodegradation of different PET variants from food containers by Ideonella sakaiensis. *Arch. Microbiol.* 204:711. https://doi.org/10.1007/s00203-022-03306-w

Wang, J., Peng, C., Dai, Y., Li, Y., Jiao, S., Ma, X., Liu, X. and Wang, L. 2022. Slower antibiotics degradation and higher resistance genes enrichment in plastisphere. *Water Res.* 222:118920. https://doi.org/10.1016/j.watres.2022.118920

Wang, L., Tong, J., Li, Y., Zhu, J., Zhang, W., Niu, L. and Zhang, H., 2021. Bacterial and fungal assemblages and functions associated with biofilms differ between diverse types of plastic debris in a freshwater system. *Environ. Res.* 196:110371. https://doi.org/10.1016/j.envres.2020.110371

Wen, B., Liu, J.H., Zhang, Y., Zhang, H.R., Gao, J.Z. and Chen, Z.Z. 2020. Community structure and functional diversity of the plastisphere in aquaculture waters: does plastic color matter?. *Sci. Total Environ.* 740:140082. https://doi.org/10.1016/j.scitotenv.2020.140082

Wright, R.J., Bosch, R., Langille, M.G., Gibson, M.I. and Christie-Oleza, J.A. 2021. A multi-OMIC characterisation of biodegradation and microbial community succession within the PET plastisphere. *Microbiome* 9:1–22. https://doi.org/10.1186/s40168-021-01054-5

Wu, X., Pan, J., Li, M., Li, Y., Bartlam, M. and Wang, Y., 2019. Selective enrichment of bacterial pathogens by microplastic biofilm. *Water Res.* 165:114979. https://doi.org/10.1016/j.watres.2019.114979

Wu, X., Liu, Z., Li, M., Bartlam, M. and Wang, Y. 2022. Integrated metagenomic and metatranscriptomic analysis reveals actively expressed antibiotic resistomes in the plastisphere. *J. Hazard. Mater.* 430:128418. https://doi.org/10.1016/j.jhazmat.2022.128418

Xia, Y., Zhang, X., Zhang, M., Chen, L., Tang, X., Sun, Y. and Li, X. 2022. Plastic materials and water sources actively select and shape wastewater plastispheres over time. *Front Environ Sci Eng.* 16:145. https://doi.org/10.1007/s11783-022-1580-1

Xu, C., Lu, J., Shen, C., Wang, J. and Li, F. 2022. Deciphering the mechanisms shaping the plastisphere antibiotic resistome on riverine microplastics. *Water Res.* 225:119192. https://doi.org/10.1016/j.watres.2022.119192

Xu, L., Li, K., Zhang, M., Guo, J., Jia, W., Bai, X., Tian, X. and Huang, Y. 2023. Plastic substrate and residual time of microplastics in the urban river shape the composition and structure of bacterial communities in plastisphere. *J. Environ. Manage.* 345:118710. https://doi.org/10.1016/j.jenvman.2023.118710

Xu, X., Wang, S., Gao, F., Li, J., Zheng, L., Sun, C., He, C., Wang, Z. and Qu, L. 2019. Marine microplastic-associated bacterial community succession in response to geography, exposure time, and plastic type in China's coastal seawaters. *Mar. Pollut. Bull.* 145:278–286. https://doi.org/10.1016/j.marpolbul.2019.05.036

Xue, N., Fang, Q., Pan, X. and Zhang, D. 2021. Distinct fungal plastisphere across different river functional zones: A watershed scale study. *Sci. Total Environ.* 752:141879. https://doi.org/10.1016/j.scitotenv.2020.141879

Yang, K., Chen, Q.L., Chen, M.L., Li, H.Z., Liao, H., Pu, Q., Zhu, Y.G. and Cui, L. 2020. Temporal dynamics of antibiotic resistome in the plastisphere during microbial colonization. *Environ. Sci. Technol.* 54:11322–11332. https://doi.org/10.1021/acs.est.0c04292

Yu, X., Zhang, Y., Tan, L., Han, C., Li, H., Zhai, L., Ma, W., Li, C. and Lu, X. 2022. Microplastisphere may induce the enrichment of antibiotic resistance genes on microplastics in aquatic environments: A review.*Environ. Pollut.* 310:119891. https://doi.org/10.1016/j.envpol.2022.119891

Zadjelovic, V., Wright, R.J., Borsetto, C., Quartey, J., Cairns, T.N., Langille, M.G., Wellington, E.M. and Christie-Oleza, J.A. 2023. Microbial hitchhikers harbouring antimicrobial-resistance genes in the riverine plastisphere. *Microbiome* 11:225. https://doi.org/10.1186/s40168-023-01662-3

Zeghal, E., Vaksmaa, A., Vielfaure, H., Boekhout, T. and Niemann, H. 2021. The potential role of marine fungi in plastic degradation–a review. *Front. Mar. Sci.* 8:738877. https://doi.org/10.3389/fmars.2021.738877

Zettler, E.R., Mincer, T.J. and Amaral-Zettler, L.A. 2013. Life in the "plastisphere": microbial communities on plastic marine debris. *Environ. Sci. Technol.* 47:7137–7146. https://doi.org/10.1021/es401288x

Zhai, X., Zhang, X.H. and Yu, M. 2023. Microbial colonization and degradation of marine microplastics in the plastisphere: A review. *Front Microbiol.* 14:1127308. https://doi.org/10.3389/fmicb.2023.1127308

Zhou, L., Sang, S., Li, J., Li, Y., Wang, D., Gan, L., Zhao, Z. and Wang, J. 2023. From waste to resource: Metagenomics uncovers the molecular ecological resources for plastic degradation in estuaries of South China. *Water Res.* 242:120270. https://doi.org/10.1016/j.watres.2023.120270

Zhou, Q., Tu, C., Liu, Y., Li, Y., Zhang, H., Vogts, A., Plewe, S., Pan, X., Luo, Y. and Waniek, J.J. 2022. Biofilm enhances the copper (II) adsorption on microplastic surfaces in coastal seawater: Simultaneous evidence from visualization and quantification. *Sci. Total Environ.* 853:158217. https://doi.org/10.1016/j.scitotenv.2022.158217

11 Microbial Biofilms and Their Role in Biodeterioration of Organic Materials

Srujana Kathi

11.1 INTRODUCTION

Biodeterioration, as described by Hueck (1968), is any undesirable change in the properties of a material caused by the vital activities of organisms (Di Carlo et al., 2022). It can be interpreted as an irreparable damage caused by microorganisms (Aliasghari et al., 2020). The damage may be severe, affecting anything from home plastics to precisely used polymers, polymers used to protect cultural treasures, and/or polymeric materials used in space applications (Flemming, 2010). Biofilms contribute to biodeterioration of cultural heritage materials, causing discoloration, dissolution, rupture, and efflorescence (Rivera et al., 2018). The sole distinction between biodeterioration and biodegradation is that biodeterioration is improper biodegradation. Therefore, biodeterioration refers to the microbial fuel breakdown in an aircrafts wing tanks, whereas biodegradation refers to the same process happening in an oil slick. Anaerobic environments are incompatible with the purely aerobic process of hydrocarbon biodegradation (Smith et al., 1991).

The results of tricky interactions between microbes in aquatic biofilms are influenced by the presence of limited resources, signifying that background levels of organic carbon control producer and decomposer reactions to food availability during biofilm formation. (Wyatt et al., 2019). Microorganisms such as phototrophs, lithotrophs, and heterotrophs rapidly colonize and coat newly exposed rock and sandstone construction surfaces, creating a biofilm that modifies the surrounding environment of the boulder surface (Gu et al., 2021). Multiomics can be used to study the biodeterioration of historical materials. The combination of current and classic biochemical approaches gives fresh insights (Liu et al., 2022). Isolated strains of *Bacillus* and *Paenibacillus* from epilithic biofilms in Lake Baikal have the possibility to produce antibacterial chemicals useful for biotechnological applications (Sukhanova et al., 2018).

The surface association of microbes also leads to adverse consequences like biofouling, biocorrosion, and the survival and spread of hazardous or pathogenic bacteria and their genetic determinants. For decades, researchers have been studying

 DOI: 10.1201/9781003487203-11

colonization processes and mechanisms, as well as significant participants among surface-related microorganisms. Specific cell-surface, cell-cell, and interpopulation interactions appear to impact the composition, structure, spatiotemporal dynamics, and roles of surface-associated microbial communities. The microbial surface association has been found to be dependent on a number of noteworthy microbial methods and mechanisms, including (i) surface, population, and community recognition and signaling; (ii) interspecies communication and interaction; and (iii) regulatory balance amid collaboration and competition (Dang et al., 2016). Sivadon (2019) highlights the prominence of biofilm formation for the degradation of particulate substrates and its ecological functions in carbon recycling and pollution removal. The interfaces amid biofilms and microplastics were the key focus of Rummel's 2017 study, which also emphasized the influence of biofilms on the fate and impacts of microplastics in aquatic ecosystems. The significance of biofilms in trophic transmission and pollutant bioaccumulation in aquatic environments was studied by Bonnineau (2020), who also highlighted the need for more study in this field.

Water and nutrients are two essential elements for biodeterioration. Fungal growth can occur with relatively low water activity if the polymer contains nutrients, such as additives, but bacteria need higher water activity (Flemming, 2010). In addition, nutrients and water temperature are two major determinants in modifying the quantity and ecological structure of macroinvertebrates in the Mohammad-Abad River, and these parameters accounted for more than 73% of the correlations between macroinvertebrate population changes. The higher the local residents' access to aquatic habitats, the larger the impacts of nutrient and water temperature changes on environmental conditions, and the greater their negative ecological impact (Aazami et al., 2020). In many aquatic habitats, biofilms—complex communities of microbiota that coexist with aquatic interfaces—are thought to be hotspots for microbial life (Weitere et al., 2018).

The proliferation and metabolic activity of bacteria trigger a variety of processes resulting in microbial biofilm-induced stone deterioration. These processes, which are not always visible to the unaided eye, recurrently modify the interior structure of the stone substrates. Fungi have a major role in the weathering of stone monuments because of their highly erosive actions through mycelial growth and the production of organic acids and colors (Figure 11.1). The pressure applied by photosynthetic bacteria within the stone causes variations in the volume of water within their gelatinous casings throughout drying and rehydration process. The precipitation of oxalates and the mobilization of calcium ions also aid in the degradation of calcareous substrata. In addition to the unique characteristics of the various microbe families, biofilm growth itself is a biodeteriogenic factor (Mascaro et al., 2021).

11.2 BIOFILM FORMATION AND COMPOSITION

A biofilm is an assemblage of bacteria that is enclosed in an extracellular polymeric substance (EPS) matrix and adhered to a substrate. The intricate process of biofilm development (Figure 11.2) is governed by both intracellular and intercellular signaling networks (Mahto et al., 2021). It is a type of microbial community that is

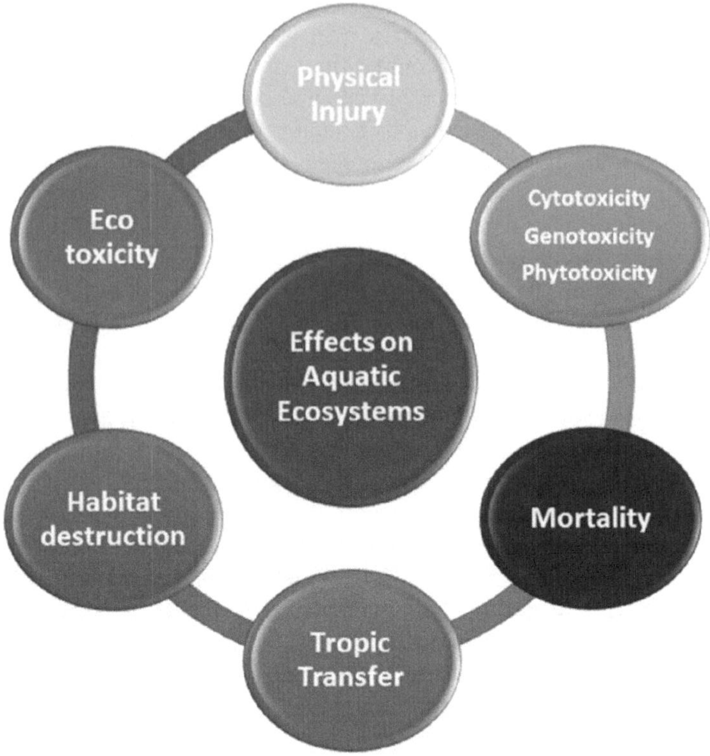

FIGURE 11.1 Effects of emerging contaminants on aquatic ecosystems.

adhered to a solid surface. It is made up of microcolonies of cells entrenched in an organic polymer matrix that is originated from microbes. Therefore, water, microbial cells, and EPS make up a biofilm. EPS is highly hydrated even if the majority of the water in them is indirectly linked to the EPS. Both charged and uncharged biopolymers are included in EPS. The biofilm matrix can contain any of the major types of biological macromolecules, comprising proteins, lipids, polysaccharides, and nucleic acids. While extracellular polysaccharides have long been thought to be the main structural elements of the matrix, it is now recognized that certain proteins known as functional amyloids strengthen the toughness of the matrix (Martino and Patrick, 2016).

The thickness of biofilms is a key factor closely associated with external parameters, including temperature, water flow, development age, and nutrient conditions (Suarez et al., 2019). Interestingly, the keystone taxa within the biofilm ecological networks exhibited strong habitat characteristics corroborated by discrete interaction patterns. Therefore, environmental filtering plays a crucial role in structuring biofilm communities demonstrating the significance of planktonic floc characteristics in the formation of biofilm communities (Xu et al., 2019). Unculturable microbes exploit mixed-species biofilm production to survive in a complex ecology where they compete with

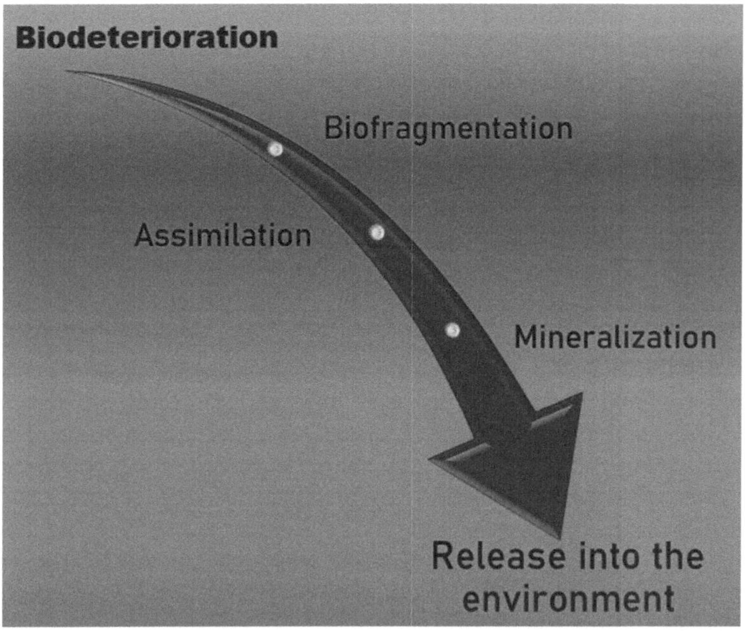

FIGURE 11.2 Process of biodeterioration of a rocky surface and mineralization.

other bacteria (Fan et al., 2020). It has been hypothesized that organic material worn from the bed acts as a "biofilm precursor," encouraging the formation of new biofilms (Chen et al., 2019). Active biodeterioration processes are characterized by biochemical activity and collaboration amid functional microorganisms in epilithic biofilms that consume inorganic resources and metabolize anthropogenic pollutants via biogeochemical cycles (Liu et al., 2020). The microbial composition of these biofilms as revealed by contemporary metagenomics and metatranscriptomics techniques include phototrophs, chemolithotrophs, and chemoorganotrophs (Liu et al., 2022b).

11.3 DEGRADATION OF ORGANIC COMPOUNDS IN BIOFILMS

The common process of biodegradation of *P. laurentii* unsaturated biofilm cells on Polyether Polyurethane (PEST)-Polyester-Polyether Polyurethane (PEUR) coatings of paints is found in high humidity environments. AFM-IR, or atomic force microscopy coupled with infrared nanospectroscopy, was important in identifying the spatially augmented hydrolysis and erosion processes. The preferential hydrolysis of ester components, which was monitored by oligomer erosion and accumulation to generate secondary microplastics rich in urethane, was used to determine the breakdown of polymers.

The immobilization of hydrolase in these areas and the ability to condense water into the polymer are the reasons for the edge position. It is also postulated that the proteinaceous EPS that sticks to the underlying polymer during cell transfer acts as a barrier against water infiltration. Over time, notwithstanding the accessibility of polyester metabolites as a carbon source, cells have become metabolically inert. The beginning of autophagy is predicted to be facilitated by nutrient-limited circumstances, particularly nitrogen starvation, which may be reflected in the observed cell shape (Barlow et al., 2020).

Biodeterioration of polymers is predominantly caused by enzyme-catalyzed chemical processes that can be caused by endoenzymes and exoenzymes. The former results in random chain breakage with a significant loss in molecular weight, whereas the latter, which has a considerably less immediate effect on the remaining polymer's molecular weight, results in removal of just terminal units, which are typically monomers, dimers, or trimers (Mohan and Krishna, 2011). The transfer of dissolved substances from the water column to the biofilm can be adjusted by the rate at which the water is flowing. The flow velocity has an impact on the accumulation and distribution of diuron in freshwater biofilms, as a decrease in flow velocity leads to a reduction in organic matter and a decrease in the overall accumulation of diuron (Chaumet et al., 2019). The performance of composites is contingent upon the conditions of their environment. When composites are subjected to aquatic environments, the presence of biological factors may expedite the deterioration of the material. Specifically, the biodeterioration of E-glass polyester composites in aquatic environments can be accelerated, resulting in the production of isophthalaldehyde as a by-product. Moreover, the toxicity of this compound leads to a decrease in microbial populations (Méndez-Sánchez et al., 2004). In most ecosystems, a significant portion of the organic carbon is not dissolved in the aqueous phase. Instead, it manifests as a collection of synthetic or natural hydrophobic and/or polymeric organic compounds. Heterotrophic bacteria that grow biofilms on these particle substrates improve their extracellular digestion and solubilization. These biofilms are present in polluted environments, animal digestive systems, and both terrestrial and aquatic natural ecosystems. They are responsible for the degradation of organic particles. They are essential parts of ecological systems because they are vital to the removal of pollutants and the recycling of carbon (Sivadon et al., 2019). Interdisciplinary research is needed to better understand flow-biofilm-sediment interactions, which impact water quality and ecosystem health (Gerbersdorf et al., 2020) (Table 11.1).

11.3.1 Hydrocarbons

Biofilm-mediated polycyclic aromatic hydrocarbon (PAH) degradation is an eco-friendly and efficient method, with factors like pH, temperature, and nutrient supplements influencing its effectiveness (Guo et al., 2022). Biofilms are effective and efficient in reducing pollution load in aquatic ecosystems due to their unique binding mechanism with pollutants and high tolerance limit (Dar & Bhat, 2019). Bacterial adhesion and biofilm formation enable the assimilation of undissolved organic carbon, playing a central role in carbon recycling and pollution removal in various ecosystems (Sivadon et al., 2020). Freshwater biofilms with aliphatic carbon and surface polarity promote the sorption of phenanthrene and ofloxacin, underlining their usefulness in managing organic contaminants in aquatic environs (Wang et al., 2019). Photosynthetic purple bacteria strains DQ41, PY2, PY6, and DG12 can

TABLE 11.1
Biofilm-Mediated Biodeterioration of Organic Molecules in Aquatic Environments

Material	Organisms	Intervention	Reference
Organic and inorganic materials	Fungi	Mycogenic biomineral production, rock and mineral bioweathering, and nutrient and element cycling are all impacted by fungi, which are essential to both organic and inorganic environmental changes.	Gadd, 2017
Pyrogenic dissolved organic matter		Lower-temperature char leachates are more biomineralized by microbials, while photodegraded leachates are more biomineralized than fresh non-photodegraded ones.	Bostick et al., 2020
Non Humic and Humic substances		Humic substances in aquatic systems can reduce bioavailability and toxicity of organic microcontaminants, potentially offering a protective effect against their impact.	Artifon et al., 2019
Crude oil	*Azomonas sp.*	Azomonas sp. effectively biodegrades crude oil in aquatic environments, with turbidity readings between 82 1.0 NTU.	Laurelta et al., 2020
Crude oil	Bacteria and Archaea Oleibacter sp., Chromatiales	Oil-contaminated mudflats show rapid biodegradation of petroleum hydrocarbons, with bacterial communities showing increased abundance in floating biofilms, supporting their role in oil degradation.	Sanni et al., 2015
Microplastic particles	bacteria, cyanobacteria and algae	Microbial biofilms can effectively detect hydrocarbon contamination in aquatic environments, with potential for use in bioelectrochemical treatment of polluted waters.	Leiser et al., 2020 Rummel et al., 2017
Microplastic	Deinococcus-thermus > Cyanobacteria > Firmicutes > Bacteroidetes are the most common phyla in biofilms occurring normally.	Periphytic biofilm with glucose enhances the biodegradation of microplastics, offering a promising approach for reducing marine litter.	Shabbir et al., 2020
Microplastics in freshwater ecosystems	*Pseudomonas monteilii, Pseudomonas mendocina Pseudomonas syringae*	Microplastic biofilms in freshwater ecosystems may serve as a vector for antibiotic resistance genes and pathogens, potentially causing environmental risks and adverse impacts on human health.	Wu et al., 2019

form biofilms and efficiently degrade naphthalene and pyrene in oil contaminated areas, potentially benefiting polluted area treatment in Vietnam (Nguyen et al., 2020). Microbial hydrocarbonoclastic biofilms at oil-water interfaces can form spontaneously, forming liquid microbial fuel cells and potentially adapting to oil spills (Ugo et al., 2021). Hydrocarbon-degrading marine bacteria can potentially degrade poly (ethylene terephthalate) in the marine environment, potentially triggering the formation of harmful microplastics (Denaro et al., 2020). Modified multi-walled carbon nanotubes (MWCNTs) can both inhibit and promote biofilm formation, with oleophilic multi-walled carbon nanotubes (oMWCNTs) showing the greatest antibiofilm effect on Gram-negative bacteria (Maksimova et al., 2022).

There has been a lot of discussion lately over the role that fungus play in the breakdown of low-density polyethylene (LDPE). Endolichenic fungi (ELF) represent a unique group of organisms possessing rich and diverse metabolite profiles and the capacity to thrive in harsh environments. It was possible for six mangrove-associated ELF species, namely *Chaetomium globosum, Neofusicoccum occulatum, Schizophyllum commune, Daldinia eschscholtzii, Phanerochaete chrysosporium,* and *Xylaria feejeensis* with rich enzymatic profiles in Negombo lagoon, Sri Lanka to biodegrade LDPE (Perera et al.,2022). With a current recycling rate of barely 1%, polypropylene (PP) has aroused serious environmental concerns due to its non-degradability. Consistent biodeterioration evidence is provided by surface morphological changes and a decrease in the carbonyl index. This work supports biological routes for the remediation of plastic waste by demonstrating that *A. terreus* and *E. album* can grow on, modify, and use PP as a carbon source with the aid of pre-treatments (Samat et al., 2023). Biosorbent-biodegrading biofilms, immobilized on eco-friendly carriers, effectively biodegrade more than 66% of oil in 10 days, offering a low-cost, low-impact solution for oil-contaminated water treatment (Catania et al., 2020). Nanoplastics can inhibit the metabolic diversity of freshwater biofilms, affecting carbon source utilization and overall microbial functions (Miao et al., 2019). Bubble biofilms, dominated by Nevskia bacteria, form stable gas bubbles at the air-air interface, enabling long-term microbial colonization and carbon dioxide recycling in harsh environments (Sjöberg et al., 2020).

11.3.2 Microplastics

Microplastics in marine environments exhibit physical, chemical, and biobehavioral behaviors, which can harm ecosystems through migration, sedimentation, accumulation, and biodegradation (Wang et al., 2016). Microbial biofilms can cause sinking of microplastics in freshwater reservoirs, potentially preventing their transport to the oceans (Leiser et al., 2020). Microplastics and biofilms interact in fluvial ecosystems, affecting biodiversity, ecosystem function, and requiring further research to better manage their interactions (Guasch et al., 2022). Microplastic pollution in riverine environments leads to reduced microbial diversity and potential pathogens, with seasonal variations, polymer types, and particle sizes affecting the microbial community (Mughini-Gras et al., 2021). Bacterial communities on microplastics show higher diversity and complexity, while fungal communities on plastic sheets show no significant alterations in diversity or structure (Wang et al., 2020). There is growing evidence that the interactions between biofilm and plastic can change the destiny and effects of microplastic pollution by altering the physical characteristics of the particles (Rummel et al., 2017). In aquatic environments, microplastic biofilms have the ability to modify the nitrogen and phosphorus cycle processes, perhaps leading to an increase in nitrogen input through biological nitrogen fixation (Chen et al., 2020). Biofilm formation in aquatic ecosystems maximizes organic matter degradation and horizontal gene transfer affecting physical and chemical fate of microplastics (Arias-Andres et al., 2019). Microplastics negatively impact ecosystem functions in marine sediments, affecting the interface between *Macomona liliana*, microphytobenthos, and mPETs (You et al., 2020). Understanding these interactions is essential to understanding plastic's movement and destiny in the seas, since microplastic debris and microorganisms interact to re-distribute plastic (Clark et al., 2016). Extended contact with silver nanoparticles in eutrophic water can alter biogeochemical cycling and lessen the self-recovery of the ecosystem (Niu et al., 2021). Biofilm organisms can cover microplastics and mask their pollution, while faecal indicator bacteria can be found in intertidal sediments, with higher concentrations in areas influenced by sewage discharges (Pazos et al., 2020). The plastisphere, a microbial community on plastic debris, spans multiple biomes and influences aquatic ecosystems, with potential for a 'core' community specific to plastic (Amaral-Zettler et al., 2020). In aquatic environments, microplastic particles may function as a distinct biological niche that promotes the survival of diseases and germs resistant to antibiotics, thus posing a serious risk to human health (Sathicq et al., 2021) (Table 11.2).

11.3.3 Wood Degradation

Wood durability in ground and aquatic environments is influenced by temperature, moisture, and biotic factors, with heartwood and sapwood showing different resistance to degradation between species, provenance, and individual trees (Marais et al., 2020). Microscopic fungi in biofilms on various substrates in the Black Sea can cause biodeterioration of natural and man-made materials, with 123 species identified, highlighting their importance in biodegradation processes (Kopytina and Bocharova, 2021). Artifacts consisting of naturally occurring biodegradable polymers, known as waterlogged archaeological wood (WAW), play a significant role in several valuable cultural heritages. Archaeological wood that has been wet and somewhat deteriorated has more hygroscopicity and hysteresis than less degraded wood, which might alter drying procedures and safe storage options for cultural property (Han et al., 2020).

Cellulose and hemicellulose in wood pulp are readily biodegradable, while lignin and ash negatively affect the biodegradability of other components (Kwon et al., 2021). Preservative treatments for fast-growing woods require non-toxic compounds to protect against biodeteriorating organisms, but traditional treatments pollute the environment (Dias and Barreiros, 2021). Climate change has an influence on historic structures, necessitating a multidisciplinary approach and monitoring panels to determine the causation of damages (Austigard and Mattsson, 2019). The Riga Cathedral (1211 CE) suffered structural deformations as a result of severe biodeterioration by microorganisms in wet foundation piles that significantly reduced the amount of

TABLE 11.2
Current Developments in the Research on Microplastics and the Growth of Microbial Biofilms in Aquatic Environments

Salient Findings	First Author	Year
Biofilm formation on microplastics can impact their environmental fate, including weathering processes, vertical transport, and potential environmental toxicity in marine ecosystems.	Tu	2020
Aquatic biofilms can function as both a sink and a source of microplastics, depending on environmental conditions, and understanding their interactions is crucial for understanding their fate in aquatic ecosystems.	Kalčíková and Bundschuh	2021
Microplastics produced from biofilms pose a serious danger to global pollution because they may absorb additional contaminants and serve as vectors for pollutants and microbes in aquatic settings.	Wang et al.,	2021
Biofilm formation on plastic surfaces can influence the fate and impacts of microplastic in aquatic environments by modifying its physical properties.	Rummel	2017
Microplastic substrates act as distinct microbial habitats, affecting community structure and microbial functions, potentially impacting aquatic ecosystems.	Miao et al.,	2019
Microplastics evolve novel microbiological niches in aquatic ecosystems, with ecological roles and environmental and human health ramifications.	Yang et al.,	2020
Microbial biofilms can cause sinking of microplastics in freshwater reservoirs, potentially preventing their transport to the oceans.	Leiser	2020
Microbial life on the surface of microplastics in natural waterways can pose environmental and public health issues because it can interact with them, use them as growth substrates, modify their characteristics, and biodegrade them.	Stabnikova et al.,	2021
Nanoplastics can disrupt basic ecological functions of freshwater biofilms, with positively charged amide-modified nanoplastics showing the highest toxicity.	Miao	2019
A microbial biofilm with a 'capture-release mechanism' effectively traps and aggregates microplastics, making their removal easier and potentially applicable to other micropollutants.	Liu	2021
Green algae can facilitate microplastic deposition by producing extracellular polymeric substances, which can increase cell density and accelerate microplastic transport to sediments.	Gopalakrishnan and Kashian	2021
Microplastic selects for consistent bacterial assemblages across diverse riverine sources, potentially impacting the long-term fate of microplastic in freshwater habitats.	Kelly et al.,	2020

Microplastics negatively impact the aquatic environment by forming biofilms, exchanging microbial genes, and acting as reservoirs for antimicrobial and metal resistance genes.	Rinanda et al., 2019	2019
The settling velocity of microplastic sheets increases with size and biofilms influence it, while fibers' settling velocity is independent of length.	Jalón-Rojas et al.,	2021
Microplastic pollution in aquatic ecosystems leads to distinct fungal communities, with parasitic and saprophytic fungi thriving in microplastic biofilms, potentially benefiting biodiversity and food web dynamics.	Kettner	2017
Microplastics provide novel habitats for aquatic microorganisms, with oligo-mesotrophic lakes exhibiting more functional richness than ambient water.	Arias-Andres	2018
The plastisphere, a microbial community on plastic debris, spans multiple biomes and influences aquatic ecosystems, with potential for a 'core' community specific to plastic.	Amaral-Zettler	2020
Weathered microplastic and macroplastic particles in aquatic environments show varying settling and rising velocities, with formulas suitable for describing pellets, fragments, and foams but less suitable for films and macroplastic particles.	Kryss Waldschläger	2020
Plastic pollution in marine ecosystems may be a reservoir for rare and understudied microbes.	Scales	2021
Polystyrene ingestion by marine organisms alters bacterial composition, potentially enhancing pathogenic biofilms and potentially affecting ecosystems.	Kesy	2016
Bio-modulated flocculation can effectively remove microplastics from river-estuary transition zones, improving our understanding of their transport, fate, and biological interactions.	Mendrik	2021
Microplastic biofilms in freshwater ecosystems may serve as a vector for antibiotic resistance genes and pathogens, potentially causing environmental risks and adverse impacts on human health.	Wu	2019

hemicellulose and cellulose sugars while leaving lignin intact (Irbe et al., 2019). Bio-based natural products show promise in preventing wood degradation and improving outdoor performance, with potential applications in coatings, paints, and varnishes (Teaca et al., 2019). Biocides containing >14% DDAC effectively protect historic building materials from biodeterioration by aerial phototrophs, effectively eliminating 'green' coatings on wood, brick, and plaster (Nowicka-Krawczyk et al., 2019). Nanostructured zinc oxide-based treatments, applied to maple wood surfaces, significantly enhance their resistance to decay and biodeterioration, offering a combination of aesthetics and protection (Weththimuni et al., 2019). Using an aqueous solution of tetra-n-butylphosphonium hydroxide, a study created a unique way to swiftly benzylate the surface of block-shaped wood, resulting in hydrophobicity and thermoplasticity with less harm to carbohydrates than with existing approaches (Abe et al., 2021). Silver nanoparticles (AgNPs) offer eco-friendly, long-lasting wood preservation and increased product durability by inhibiting fungal growth and extending the durability of wood and textiles (Salem, 2021). Wood quality can be improved by structural engineering, chemical, and thermal processes for a variety of uses in energy storage, environmental remediation, and high-performance building materials (Chen et al., 2020). *Bacillus megaterium* UMCV1 is as a promising biotechnological option to control decay fungi in wood-based materials (García-Ortiz et al., 2019).

11.4 MICROBIAL COMMUNITIES IN AQUATIC BIOFILMS

The eco-physiological role of biofilms is analyzed by the morphology of the microbial community and its activities (Liu et al., 2022b). During cycles of drying and rehydration, the pressure exerted by photosynthetic bacteria within the stone causes changes in the water volumes within their gelatinous sheaths. The breakdown of calcareous substrata is also facilitated by the precipitation of oxalates and the organization of calcium ions. In addition to the distinctive qualities of the many families of microorganisms, biofilm development itself is a biodeteriogenic factor (Mascaro et al., 2021). The significance of connected bacteria and algae is widely recognized, but not much is agreed on about the function of the incredibly prevalent biofilm-dwelling micro grazers, such as tiny metazoans and heterotrophic protists (Weitere et al., 2018).

Reverse osmosis systems' membrane colonization and degradation by microorganisms were investigated. A total of over 400 clones, 17 filamentous fungi, and 37 bacterial isolates were gathered and examined (Figure 11.3). The most prevalent bacterial genera found were *Shinella, Microbacterium, Rhizobium, Acidovorax,* and *Sphingobium* (Belgini et al., 2018). Aspergillus and the black fungus species *Aeminium ludgeri* only indicated a somewhat overemphasized development in low nutrient settings, and all species generated notable aesthetic modifications. Penicillium growth was spread over wider regions under rich micronutrient environments (Trovão et al., 2021). When it comes to biodeterioration, bacteria in biofilms have many advantages. These include being better able to withstand toxic pollutants within the biofilm, producing EPS that aids in the appropriation of pollutants, having higher cell densities with a larger genetic resource pool, being competent to stick to a

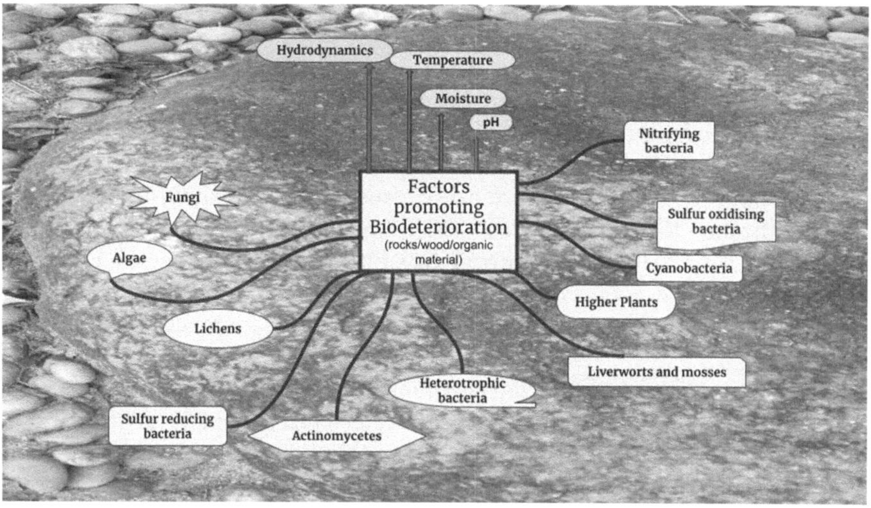

FIGURE 11.3 Microbial communities in aquatic biofilms involved in biodeterioration.

variety of substrates, and having a metabolic heterogeneity (Mahto et al., 2021). The KEGG functional profiles predicted by Tax4Fun indicated that the biofilms on MPs had an enrichment in the metabolic pathways of aspartate, alanine, glutamate, and biotin. However, the increased complexity of biofilms associated with plastic sheets may confer greater resistance to environmental perturbations (Wang et al., 2020). Fungi are potent biodeteriogens, and when they colonize stone monuments, they can cause severe physical, chemical, and aesthetical changes (Trovão et al., 2019).

The microalga *Haematococcus lacustris* is well known for its exceptional capacity to endure in a variety of unfavourable environments, especially in the White Sea's supralittoral temporal rock ponds. In supralittoral rock ponds, *H lacustris* lives and naturally develops biofilms that include multilayer cyanobacterial and fermentative microbial mats (Kublanovskaya et al., 2020). Lichens are recognized as the forerunners in the process of decomposing rocks. Additionally, fungi and sulfur-oxidizing bacteria play a role in the breakdown of sandstone by cycling sulfur and initiating salt attack, which causes the stone to become defoliated and crack while archaea and bacteria oxidize ammonia (Gu et al., 2021). Like cyanobacteria, lichens are poikilohydric organisms that can alter their metabolism in response to water availability. Lichens are visually beautiful symbiotic relationships between fungus and phototrophic microorganisms (cyanobacteria or algae) and related non-photosynthetic bacteria. Nonetheless, their tiny roots, known as rhizines, pierce the stone and do evident harm. They also generate acids, primarily oxalic acid, which is referred to as "lichenic acids." While siliceous rocks may also be impacted, carbonate stones are more vulnerable to this (Gaylarde et al., 2020). Microbioeroders are Euendoliths that aggressively penetrate the inside of rocks, leaving their imprints on the minerals by the biochemical disintegration of calcium pumps, respiratory carbonic acid, and/or enzymes (Cámara et al., 2017).

11.5 MICROBIAL DIVERSITY IN DIFFERENT AQUATIC ECOSYSTEMS

A considerable proportion of organic carbon present in most ecosystems is not soluble in water. Instead, it manifests as particles consisting of a collection of natural or synthetic hydrophobic and/or polymeric organic molecules. Heterotrophic bacteria that grow biofilms on these particle substrates use extracellular digestion and solubilization to break them down. Biofilms that degrade organic particles are common and have been found in animal digestive tracts, contaminated and human-caused habitats, and aquatic and terrestrial natural ecosystems. Moreover, they serve essential ecological roles in pollution removal and carbon recycling (Sivadon et al., 2019).

Within freshwater environments, biofilms have the ability to gather metals and organic pollutants that are carried by water currents and/or deposited onto surfaces (Bonnineau et al., 2020). The structure of the biofilm bacterial population serves as an effective and sensitive indication of environmental changes in freshwater habitats. It is also a useful tool for monitoring water quality in enclosed storm water networks (Ancion et al., 2014). Isolated strains of *Bacillus* and *Paenibacillus* from epilithic biofilms in Lake Baikal have the possibility to create antibacterial chemicals and might be useful for biotechnological applications (Sukhanova et al., 2018). Benthic microbial biofilms that are adherent to rocks in oligotrophic lakes are known as epilithic biofilms, and they play a significant role in the carbon cycle and ecosystem primary production. Phototrophic, chemotrophic, anaerobic, and aerobic microorganisms are the colonizers of biofilms; importantly the two most important bacterial taxa are Cyanobacteria and Bacteroidetes, followed by Verrucomicrobia, Chlorobi, Planctomycetes, and Alpha-, Beta-, Gamma-, and Deltaproteobacteria (Bartrons et al., 2012). The primary cause of the eco-environmental deterioration in the headwater streams of the Yellow River is probably the combination of increasingly disruptive human activities and fragile natural conditions (Liu et al., 2019). Biofilms are vital locations for organic matter breakdown and nutrient cycling in streams. They can consume organic molecules exhaled by actively developing algae, whereas microbial degradation of non-living organic materials produces inorganic chemicals favorable to algal formation. Microbial production is the foundation of many stream food webs, and complex interactions are involved in the transfer of carbon and other materials from microbial communities to higher trophic levels. Higher trophic levels can access this energy source thanks to the assistance of bacteria and fungi in breaking down organic detritus such as leaves that fall in the autumn. The microbial tissue in biofilms is consumed by a wide range of micro-consumers, such as protozoans and the microscopic metazoans known as meiofauna. Within biofilms and plankton assemblages, a meiofauna-based food web may be present and serve as a food source (Allan et al., 2021). Buried pipelines are critical for delivering drinkable water all over the world. These pipes, especially those made of carbon steel, frequently leak or break due to increased localized corrosion brought on by soil microorganisms (Spark et al., 2020).

11.6 ECOLOGICAL IMPACTS OF BIODETERIORATION ON AQUATIC ECOSYSTEMS

Biodeterioration impacts cultural heritage through discoloration, dissolution, and rupture affect human health by contributing to sick building syndrome, and influence residential development and concrete constructions (Rivera et al., 2018). Heavy metals, nanoparticles, anthropogenic disturbance, and climate change all have a deleterious influence on aquatic ecosystems, impacting bioindicators, biodiversity, and water quality. Biofilms on river stones show potential as a biosorbent for copper (II) pollution treatment, with a maximum absorption capacity of 2.14 mg/wet-g of biofilm (Kurniawan et al., 2022). Microbial biofilms on material surfaces influence their characteristics, endurance, and performance, hence impacting the integrity of substructure materials (Gu et al., 2018). Aquatic microbiomes are critical to the health and ecological state of aquatic ecosystems, with important processes such as primary production, nitrogen cycle, and nutrient metabolism being preserved throughout aquatic settings (Sehnal et al., 2021). Freshwater ecosystems show more antagonistic responses to multiple stressors than marine systems, possibly due to greater environmental variability and potential for acclimation and co-adaptation (Jackson et al., 2016). Moreover, due to high population densities and genetic exchange across bacterial species, biofilms in aquatic habitats have the potential to become repositories of antibiotic resistance, which might aid in the development of antibiotic resistance (Balcázar et al., 2015). In marine and freshwater environments, harmful algal blooms (HABs) are a co-stressor caused by climate change. As such, their effects are amplified, necessitating integrated management measures to safeguard human health, fisheries, aquaculture, and aquatic ecosystems (Griffith and Gobler, 2020). Antibiotics are the most significant because of their chemical variety and greater quantities in the environment, and they may impact cyanobacteria ecology, thereby producing changes in ecosystem dynamics (Costa et al., 2021).

Climate change and anthropogenic-caused hydric deficits may have a profound impact on microbial populations, possibly affecting ecosystem functioning (Iniesto et al., 2022). Microbial biofilms in aquatic ecosystems accumulate nutrient ions like NH_{4+} and NO_3, which may impact the seasonal dynamics of nutrients in aquatic ecosystems (Kurniawan and Yamamoto, 2019). These biofilms can accumulate metals and organic contaminants, influencing contaminant fate in aquatic ecosystems (Bonnineau et al., 2020). Biofilm formation increases the impact of microplastics (MPs) on the transport and destiny of trace metals in freshwater (Guan et al., 2020). In addition to amplifying the role of MPs as vectors for heavy metal migration in freshwater, biofilms significantly increase the toxicity of MPs in combination, potentially posing additional ecological concerns to freshwater ecosystems (Qi et al., 2021). The fact that permanent stream biofilms are disrupted during the non-flow period suggests that transitory stream biofilms will be more resilient when climate change causes the hydroperiod to decrease (Timoner et al., 2020). Potential human pathogens, such as *Klebsiella pneumoniae* MBB9, can form biofilms in aquatic ecosystems, potentially altering ecosystem composition (Alotaibi, 2020). In maritime environments, microbial biofilms can serve as prospective candidates for bioremediation as well as sentinels of environmental changes (Caruso, 2020). Omics technology may be used to detect the biodeterioration of cultural heritage materials such as stone, cellulose-based substrates, and parchment. This allows for the identification of the biodegradation's fingerprint as well as its biological sources (Marvasi et al., 2019).

The established benefits of nitrogen enrichment on benthic biofilm communities may be overshadowed by the intense rains and increased runoff output in these tropical catchments that are influenced by agriculture. This study suggests that there may be changes in benthic biofilm communities in tropical streams due to predicted changes in temperature and land use, which could have functional implications for aquatic food webs in these habitats (Taniwaki et al., 2019). Microbes contribute to the deterioration of buildings, endangering public health through mycotoxins, and necessitating enhanced cleaning and air conditioning systems (Ahmed et al., 2017).

11.7 MARINE AND FRESHWATER INFRASTRUCTURE DAMAGE AND CULTURAL HERITAGE

Microbial biodeterioration of cultural heritage leads to physical weakening and discoloration of stones, requiring effective conservation strategies to preserve original integrity (Negi and Sarethy, 2019). In the life cycle of infrastructure systems, deterioration is a crucial component. Living animals' actions have been shown to be major among all causes of deterioration (Sánchez-Silva and Rosowsky, 2008). Moreover, it has a substantial impact on structural performance and has the potential to significantly shorten its lifespan (Bastidas-Arteaga et al., 2008). In the maritime environment, the biodeterioration of concrete starts when microorganisms stick to its surface and create a biofilm. Because fresh concrete is alkaline, microbes cannot thrive there. However, when wet and leached by the sea, which has a pH typically between 7.5 and 9, the pH quickly drops, allowing for microbial adherence and metabolic activity (Gaylarde et al., 2023). In saltwater, biofilm formation on material surfaces is followed by the biocorrosion phenomena. Biofouling is a threat to all manmade facilities, but especially to timber ones (Rao and Kuppusamy, 2022). Concrete sewer biodeterioration causes disintegration and substantial damage, with sulfur-oxidizing bacteria being the major culprit (Hudon et al., 2011).

11.7.1 Impacts on Cultural Heritage

The process of biodeterioration materializes when hazardous bacteria, fungus, and other microbes proliferate on historically significant structures, leading to their degradation (Rinanti et al., 2023). The deterioration of historic structures and ancient sites due to biopatina, biocrust, or biofouling causes changes in the stone matrix's aesthetics and initiates degradation processes (Becerra et al., 2019). Therefore, this is a global issue, with China and Japan being the most productive authors, and natural biocides playing a significant role in combating fungi and bacteria (Rinanti et al., 2023). Microorganisms classified as chemoorganotrophic, chemolithotrophic, or photolithotrophic may predominate, contingent upon the substance's chemical composition and surrounding environmental factors. However, it is central to concurrently recognize the prevailing microbial populations and any potential damage-causing metabolic activities. While culture techniques are helpful in finding biodegradogenic agents, they are not adequate for examining the entire range and abundance of microorganisms. Because of their complementarity, it is therefore preferable to combine the two approaches, and the outcomes may provide more understanding (Branysova et al., 2022). Biodeterioration of historic buildings, such as Persepolis, is instigated by biological agents like fungi, bacteria, lichens, plants, and animals, causing irreparable damage to the stone surface (Sohrabi et al., 2020). Changing water regime and increased CO_2 concentration can lead to more intense biodeterioration of stone cultural heritage, affecting conservation and prevention policies (Prieto et al., 2020).

For a better understanding of EPS's involvement in monument biodegradation, research on the cyanobacteria that form biofilms most frequently, including *Leptolyngbya*, has become quite interesting. Changes in surface color, the breakdown of minerals by biogenic acids, an increase in water retention, the mobilization and crystallization of metallic cations, salt efflorescence, and the penetration of fungal hyphae and other endolithic organisms are all probable reasons (Mascaro et al., 2020). Epilithic microorganisms, including cyanobacteria, green algae, and pink yeasts, can modify the mechanical and visual potentials of stone monuments by means of metabolic processes that control the microbiome. Air pollution, particularly sulfurous and/or nitrogenous substances, which chemolithotrophs and chemoorganotrophs use as nourishment, has a detrimental effect on stone monuments. Selective community members like nitrifiers and sulfur-oxidizers, need to be closely monitored since they can create acids that dissolve minerals and are corrosive, such nitric acid and sulfuric acid. Furthermore, the existence of halophiles, such as those belonging to the Haloarchaea, might intensify the biodeterioration by causing salt efflorescence (Liu et al., 2022b). Agrochemicals offer a rising concern to aquatic ecosystems, impacting macroinvertebrates, due to their frequent penetration into water bodies and more toxicity (Mello et al., 2018). Fluctuations in functional features of freshwater and marine fishes can occur from habitat disturbances, influencing community dynamics and essential ecological processes (Kelley et al., 2018). Ecotoxicological assays are supplementary methodologies that directly address biological impacts and give a connection to chemical exposure, allowing for a more complete valuation of chemical stressors on marine environments (Schuijt et al., 2021). Contamination in aquatic environments degrades water quality and spreads infectious illnesses, resulting in poor health consequences and death (Bashir et al., 2020). Environmental DNA (eDNA) techniques are useful for assessing the effects of chemical contaminants on biodiversity, community structure, and ecosystem function in the field (Zhang, 2019).

Abbreviations used in the illustration: FM –Fluorescence Microscopy; IC-Ion Chromatography; MT-Micro Tomography; GC-MS-Gas Chromatography-Mass Spectrometry; HPLC-MS-High Performance Liquid Chromatography-Mass Spectrometry; SEM(-EDS)- Scanning Electron Microscopy (Energy-Dispersive X-ray Spectroscopy); FT-IR-Fourier Transform Coupled Infrared Spectroscopy; TEM-Transmission Electron Microscopy; RS-Raman Spectroscopy; XRF-X-ray Fluorescence; XRD-X-ray Diffraction).

11.8 MITIGATION AND MANAGEMENT STRATEGIES

Water-quality management plays a decisive role in mitigating biodeterioration. Biodeterioration in aquatic environments poses substantial challenges to the maintenance and preservation of these fragile ecosystems. The degradation caused by microorganisms, such as bacteria, fungi, and algae, can result in fouling of surfaces, corrosion of infrastructure, and depletion of oxygen levels. To mitigate and manage this biodeterioration effectively, various strategies have been developed. Recent developments in multiomics and culture-independent high-throughput sequencing have produced multifaceted results for cultural heritage microbiology. These instruments can find metabolically active bacteria, reveal community functional activity, and greatly enhance taxonomic and functional profiling (Liu et al., 2020). Omics technology can detect metabolites and microbial populations to offer a fingerprint of biodeterioration and its biological origins (Marvasi et al., 2019).

By halting fungal deterioration, natural substances like chitosan, tannins, and essential oils may be able to prolong the life of wood goods (Broda, 2020). For conservation purposes, it has been determined that the eco-friendly, self-healing, and extremely durable nature of these bio-binders makes Microbially Induced Calcium Carbonate Precipitation (MICP) technology more suitable (Zamani et al., 2017). To effectively manage biodeterioration in aquatic environments, monitoring tools and early detection techniques are essential. Molecular tools allow for rapid identification of specific pathogens responsible for deterioration whereas sensor technologies provide real-time data on water quality parameters indicating potential risks. Zinc oxide-based nanomaterials can help as a remedy to biodeterioration in historic buildings by demonstrating high antibacterial and antibiofilm activities on various substrates (Schifano et al., 2020). Biorestoration, using microorganisms as potential tools for restoration, offers a promising alternative to conventional restorative methods for Cultural Heritage items (Soffritti et al., 2019). Microbial weakening of stone monuments and buildings adds to a damage of world cultural heritage, and

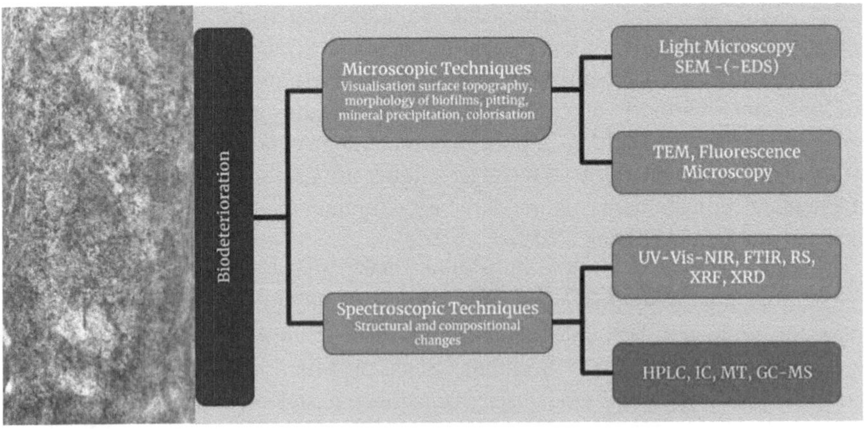

FIGURE 11.4 Various approaches used to investigate natural biodeterioration.

accepting these processes can help develop sustainable conservation strategies (Liu et al., 2020). Inorganic nanoparticles, such as silver, titanium, and zinc oxide, offer a promising alternative to organic biocides for protecting cultural heritage from biodeterioration (Kakakhel et al., 2019). Since deep eutectic solvents (DESs) are non-toxic to the environment or the operator, they have potential as biocidal agents for treating surfaces containing biodegradable cultural heritage materials (Macchia et al., 2022). MgB_2 powders show potential as eco-friendly, efficient tools to combat fungal biodeterioration of heritage buildings and objects, with PVZ and CERAC being more effective against stone and wood materials (Gheorghe et al., 2021). *Trichoderma* spp., a potential sustainable wood protection strategy, shows potential against wood decay basidiomycetes, offering a sustainable integrated wood protection strategy with chromium-free preservatives (Ribera and Schwarze, 2020).

11.9 CONCLUSIONS

Biodeterioration is an irrevocable damage caused by microorganisms, affecting materials like household plastics, polymers, and cultural heritage materials. It can cause discoloration, dissolution, rupture, and efflorescence. Biofilms, formed by microorganisms, can cause harmful effects like biofouling, biocorrosion, and the perseverance of detrimental or pathogenic microorganisms. The presence of limiting resources and microbial interactions in aquatic biofilms can influence the outcomes

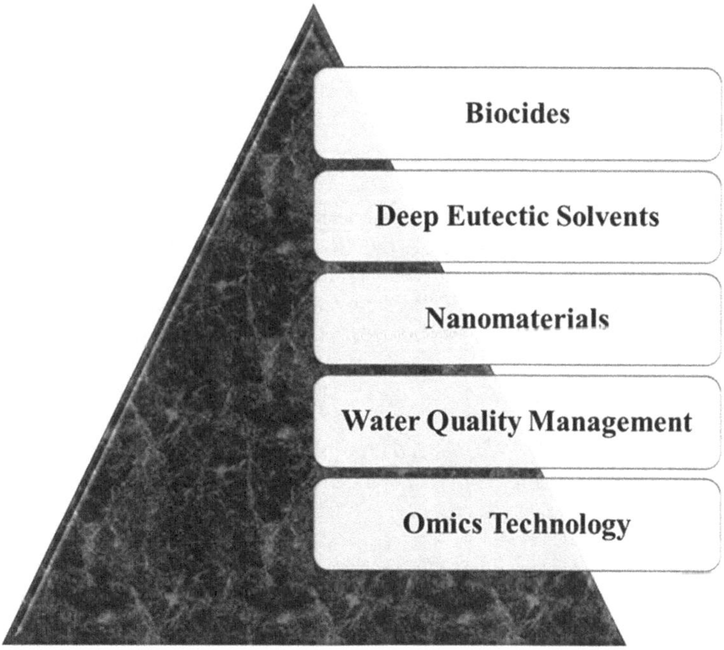

FIGURE 11.5 Mitigation strategies to sustainably handle biodeterioration.

of biodeterioration. Nutrients and water are critical conditions for biodeterioration, with nutrient and water temperature influencing macroinvertebrate population changes. Biofilm-induced stone degradation is a complex process involving microbial development and metabolic activities, altering the internal structure of stone substrates. Microbial biofilms in aquatic habitats can become repositories of antibiotic resistance. Microbial communities can be profoundly impacted by climate change and anthropogenic-induced hydric deficiency, which may influence how ecosystems operate. Understanding the mechanisms by which biofilms contribute to biodeterioration can inform strategies for preventing or mitigating their impact on various industries, such as construction, maritime, and cultural heritage preservation. Biofilm-mediated polycyclic aromatic hydrocarbon (PAH) degradation is an eco-friendly and efficient method for reducing pollution in aquatic ecosystems. Bacterial adhesion and biofilm formation enable carbon recycling and pollution removal. Photosynthetic purple bacteria can form biofilms to degrade naphthalene and pyrene in oil-contaminated areas. Endolichenic fungi can biodegrade LDPE and polypropylene, while nanoplastics can inhibit biofilm formation and affect carbon source utilization. Further, studies in this field are essential to develop effective interventions and management strategies. Microbial biofilms' function in the biodeterioration of organic materials is significant and warrants further research. Further studies in this field are essential to develop effective interventions and management strategies. Water-quality management is crucial in mitigating biodeterioration in aquatic environments caused by microorganisms like bacteria, fungi, and algae. Technological developments in multiomics and culture-independent high-throughput sequencing have enhanced taxonomic and functional profiling, identified metabolically active bacteria, and revealed functional activity within communities. Ecotoxicological assays and environmental DNA techniques can help to assess the biological impacts of chemical contaminants on aquatic ecosystems and biodiversity.

REFERENCES

Ahmed, T., M. Usman and M. Scholz. "Biodeterioration of Buildings and Public Health Implications Caused by Indoor Air Pollution." *Indoor and Built Environment*, 27 (2017): 752–765. https://doi.org/10.1177/1420326X17690912.

Aliasghari, A. and P. Mohammadi. "Microorganisms and Deterioration of Artistic and Historical Substrate of Cultural Heritage." (2020). https://doi.org/10.6084/M9.FIGSHARE.12845831.V2.

Allan, J., M. M. Castillo and K. Capps. "Stream Microbial Ecology." *Stream Ecology* (2021). https://doi.org/10.1007/978-3-030-61286-3_8.

Alotaibi, G. F. "Occurrence of Potentially Pathogenic Bacteria in Epilithic Biofilm Forming Bacteria isolated from Porter Brook River-stones, Sheffield, UK." *Saudi Journal of Biological Sciences*, 27 (2020): 3405–3414. https://doi.org/10.1016/j.sjbs.2020.09.030.

Amaral-Zettler, L., E. Zettler and T. Mincer. "Ecology of the Plastisphere." *Nature Reviews Microbiology*, 18 (2020): 139–151. https://doi.org/10.1038/s41579-019-0308-0.

Ancion, P., G. Lear, M. Neale, Kelly Roberts and G. Lewis. "Using Biofilm as a Novel Approach to Assess Storm Water Treatment Efficacy." *Water Research*, 49 (2014): 406–415. https://doi.org/10.1016/j.watres.2013.10.023.

Arias-Andres, M., Marie Therese Kettner, T. Miki and H. Grossart. "Microplastics: New Substrates for Heterotrophic Activity Contribute to Altering Organic Matter Cycles in Aquatic Ecosystems." *The Science of the Total Environment*, 635 (2018): 1152–1159. https://doi.org/10.1016/j.scitotenv.2018.04.199.

Arias-Andres, M., Keilor Rojas-Jimenez and H. Grossart. "Collateral Effects of Microplastic Pollution on Aquatic Microorganisms: An Ecozlogical Perspective." *TrAC Trends in Analytical Chemistry* (2019). https://doi.org/10.1016/J.TRAC.2018.11.041.

Artifon, V., E. Zanardi-Lamardo and G. Fillmann. "Aquatic Organic Matter: Classification and Interaction with Organic Microcontaminants." *The Science of the Total Environment*, 649 (2019): 1620–1635. https://doi.org/10.1016/j.scitotenv.2018.08.385.

Austigard, M. S, and J. Mattsson. "Monitoring Climate Change Related Biodeterioration of Protected Historic Buildings." *International Journal of Building Pathology and Adaptation* (2019). https://doi.org/10.1108/ijbpa-11-2018-0094.

Balcázar, J., J. Subirats and C. Borrego. "The Role of Biofilms as Environmental Reservoirs of Antibiotic Resistance." *Frontiers in Microbiology*, 6 (2015). https://doi.org/10.3389/fmicb.2015.01216.

Barlow, D. E., J. C. Biffinger, L. Estrella, Q. Lu, C.-S. Hung, L. J. Nadeau, A. L. Crouch, J. N. Russell Jr and W. J. Crookes-Goodson. "Edge-Localized Biodeterioration and Secondary Microplastic Formation by Papiliotrema Laurentii Unsaturated Biofilm Cells on Polyurethane Films." *Langmuir* 36, no. 6 (2020): 1596–1607. https://doi.org/10.1021/acs.langmuir.9b03421

Bartrons, M., J. Catalan and E.O. Casamayor. "High Bacterial Diversity in Epilithic Biofilms of Oligotrophic Mountain Lakes." *Microbial Ecology*, 64 (2012): 860–869. https://doi.org/10.1007/s00248-012-0072-4

Bashir, I., F. Lone, R. A. Bhat, S. A. Mir, Z. Dar and Shakeel Ahmad Dar. "Concerns and Threats of Contamination on Aquatic Ecosystems." *Bioremediation and Biotechnology* (2020): 1–26. https://doi.org/10.1007/978-3-030-35691-0_1.

Bastidas-Arteaga, E., M. Sánchez-Silva, A. Chateauneuf and Moema Ribas Silva. "Coupled Reliability Model of Biodeterioration, Chloride Ingress and Cracking for Reinforced Concrete Structures." *Structural Safety*, 30 (2008): 110–129. https://doi.org/10.1016/J.STRUSAFE.2006.09.001.

Becerra, J., M. Mateo, P. Ortiz, G. Nicolas and A. P. Zaderenko. "Evaluation of the Applicability of Nano-Biocide Treatments on Limestones Used in Cultural Heritage." *Journal of Cultural Heritage* (2019). https://doi.org/10.1016/J.CULHER.2019.02.010.

Belgini, D. R., V. Siqueira, D. M. Oliveira, S. G. Fonseca, V. Piccin-Santos, R. Dias, L. Quartaroli, R. Souza, A. Torres, M. Sousa, C. M. Silva, C. C. Silva, S. D. de Paula and V. M. Oliveira. "Integrated Diversity Analysis of the Microbial Community in a Reverse Osmosis System from a Brazilian Oil Refinery." *Systematic and Applied Microbiology*, 41 No. 5 (2018): 473–486. https://doi.org/10.1016/j.syapm.2018.03.007.

Bonnineau, C., J. Artigas, B. Chaumet, A. Dabrin, J. Faburé, B. Ferrari, J. Lebrun, C. Margoum, N. Mazzella, C. Miège, S. Morin, E. Uher, M. Babut and S. Pesce. "Role of Biofilms in Contaminant Bioaccumulation and Trophic Transfer in Aquatic Ecosystems: Current State of Knowledge and Future Challenges." *Reviews of Environmental Contamination and Toxicology* (2020). https://doi.org/10.1007/398_2019_39.

Bostick, K. W., A. Zimmerman, A. Goranov, S. Mitra, P. Hatcher and A. Wozniak. "Biolability of Fresh and Photodegraded Pyrogenic Dissolved Organic Matter From Laboratory-Prepared Chars." *Journal of Geophysical Research: Biogeosciences*, 126 (2020). https://doi.org/10.1029/2020JG005981.

Branysova, T., K. Demnerova, M. Durovic, and H. Stiborova. "Microbial Biodeterioration of Cultural Heritage and Identification of the Active Agents Over the Last Two Decades."

Journal of Cultural Heritage, 55 (2022): 245–260. https://doi.org/10.1016/j.culher.2022.03.013

Broda, M.. "Natural Compounds for Wood Protection against Fungi—A Review." *Molecules*, 25 (2020). https://doi.org/10.3390/molecules25153538.

Cámara, B., M. Á. de Buergo, M. Bethencourt, T. Fernández-Montblanc, M. F. La Russa, M Ricca and R. Fort. "Biodeterioration of Marble in an Underwater Environment." *Science of the Total Environment*, 609 (2017): 109–122. https://doi.org/10.1016/j.scitotenv.2017.07.103

Carmen Alice Teaca, D. R., F. Mustaţă, T. Rusu, L. Rosu, I. Rosca and C. Varganici. "Natural Bio-Based Products for Wood Coating and Protection Against Degradation: A Review." *BioResources* (2019). https://doi.org/10.15376/BIORES.14.2.TEACA.

Caruso, G. "Microbial Colonization in Marine Environments: Overview of Current Knowledge and Emerging Research Topics." *Journal of Marine Science and Engineering*, 8 (2020): 78. https://doi.org/10.3390/jmse8020078.

Catania, V., F. Lopresti, S. Cappello, R. Scaffaro and P. Quatrini.. "Innovative, Ecofriendly Biosorbent-Biodegrading Biofilms for Bioremediation of Oil- Contaminated Water." *New Biotechnology* (2020). https://doi.org/10.1016/j.nbt.2020.04.001.

Chaumet, B., S. Morin, O. Hourtané, J. Artigas, B. Delest, M. Éon and N. Mazzella. "Flow Conditions Influence Diuron Toxicokinetics and Toxicodynamics in Freshwater Biofilms." *The Science of the Total Environment,* 652 (2019): 1242–1251. https://doi.org/10.1016/j.scitotenv.2018.10.265.

Chen, C., Y. Kuang, S. Zhu, I. Burgert, T. Keplinger, A. S. Gong, T. Li, L. Berglund, S. Eichhorn and L. Hu. "Structure–Property–Function Relationships of Natural and Engineered Wood." *Nature Reviews Materials* (2020): 5(9), 1–25. https://doi.org/10.1038/s41578-020-0195-z.

Chen, X., C. Zhang, D. Paterson, I. Townend, C. Jin, Zeng Zhou, Z. Gong and Q. Feng. "The Effect of Cyclic Variation of Shear Stress on Non-Cohesive Sediment Stabilization by Microbial Biofilms: The Role of 'Biofilm Precursors'." *Earth Surface Processes and Landforms*, 44 (2019): 1471–1481. https://doi.org/10.1002/esp.4573.

Clark, J., M. Cole, P. Lindeque, Elaine S. Fileman, J. Blackford, C. Lewis, T. Lenton and T. Galloway. "Marine Microplastic Debris: A Targeted Plan for Understanding and Quantifying Interactions with Marine Life." *Frontiers in Ecology and the Environment*, 14 (2016): 317–324. https://doi.org/10.1002/FEE.1297.

D'Ugo, E., L. Bertuccini, F. Spadaro, R. Giuseppetti, F. Iosi, F. Santavenere, F. Giuliani, M. Gricia, A. Rodomonte, N. Lovecchio, Arghya Mukherjee, P. Bucci, M. Bruno, E. Stellacci, A. Bernardo and F. Magurano. "Electrogenic and hydrocarbonoclastic biofilm at the oil-water interface as microbial responses to oil spill." *Water Research*, 197 (2021): 117092 . https://doi.org/10.1016/j.watres.2021.117092.

Dang, H. and C. R. Lovell. "Microbial Surface Colonization and Biofilm Development in Marine Environments." *Microbiology and Molecular Biology Reviews* 80, no. 1 (2016): 91–138. https://doi.org/10.1128/mmbr.00037-15

Dar, S. A. and R. A. Bhat. "Aquatic Pollution Stress and Role of Biofilms as Environment Cleanup Technology." *Fresh Water Pollution Dynamics and Remediation* (2019). https://doi.org/10.1007/978-981-13-8277-2_16.

Denaro, R., F. Aulenta, F. Crisafi, F. Di Pippo, C. Cruz Viggi, B. Matturro, P. Tomei, F. Smedile, A. Martinelli, V. Di Lisio, C. Venezia and S. Rossetti. "Marine Hydrocarbon-Degrading Bacteria Breakdown Poly(Ethylene Terephthalate) (PET)." *The Science of the Total Environment*, 749 (2020): 141608. https://doi.org/10.1016/j.scitotenv.2020.141608.

Di Carlo, E., G. Barresi and F. Palla. "Biodeterioration." In *Biotechnology and Conservation of Cultural Heritage*, pp. 1–30. Cham: Springer International Publishing, 2022.

Dias, K. and R. Barreiros. "Preservative Treatments on Wood and Their Effects on Metal Fasteners." *Engineered Wood Products for Construction [Working Title]* (2021). https://doi.org/10.5772/intechopen.98467.

dos Santos Costa, R., G. Quadra, Helena de Oliveira Souza, V. S. do Amaral and J. Navoni. "The Link Between Pharmaceuticals and Cyanobacteria: A Review Regarding Ecotoxicological, Ecological, and Sanitary Aspects." *Environmental Science and Pollution Research*, 28 (2021): 41638–41650. https://doi.org/10.1007/s11356-021-14698-5.

Fan, Y.-H., X. Huang, J. Chen and B. Han. "Formation of a Mixed-Species Biofilm Is a Survival Strategy for Unculturable Lactic Acid Bacteria and Saccharomyces cerevisiae in Daqu, a Chinese Traditional Fermentation Starter." *Frontiers in Microbiology*, 11 (2020). https://doi.org/10.3389/fmicb.2020.00138.

Flemming, H-C. "Biodeterioration of Synthetic Materials–A Brief Review." *Materials and Corrosion,* 61, no. 12 (2010): 986–992. https://doi.org/10.1002/maco.201005837

Gadd, G.. "The Geomycology of Elemental Cycling and Transformations in the Environment." *Microbiology Spectrum,* 5 (1) (2017). https://doi.org/10.1128/microbiolspec.FUNK-0010-2016.

García-Ortiz, V., C. Hernández-Soberano, M. Martínez-Pacheco, E. Ambriz-Parra and C. V. Becerra. "Protective Effect on Wood by Metabolic Extracts from Plant Growth-Promoting Rhizobacteria Against Decay Fungi." *Revista Argentina de Microbiologia* (2019). https://doi.org/10.1016/j.ram.2019.04.003.

Gaylarde, C. C., and B. O. Ortega-Morales. "Biodeterioration and Chemical Corrosion of Concrete in the Marine Environment: Too Complex for Prediction" *Microorganisms* 11, no. 10 (2023): 2438. https://doi.org/10.3390/microorganisms11102438.

Gaylarde, C., Christine. "Influence of Environment on Microbial Colonization of Historic Stone Buildings with Emphasis on Cyanobacteria" *Heritage* 3, no. 4 (2020): 1469–1482. https://doi.org/10.3390/heritage3040081

Gerbersdorf, S., K. Koca, D. de Beer, A. Chennu, C. Noss, Ute Risse-Buhl, M. Weitere, O. Eiff, M. Wagner, J. Aberle, M. Schweikert and K. Terheiden. "Exploring Flow-Biofilm-Sediment Interactions: Assessment of Current Status and Future Challenges." *Water Research*, 185 (2020): 116182. https://doi.org/10.1016/j.watres.2020.116182.

Gheorghe, I., I. Avram, Viorica Matis Corbu, L. Marutescu, M. Popa, Irina Balotescu, I. Blăjan, V. Mateescu, D. Zaharia, Andreea Ştefania Dumbravă, Octavia Emilia Zetu, Ionut Pecete, V. Cristea, D. Batalu, M. Grigoroscuta, M. Burdusel, G. Aldica, P. Badica, A. Datcu, N. Ianovici, C. Bleotu, V. Lazăr, L. Diţu and M. Chifiriuc. "In Vitro Evaluation of MgB2 Powders as Novel Tools to Fight Fungal Biodeterioration of Heritage Buildings and Objects." 7 (2021). https://doi.org/10.3389/fmats.2020.601059.

Gopalakrishnan, K. and D. Kashian. "Extracellular Polymeric Substances in Green Alga Facilitate Microplastic Deposition." *Chemosphere*, 286, (2021): 131814. https://doi.org/10.1016/j.chemosphere.2021.131814

Griffith, A. W. and C. Gobler. "Harmful Algal Blooms: A Climate Change Co-Stressor in Marine and Freshwater Ecosystems." *Harmful Algae* 91 (2020): 101590. https://doi.org/10.1016/J.HAL.2019.03.008.

Gu, J. "Microbial Biofilms, Fouling, Corrosion, and Biodeterioration of Materials." (2018): 273–298. https://doi.org/10.1016/B978-0-323-52472-8.00014-9.

Gu, J.-D., and Y. Katayama. "Microbiota and Biochemical Processes Involved in Biodeterioration of Cultural Heritage and Protection." *Microorganisms in the Deterioration and Preservation of Cultural Heritage* (2021). https://doi.org/10.1007/978-3-030-69411-1_2.

Guan, J., K. Qi, J. Wang, W. Wang, Z. Wang, N. Lu and J. Qu. "Microplastics as an Emerging Anthropogenic Vector of Trace Metals in Freshwater: Significance of Biofilms and Comparison with Natural Substrates." *Water Research*, 184 (2020):116205. https://doi.org/10.1016/j.watres.2020.116205.

Guasch, H., S. Bernal, D. Bruno, B. C. Almroth, J. Cochero, N. Corcoll, Delfina Cornejo, E. Gacia, A. Kroll, I. Lavoie, J. Ledesma, A. Lupon, H. Margenat, S. Morin, Enrique Navarro, M. Ribot, T. Riis, M. Schmitt-Jansen, A. Tlili and E. Martí. "Interactions between Microplastics and Benthic Biofilms in Fluvial Ecosystems: Knowledge Gaps and Future Trends." *Freshwater Science*, 41 (2022): 442–458. https://doi.org/10.1086/721472.

Han, L., J. Guo, K. Wang, P. Grönquist, R. Li, X.-L. Tian and Y. Yin. "Hygroscopicity of Waterlogged Archaeological Wood from Xiaobaijiao No.1 Shipwreck Related to Its Deterioration State." *Polymers*, 12 (2020). https://doi.org/10.3390/polym12040834.

Hudon, E., S. Mirza and D. Frigon. "Biodeterioration of Concrete Sewer Pipes: State of the Art and Research Needs." *Journal of Pipeline Systems Engineering and Practice*, 2 (2011): 42–52. https://doi.org/10.1061/(ASCE)PS.1949-1204.0000072.

Iniesto, M., D. Moreira, K. Benzerara, G. Reboul, P. Bertolino, R. Tavera and P. López-García. "Planktonic Microbial Communities from Microbialite-Bearing Lakes Sampled Along a Salinity-Alkalinity Gradient." *Limnology and Oceanography*, 67 (2022). https://doi.org/10.1002/lno.12233.

Irbe, I., Oskars Bikovens, V. Fridrihsone and M. Dzenis. "Impact of Biodeterioration on Structure and Composition of Waterlogged Foundation Piles from Riga Cathedral (1211 CE), Latvia." *Journal of Archaeological Science: Reports* (2019). https://doi.org/10.1016/J.JASREP.2018.11.002.

Jackson, M., C. J. G. Loewen, R. Vinebrooke and C. Chimimba. "Net Effects of Multiple Stressors in Freshwater Ecosystems: A Meta-Analysis." *Global Change Biology*, 22 (2016). https://doi.org/10.1111/gcb.13028.

Jalón-Rojas, I., A. Roméro-Ramirez, K. Fauquembergue, L. Rossignol, B. Morin and J. Cachot. "Experimental Assessment of Settling Velocity of Pristine and Biofouled Microplastics." (2021). https://doi.org/10.5194/EGUSPHERE-EGU21-15053.

Kakakhel, M. A., F. Wu, J.-D. Gu, H. Feng, K. Shah and W. Wang. "Controlling Biodeterioration of Cultural Heritage Objects with Biocides: A Review." *International Biodeterioration & Biodegradation* (2019). https://doi.org/10.1016/J.IBIOD.2019.104721.

Kalčíková, G. and M. Bundschuh. "Aquatic Biofilms—Sink or Source of Microplastics? A Critical Reflection on Current Knowledge." *Environmental Toxicology and Chemistry*, 41 (2021). https://doi.org/10.1002/etc.5195.

Kelley, J., P. Grierson, S. Collin and P. Davies. "Habitat Disruption and the Identification and Management of Functional Trait Changes." *Fish and Fisheries*, 19 (2018): 716–728. https://doi.org/10.1111/FAF.12284.

Kelly, J., M. G. London, N. Oforji, A. Ogunsola and T. Hoellein. "Microplastic Selects for Convergent Microbiomes from Distinct Riverine Sources." *Freshwater Science*, 39 (2020): 281–291. https://doi.org/10.1086/708934.

Kettner, M. T., K. Rojas-Jimenez, S. Oberbeckmann, M. Labrenz and H. Grossart. "Microplastics Alter Composition of Fungal Communities in Aquatic Ecosystems." *Environmental Microbiology*, 19 (2017): 4447–4459. https://doi.org/10.1111/1462-2920.13891.

Kopytina, N. and E. Bocharova. "Fouling Communities of Microscopic Fungi on Various Substrates of the Black Sea." *Biosystems Diversity* (2021). https://doi.org/10.15421/012144.

Kryss Waldschläger, M. B, A. G. Win Colton Cowger and H. Schüttrumpf. "Settling and Rising Velocities of Environmentally Weathered Micro- and Macroplastic Particles." *Environmental Research* (2020): 110192. https://doi.org/10.1016/j.envres.2020.110192.

Kublanovskaya, A., O. Baulina, K. Chekanov and E. Lobakova. "The Microalga Haematococcus Lacustris (Chlorophyceae) Forms Natural Biofilms in Supralittoral White Sea Coastal Rock Ponds." *Planta* 252 (2020). https://doi.org/10.1007/s00425-020-03438-7.

Kurniawan, A., and T. Yamamoto. "Accumulation of NH_{4+} and NO_{3-} Inside Biofilms of Natural Microbial Consortia: Implication on Nutrients Seasonal Dynamic in Aquatic Ecosystems." *International Journal of Microbiology* 2019 (2019). https://doi.org/10.1155/2019/6473690.

Kurniawan, A., S. M. Ulfa and C. Chamidah. "The Biosorption of Copper(II) Using a Natural Biofilm Formed on the Stones from the Metro River, Malang City, Indonesia." *International Journal of Microbiology* 2022 (2022). https://doi.org/10.1155/2022/9975333.

Kwon, S., M. Zambrano, J. Pawlak and R. Venditti. "Effect of Lignocellulosic Fiber Composition on the Aquatic Biodegradation of Wood Pulps and the Isolated Cellulose, Hemicellulose and Lignin Components: Kinetic Modelling of the Biodegradation Process." *Cellulose*, 28 (2021): 2863–2877. https://doi.org/10.1007/s10570-021-03680-6.

Leiser, R., G.-M. Wu, T. Neu and K. Wendt-Potthoff. "Biofouling, Metal Sorption and Aggregation are Related to Sinking of Microplastics in a Stratified Reservoir." *Water Research*, 176 (2020): 115748. https://doi.org/10.1016/j.watres.2020.115748.

Lingzhan Miao, J. H., Guoxiang You, Zhilin Liu, Songqi Liu, Tengfei Li, Yujuan Mo, Song Guo and Hao Qu. "Acute Effects of Nanoplastics and Microplastics on Periphytic Biofilms Depending on Particle Size, Concentration and Surface Modification." *Environmental Pollution*, 255, no. Pt 2 (2019): 113300. https://doi.org/10.1016/j.envpol.2019.113300.

Liu, S. Y., M. Leung, J. K. Fang and S. Chua. "Engineering a Microbial 'Trap and Release' Mechanism for Microplastics Removal." *Chemical Engineering Journal*, 404 (2021): 127079. https://doi.org/10.1016/J.CEJ.2020.127079.

Liu, W., M. Xu, N. Zhao, X. Zhou, B. Pan, S. Tian and F. Lei. "River Health Assessment of the Yellow River Source Region, Qinghai-Tibetan Plateau, China, Based on Tolerance Values of Macroinvertebrates." *Environmental Science and Pollution Research*, 26 (2019): 10251–10262. https://doi.org/10.1007/s11356-018-04110-0.

Liu, X., R. J. Koestler, T. Warscheid, Y. Katayama and J.-D. Gu. "Microbial Deterioration and Sustainable Conservation of Stone Monuments and Buildings." *Nature Sustainability*, 3, no. 12 (2020): 991–1004. https://doi.org/10.1038/s41893-020-00602-5

Liu, X., R. Koestler, T. Warscheid, Y. Katayama and J. Gu. "Microbial Deterioration and Sustainable Conservation of Stone Monuments and Buildings." *Nature Sustainability*, 3 (2020): 991–1004. https://doi.org/10.1038/s41893-020-00602-5.

Liu, X., R.J. Koestler, T. Warscheid, *et al.* "Microbial Deterioration and Sustainable Conservation of Stone Monuments and Buildings." *Nature Sustainability*, **3** (2020): 991–1004. https://doi.org/10.1038/s41893-020-00602-5

Liu, X., Y. Qian, Y. Wang, F. Wu, W. Wang and J.-D. Gu. "Innovative Approaches for the Processes Involved in Microbial Biodeterioration of Cultural Heritage Materials." *Current Opinion in Biotechnology*, 75 (2022a): 102716. https://doi.org/10.1016/j.copbio.2022.102716

Liu, X., Y. Qian, F. Wu, Y. Wang, W. Wang and J.-D. Gu. "Biofilms on Stone Monuments: Biodeterioration or Bioprotection?." *Trends in Microbiology* (2022b). https://doi.org/10.1016/j.tim.2022.05.012

Macchia, A., R. Strangis, S. De Angelis, M. Cersosimo, A. Docci, M. Ricca, B. Gabriele, R. Mancuso and M. L. La Russa. "Deep Eutectic Solvents (DESs): Preliminary Results for

Their Use Such as Biocides in the Building Cultural Heritage." *Materials* 15 (2022). https://doi.org/10.3390/ma15114005.

Mahto, K. U., S. Kumari and S. Das. "Unraveling the Complex Regulatory Networks in Biofilm Formation in Bacteria and Relevance of Biofilms in Environmental Remediation." *Critical Reviews in Biochemistry and Molecular Biology*, 57 (2021): 305–332. https://doi.org/10.1080/10409238.2021.2015747.

Maksimova, Y., Y. E. Bykova and A. Maksimov. "Functionalization of Multi-Walled Carbon Nanotubes Changes Their Antibiofilm and Probiofilm Effects on Environmental Bacteria." *Microorganisms*, 10 (2022). https://doi.org/10.3390/microorganisms10081627.

Marais, B., C. Brischke and H. Militz. "Wood Durability in Terrestrial and Aquatic Environments – A Review of Biotic and Abiotic Influence Factors." *Wood Material Science & Engineering* 17 (2020): 82–105. https://doi.org/10.1080/17480272.2020.1779810.

Martino, P. D. "What About Biofilms on the Surface of Stone Monuments?" *The Open Conference Proceedings Journal*, 7, no. 1 (2016). http://dx.doi.org/10.2174/2210289201607020014

Marvasi, M., D. Cavalieri, G. Mastromei, A. Casaccia and B. Perito. "Omics Technologies for an In-Depth Investigation of Biodeterioration of Cultural Heritage." *International Biodeterioration & Biodegradation* (2019). https://doi.org/10.1016/j.ibiod.2019.104736.

Mascaro, M. E., G. Pellegrino and A. M. Palermo. "Analysis of Biodeteriogens on Architectural Heritage. An Approach of Applied Botany on a Gothic Building in Southern Italy." *Sustainability,* 14, no. 1 (2021): 34. https://doi.org/10.3390/su14010034

Mello, J. L. S., B. Cardoso, V. Colombo-Corbi and J. Corbi. "Effects of Agrochemicals on Freshwater Macroinvertebrates: Challenges and Perspectives from Southeastern Brazil." *Limnology and Oceanography Bulletin*, 28 (2018). https://doi.org/10.1002/lob.10284.

Méndez-Sánchez, N., T. Cutright and P. Qiao. "Accelerated Weathering and Biodegradation of E-Glass Polyester Composites." *International Biodeterioration & Biodegradation*, 54 (2004): 289–296. https://doi.org/10.1016/J.IBIOD.2004.03.018.

Miao, L., P. Wang, J. Hou, Y. Yao, Z. Liu, S. Liu and T. Li. "Distinct Community Structure and Microbial Functions of Biofilms Colonizing Microplastics." *The Science of the Total Environment*, 650 (2019): 2395–2402. https://doi.org/10.1016/j.scitotenv.2018.09.378.

Miao, L., S. Guo, Z. Liu, S. Liu, G. You, H. Qu and J. Hou. "Effects of Nanoplastics on Freshwater Biofilm Microbial Metabolic Functions as Determined by BIOLOG ECO Microplates." *International Journal of Environmental Research and Public Health*, 16 (2019). https://doi.org/10.3390/ijerph16234639.

Minh Nguyet, N., H. Hà, D. Van Quyen, N. H. Tra and L. Công. "Degradation of Naphthalene and Pyrene by Several Biofilm-Forming Photosynthesis Purple Bacterial Strains." 18 (2020): 561–570. https://doi.org/10.15625/1811-4989/18/3/15322.

Mitsuru Abe, M. S., T. Miki and M. Nishida. "Surface Hydrophobization of Block-Shaped Wood with Rapid Benzylation." *Forests* (2021). https://doi.org/10.3390/f12081028.

Mohan, K. "Microbial Deterioration and Degradation of Polymeric Materials." *Journal of Biochemical Technology,* 2, no. 4 (2011): 210–215.

Mughini-Gras, L., R. V. D. van der Plaats, P. van der Wielen, P. S. Bauerlein and A. M. de Roda Husman. "Riverine Microplastic and Microbial Community Compositions: A Field Study in the Netherlands.." *Water Research*, 192 (2021): 116852 . https://doi.org/10.1016/j.watres.2021.116852.

Negi, A., and I. Sarethy. "Microbial Biodeterioration of Cultural Heritage: Events, Colonization, and Analyses." *Microbial Ecology* (2019): 1–16. https://doi.org/10.1007/s00248-019-01366-y.

Niu, L., Jiaxin Hu, Yi Li, Chao Wang, Wenlong Zhang, Qing Hu, Longfei Wang and Huanjun Zhang. "Effects of Long-Term Exposure To Silver Nanoparticles on the Structure and Function of Microplastic Biofilms in Eutrophic Water." *Environmental Research* (2021): 112182. https://doi.org/10.1016/j.envres.2021.112182.

Nowicka-Krawczyk, P. J. Żelazna-Wieczorek, A. Koziróg, A. Otlewska, K. Rajkowska, M. Piotrowska, B. Gutarowska and B. Brycki. "Multistep Approach to Control Microbial Fouling of Historic Building Materials by Aerial Phototrophs." *Biofouling*, 35 (2019): 284–298. https://doi.org/10.1080/08927014.2019.1598396.

Pazos, R. S., J. C. Suárez and N. Gómez. "Study of the Plastisphere: Biofilm Development and Presence of Faecal Indicator Bacteria on Microplastics from the Río de la Plata Estuary." *Ecosistemas* (2020). https://doi.org/10.7818/ecos.2069.

Perera, T. W. N. K., W. R. H. Weerasinghe, R. N. Attanayake and P. A. Paranagama. "Biodeterioration of Low-Density Polyethylene by Mangrove-Associated Endolichenic Fungi and Their Enzymatic Regimes." *Letters in Applied Microbiology*, 75, no. 6 (2022): 1526–1537. https://doi.org/10.1111/lam.13819

Prieto, B., D. Vázquez-Nion, E. Fuentes and A. G. Durán-Román. "Response of Subaerial Biofilms Growing on Stone-Built Cultural Heritage to Changing Water Regime and CO_2 Conditions." *International Biodeterioration & Biodegradation,* 148 (2020): 104882. https://doi.org/10.1016/j.ibiod.2019.104882.

Qi, K., N. Lu, S. Zhang, W. Wang, Z. Wang and J. Guan. "Uptake of Pb(II) onto Microplastic-Associated Biofilms in Freshwater: Adsorption and Combined Toxicity in Comparison to Natural Solid Substrates." *Journal of Hazardous Materials*, 411 (2021): 125115. https://doi.org/10.1016/j.jhazmat.2021.125115.

Rao, M. V., and V. Kuppusamy. "Wood Biodeterioration in Marine Environment." In: Sundararaj, R. (eds) *Science of Wood Degradation and its Protection*. Springer, Singapore, 2022. https://doi.org/10.1007/978-981-16-8797-6_12.

Ribera, J., and F. Schwarze. "Biological Control of Wood Decay Basidiomycetes Using Trichoderma spp." (2020): 171–184. https://doi.org/10.1016/b978-0-12-819453-9.00008-8.

Rinanda, T., R. Fathima and R. Ramadhani. "Microbiological Perspectives on the Effects of Microplastics on the Aquatic Environment." *IOP Conference Series: Earth and Environmental Science*, 348 (2019). https://doi.org/10.1088/1755-1315/348/1/012048.

Rinanti, A., A. Minarti, M. Fachrul and T. Sunaryo. "A Bibliometric Analysis of Current Status on Biodeterioration of Cultural Heritage during 2018-2022." *Research Journal of Biotechnology* (2023). https://doi.org/10.25303/1803rjbt24038.

Rivera, L., A. P. Ramos, J. I. Castañeda Sánchez and M. Elisa Drago Serrano. "Origin and Control Strategies of Biofilms in the Cultural Heritage." *Antimicrobials, Antibiotic Resistance, Antibiofilm Strategies and Activity Methods* (2018). https://doi.org/10.5772/INTECHOPEN.79617.

Rummel, C. D., A. Jahnke, E. Gorokhova, D. Kühnel and M. Schmitt-Jansen. "Impacts of Biofilm Formation on the Fate and Potential Effects of Microplastic in the Aquatic Environment." *Environmental Science & Technology Letters* 4, no. 7 (2017): 258–267.

Rummel, C., A. Jahnke, E. Gorokhova, D. Kühnel and M. Schmitt-Jansen. "Impacts of Biofilm Formation on the Fate and Potential Effects of Microplastic in the Aquatic Environment." *Environmental Science and Technology Letters*, 4 (2017): 258–267. https://doi.org/10.1021/ACS.ESTLETT.7B00164.

Salem, M. "Silver Nanoparticle Applications in Wood, Wood-Based Panels, and Textiles." *Silver Nanomaterials for Agri-Food Applications* (2021). https://doi.org/10.1016/b978-0-12-823528-7.00006-8.

Samat, A. F., D. Carter and A. Abbas. "Biodeterioration of Pre-Treated Polypropylene by *Aspergillus Terreus* and *Engyodontium Album*. *npj Materials Degradation,* 7 (2023): 28. https://doi.org/10.1038/s41529-023-00342-9.

Sánchez-Silva, M and D. Rosowsky. "Biodeterioration of Construction Materials: State of the Art and Future Challenges." *Journal of Materials in Civil Engineering*, 20 (2008): 352–365. https://doi.org/10.1061/(ASCE)0899-1561(2008)20:5(352).

Sanni, G. O., F. Coulon and T. McGenity. "Dynamics and distribution of bacterial and archaeal communities in oil-contaminated temperate coastal mudflat mesocosms." *Environmental Science and Pollution Research*, 22 (2015): 15230–15247. https://doi.org/10.1007/s11356-015-4313-1.

Sathicq, M. B., Raffaella Sabatino, Gianluca Corno and A. Di Cesare. "Are Microplastic Particles a Hotspot for the Spread and the Persistence of Antibiotic Resistance in Aquatic Systems?" *Environmental Pollution*, 279 (2021): 116896 . https://doi.org/10.1016/j.envpol.2021.116896.

Schifano, E., D. Cavallini, G. De Bellis, M. Bracciale, A. C. Felici, M. Santarelli, M. S. Sarto and D. Uccelletti. "Antibacterial Effect of Zinc Oxide-Based Nanomaterials on Environmental Biodeteriogens Affecting Historical Buildings." *Nanomaterials*, 10 (2020). https://doi.org/10.3390/nano10020335.

Schuijt, L., F.-J. Peng, S. J. P. Van den Berg, M. Dingemans and P. J. Van den Brink. "(Eco)toxicological Tests for Assessing Impacts of Chemical Stress to Aquatic Ecosystems: Facts, Challenges, and Future." *The Science of the Total Environment*, 795 (2021): 148776. https://doi.org/10.1016/j.scitotenv.2021.148776.

Sehnal, L., E. Brammer-Robbins, A. M. Wormington, L. Bláha, J. Bisesi, I. Larkin, C. Martyniuk, M. Simonin and O. Adamovsky. "Microbiome Composition and Function in Aquatic Vertebrates: Small Organisms Making Big Impacts on Aquatic Animal Health." *Frontiers in Microbiology*, 12 (2021). https://doi.org/10.3389/fmicb.2021.567408.

Shabbir, S., M. Faheem, Naeem Ali, P. Kerr, L. Wang, S. Kuppusamy and Y. Li. "Periphytic Biofilm: An Innovative Approach for Biodegradation of Microplastics." *The Science of the Total Environment*, 717 (2020): 137064 . https://doi.org/10.1016/j.scitotenv.2020.137064.

Sivadon, P., C. Barnier, L. Urios and R. Grimaud. "Biofilm Formation as a Microbial Strategy to Assimilate Particulate Substrates." *Environmental Microbiology Reports,* 11, no. 6 (2019): 749–764. https://doi.org/10.1111/1758-2229.12785

Sjöberg, S., Courtney W. Stairs, B. Allard, R. Hallberg, Felix Homa, T. Martin, T. Ettema and C. Dupraz. "Bubble Biofilm: Bacterial Colonization of Air-Air Interface." *Biofilm*, 2 (2020). https://doi.org/10.1016/j.bioflm.2020.100030.

Smith, R. N. Biodeterioration of Fuels. In: Betts, W.B. (eds) *Biodegradation. Springer Series in Applied Biology*. Springer, London, 1991. https://doi.org/10.1007/978-1-4471-3470-1_3.

Soffritti, I., M. D'Accolti, L. Lanzoni, Antonella Volta, M. Bisi, S. Mazzacane and E. Caselli. "The Potential Use of Microorganisms as Restorative Agents: An Update." *Sustainability* (2019). https://doi.org/10.3390/SU11143853.

Sohrabi, M., Mahsa Esmaeillou, H. Fadaei, Mohammad Hossein Talebian and N. Noohi. "The Field Monitoring of Influential Biodeteriogenic Agents on the Historic Rock Surfaces in Persepolis-UNESCO World Heritage Site." *Journal of Research on Archaeometry* (2020). https://doi.org/10.29252/jra.6.1.175.

Spark, A., K. Wang, I. Cole, D. Law and L. Ward. "Microbiologically Influenced Corrosion: A Review of the Studies Conducted on Buried Pipelines." *Corrosion Reviews* 38, no. 3 (2020): 231–262.

Stabnikova, P, V. Stabnikov, A. Marinin, M. Kļaviņš, L. Klavins and Ashok Vaseashta. "Microbial Life on the Surface of Microplastics in Natural Waters." *Applied Sciences* (2021). https://doi.org/10.3390/app112411692.

Suarez, C., M. Piculell, O. Modin, S. Langenheder, F. Persson and M. Hermansson. "Thickness Determines Microbial Community Structure and Function in Nitrifying Biofilms Via Deterministic Assembly." *Scientific Reports*, 9 (2019). https://doi.org/10.1038/s41598-019-41542-1.

Sukhanova, E., E. Zimens, O. Kaluzhnaya, V. Parfenovaand O. Belykh. "Epilithic Biofilms in Lake Baikal: Screening and Diversity of PKS and NRPS Genes in the Genomes of Heterotrophic Bacteria." *Polish Journal of Microbiology* 67, no. 4 (2018): 501–516. https://doi.org/10.21307/pjm-2018-060.

Taniwaki, R. H., C. Matthaei, Tatima K. M. Cardoso, S. Ferraz, L. Martinelli and J. Piggott. "The Effect of Agriculture on the Seasonal Dynamics and Functional Diversity of Benthic Biofilm in Tropical Headwater Streams." *Biotropica*, 51 (2019): 18–27. https://doi.org/10.1111/btp.12617.

Timoner, X., Miriam Colls, Samia M. Salomón, F. Oliva, V. Acuña and S. Sabater. "Does Biofilm Origin Matter? Biofilm Responses to Non-Flow Period in Permanent and Temporary Streams." *Freshwater Biology*, 65 (2020): 514–523. https://doi.org/10.1111/fwb.13447.

Trovão, J., A. Portugal, F. Soares, D. S. Paiva, N. Mesquita, C. Coelho, A. C. Pinheiro, L. Catarino, F. Gil and I. Tiago. "Fungal Diversity and Distribution Across Distinct Biodeterioration phenomena in Limestone Walls of the Old Cathedral of Coimbra, UNESCO World Heritage Site." *International Biodeterioration & Biodegradation*, 142 (2019): 91–102. https://doi.org/10.1016/j.ibiod.2019.05.008.

Trovão, J., F. Soares, I. Tiago, L. Catarino, A. Portugal and Francisco Gil. "A Contribution to Understand the Portuguese Emblematic Ançã Limestone Bioreceptivity to Fungal Colonization and Biodeterioration." *Journal of Cultural Heritage* (2021). https://doi.org/10.1016/J.CULHER.2021.04.003.

Tudararo-Aherobo Laurelta, O. S., A. Tega and S. Avwerosuoghene. "Biotreatability of Crude Oil Polluted Aquatic Environment Using Indigenous Hydrocarbon Utilizing Bacteria." *Asian Journal of Environment & Ecology* (2020). https://doi.org/10.9734/ajee/2020/v12i430175.

Wang, J., X. Guo and J. Xue. "Biofilm-Developed Microplastics As Vectors of Pollutants in Aquatic Environments." *Environmental Science & Technology* (2021). https://doi.org/10.1021/acs.est.1c04466.

Wang, J., Z. Tan, J. Peng, Q. Qiu and M. Li. "The Behaviors of Microplastics in the Marine Environment." *Marine Environmental Research*, 113 (2016): 7–17. https://doi.org/10.1016/j.marenvres.2015.10.014.

Wang, L., J. Tong, Y. Li, J. Zhu, W. Zhang, L. Niu and H. Zhang. "Bacterial and Fungal Assemblages and Functions Associated with Biofilms Differ Between Diverse Types of Plastic Debris in a Freshwater System." *Environmental Research* (2020): 110371. https://doi.org/10.1016/j.envres.2020.110371.

Wang, L., X. Hua, L. Zhang, N. Song, D. Dong and Z. Guo. "Influence of Organic Carbon Fractions of Freshwater Biofilms on the Sorption for Phenanthrene and Ofloxacin: The Important Role of Aliphatic Carbons." *The Science of the Total Environment*, 685 (2019): 818–826 . https://doi.org/10.1016/j.scitotenv.2019.06.203.

Weitere, M., M. Erken, N. Majdi, H. Arndt, H. Norf, M. Reinshagen, W. Traunspurger, A. Walterscheid and J. K. Wey. "The Food Web Perspective on Aquatic Biofilms." *Ecological Monographs* 88, no. 4 (2018): 543–559. https://doi.org/10.1002/ecm.1315

Weththimuni, M., D. Capsoni, M. Malagodi and M. Licchelli. "Improving Wood Resistance to Decay by Nanostructured ZnO-Based Treatments." *Journal of Nanomaterials* (2019). https://doi.org/10.1155/2019/6715756.

Wu, X., J. Pan, M. Li, Y. Li, M. Bartlam and Y. Wang. "Selective Enrichment of Bacterial Pathogens by Microplastic Biofilm." *Water Research*, 165 (2019):114979. https://doi.org/10.1016/j.watres.2019.114979.

Wyatt, K., R. Seballos, M. Shoemaker, S. Brown, S. Chandra, K. Kuehn, A. Rober and S. Sadro. "Resource Constraints Highlight Complex Microbial Interactions During Lake Biofilm Development." *Journal of Ecology*, 107 (2019): 2737–2746. https://doi.org/10.1111/1365-2745.13223.

Xu, R., S. Zhang and F. Meng. "Large-sized Planktonic Bioaggregates Possess High Biofilm Formation Potentials: Bacterial Succession and Assembly in the Biofilm Metacommunity." *Water Research*, 170 (2019): 115307. https://doi.org/10.1016/j.watres.2019.115307.

Yang, Y., W. Liu, Z. Zhang, H. Grossart and G. Gadd. "Microplastics Provide New Microbial Niches in Aquatic Environments." *Applied Microbiology and Biotechnology*, 104 (2020): 6501–6511. https://doi.org/10.1007/s00253-020-10704-x.

Yao, W., C.-D. Li, H. Zhan, J.-Q. Zhou, R. Criss, S. Xiong and X. Jiang. "Multiscale Study of Physical and Mechanical Properties of Sandstone in Three Gorges Reservoir Region Subjected to Cyclic Wetting–Drying of Yangtze River Water." *Rock Mechanics and Rock Engineering*, 53 (2020): 2215–2231. https://doi.org/10.1007/s00603-019-02037-7.

Yuxian You, S. T., and J. A. Hope. "The Impacts of Polyethylene Terephthalate Microplastics (mPETs) on Ecosystem Functionality in Marine Sediment." *Marine Pollution Bulletin*, 160 (2020): 11624. https://doi.org/10.1016/j.marpolbul.2020.111624.

Zamani, N., M. Ghezelsofla, A. M. Ahadi and M. Zamani. "Application of Microbial Biotechnology in Conservation and Restoration of Stone Monument." *Journal of Applied Biotechnology Reports,* 4, no. 2 (2017): 587–592.

Zhang, X. "Environmental DNA Shaping a New Era of Ecotoxicological Research." *Environmental Science & Technology*, 53 10 (2019): 5605–5612. https://doi.org/10.1021/acs.est.8b06631.

12 Biofilms in Aquatic Ecosystems and Their Role in Antibiotic Resistance

Mahima S. Mohan and Siddhardha Busi

12.1 BIOFILM: A BRIEF INTRODUCTION

Biofilm is defined as an interdependent community-based subsistence of microorganisms, derived from a single or multiple species (Davey and O'toole, 2000). Biofilm consists of a self-synthesized matrix comprised of extracellular polymeric substance (EPS) that includes polysaccharides, proteins, nucleic acids, lipids, amyloids, membrane vesicles, and microbial derived humic-like refractory molecules. EPS provides physical strength to the biofilm structure, helps in attachment, and forms a three-dimensional architecture, where the biofilm cells are interconnected and immobilized within the matrix. Moreover, the extracellular enzymes produced by the biofilm aid in the digestion of biopolymers. The matrix shields the biofilm cells from antimicrobial agents, heavy metals, desiccation, ultraviolet radiation, host immunity, and so on. In addition, the matrix keeps the biofilm cells in close proximity and allows interactions such as quorum sensing, horizontal gene transfer, and synergistic microconsortium (Flemming and Wingender, 2010; Flemming et al., 2023). Biofilm formation is stimulated and controlled by quorum sensing, environmental conditions, secondary messengers, nutrient availability, etc. (Rather et al., 2021). Quorum sensing (QS) is a signaling process, where microorganisms communicate with each other based on the concentration of a signaling molecule called an autoinducer (AI). These molecules aid in the sustenance of microbial diversity and microbial community and maintain cells under stress conditions (Mishra et al., 2022).

The capacity of biofilm complex to mature into distinct forms such as films, micro- and macro-colonies, pellicles, mushroom-like structures, columns, and aggregates are mediated by the microbial community interactions and environmental conditions. Biofilm assembly and dynamics are mediated by secondary messengers, AIs or small RNAs, which have been observed in both clinical and environmental biofilms (Flemming et al., 2023).

Biofilm formation occurs through sequential steps. At first, the planktonic cells attach to a substrate, which is the most crucial process in biofilm formation. There

DOI: 10.1201/9781003487203-12

are two types of adhesion: reversible and irreversible adhesion. Reversible adhesion is mediated via flagella, pili, and fimbriae, while irreversible adhesion is an EPS-facilitated adhesion process. Followed by irreversible attachment, the cells undergo cell division and assemble over the previously attached cells to form microcolonies. Further, the cell proliferation increases and the cells produce an enormous number of EPS to form a 3D structured framework. The EPS maintains the biofilm architecture, signaling molecules, enzymes, and metabolic products within the biofilm. Once the biofilm matures, dispersion happens to start a new lifecycle. Active dispersal of biofilm depends on the motility or EPS degradation, whereas passive dispersal is based on physical forces such as shearing force. The former method is affected by deviations in the environmental settings such as temperature, starvation, lack of oxygen, and accumulation of metabolites. In this stage, the genes associated with motility (flagella synthesis) and EPS degradation are upregulated. The expression of genes involved in the EPS synthesis and fimbriae synthesis (attachment) are downregulated. Additionally, c-di-GMP level decreases lead to the transformation of biofilm cells into its planktonic forms (Toyofuku et al., 2016; Fu et al., 2021).

Biofilms are able to colonize in both biotic and abiotic surfaces. This poses severe problems to industries leading to economic loss and declines in product quality. Medical, food, water, marine, dairy processing plants, etc., are also widely affected leading to pipe blockages, waterlogging, and obstruction of the functioning of the plants. Studies have reported that ~65–80% of human infections are biofilm associated; microbes can form biofilm on medical devices such as surgical devices, catheters, analytical and diagnostic equipment (Vishwakarma, 2020).

12.2 AQUATIC BIOFILMS

Aquatic biofilms are the microbial communities of bacteria, algae, protozoa, and fungi formed on a wet surface that are encapsulated in a matrix made up of extracellular polymeric substance (EPS) (Rivera et al., 2021). Aquatic systems offer different biotic and abiotic surfaces for the colonization of biofilm communities. The structure and function of biofilm depend on the nature of the surface where it gets attached (Besemer, 2015). Biofilms are omnipresent in aquatic systems, which maintain the stability of biological and ecological processes (Guo et al., 2017). Biofilm offers a wide array of advantages to the enclosed microbial communities including resistance to stress factors, nutrient availability, metabolic products, and horizontal gene transfer (HGT) (Chao et al., 2015).

In fast flowing rivers, the periphytic biofilms (formed on plant) enhance water retention in some areas, which leads to prolonged exchange of organic materials and capture limiting nutrients such as nitrogen and phosphorus. Moreover, biofilm limits erosion of sediments with the flow of water. In polluted water, periphyton reduces the amount of phosphorous and nitrogen responsible for the eutrophication and acidification of water bodies. Biofilms are reported to adsorb and degrade microcystin and maintain water quality, human and animal health (Sentenac et al., 2022).

In intense light environment, biofilms are occupied by photoautotrophs such as photosynthetic organisms, algae, diatoms, and blue-green algae, while in low light environment, heterotrophs such as bacteria are predominant.Biofilms are the major reservoir of carbon along with macroalgae and higher aquatic plants. Biofilm also act as a major nutrient source by the release of leached exudates, senescent material, and live cells. They become food sources for crustaceans, fish, molluscs, and insects (Burns and Ryder, 2001).

In freshwater systems, the prevalent bacterial community generally includes *Proteobacteria,* cyanobacteria, and *Bacteroidetes.* In lakes, streams, and rivers the dominant bacterial communities belong to β-*Proteobacteria*, whereas α-*Proteobacteria* is predominant in marine systems, which can be even more abundant than β-*Proteobacteria* in freshwater biofilms. Some members of α-*Proteobacteria* degrades humic substances and able to form grazing-restistant phenotypes make them to survive in freshwater biofilms. *Bacteroidetes* present in the biofilms are capable of degrading suspended particles in water. Other taxonomical groups typically present in biofilm communities are *Acidobacteria* (low pH), *Firmicutes, Actinobacteria, Verrucomicrobia, γ* and *δ-Proteobacteria*, *Planctomycetes,* etc. (Besemer, 2015).

Pollution of water bodies is a threat to the environment as it decreases the water quality. It can negatively influence the biofilm biomass, growth, composition, and functions. However, biofilm captures the nutrients and pollutants from the water bodies and detoxifies the environment (Sentenac et al., 2022). Aquatic biofilms can be used to study pollution in the aquatic environment due to anthropogenic activities, as they respond rapidly to environmental stress. Biofilms have multiple cationic and anionic absorption sites and hydrophobic/lipophilic sites to which contaminants can bind. Hence, biofilms can bioaccumulate contaminants and be used as a tool in monitoring aquatic environment pollution (Fernandes et al., 2020).

12.3 BIOFILM AND ANTIBIOTIC RESISTANCE

Since its discovery, antibiotics have saved millions of people worldwide from bacterial infections. The prevalent use of antibiotics leads to antibiotic resistance and hence emergence of antibiotic-restistant bacteria (ARB). According to the World Health Organization, the development of antibiotic resistance (AR) is regarded as one among the three biggest threats to human health in the twenty-first century (Lorenzo et al., 2018)

The Center for Disease Control and Prevention (CDC) revealed that >2.8 million antimicrobial resistance infections arise worldwide and around 35,000 or more humans die of such microbial infections each year in the United States (Reddy et al., 2022). In India, the evolution of drug resistance has rapidly increased due to the use of antibiotics in treating human and animal infections. Infectious diseases are the major cause of death in India which demands the usage of alarmingly high levels of antibiotics. Moreover, India is the third largest country to produce pharmaceutical drugs. Effluents from pharmaceutical industries, hospitals, and domestic wastewater without treatment discharged into water bodies have become a prime cause of water contamination in aquatic ecosystems (Das et al., 2023).

Exposure of antibiotics at low concentrations can induce the progression and maintenance of resistance genes in microbial communities. These antibiotic-resistant genes (ARGs) are disseminated to other cells or species through HGT (Aubertheau et al., 2017). Lundström et al. revealed that exposure to tetracycline augmented the richness of tetracycline-resistant bacteria and *tet* (*tet*A and *tet*G- resistant genes even at lowest concentration (1 μg/L). The microbial structure of the biofilm was also modified upon exposure to varying concentrations of tetracycline. The drugs released into the aquatic niches induce selection pressure, which enriches ARGs and resistant bacteria and creates great risk to public health (Lundström et al., 2016).

The capacity of biofilm to survive under diverse physiological conditions allows it to become extremely tolerant to antimicrobial agents (~ >1000 times) and host immune systems. When compared to planktonic forms, biofilm has a greater ability to undergo horizontal gene transfer and has a significantly higher rate of mutation. With the assistance of virulence factors and ARGs, biofilm forms can easily transfer resistance genes from a resistant bacterium to an antibiotic-susceptible one and thereby acquire resistance towards antibiotics. Moreover, the innate properties of biofilm such as structure and physiological characters help to overcome the stress due to antimicrobials. The defensive characteristics of biofilm have not yet been clearly studied, but some of the possible reasons are: mutation in the resistance genes such as efflux pumps, slower growth, low metabolism, persister cells, antibiotic targeting enzymes, poor diffusion of antibiotics into the matrix, stress response mechanism, modified microenvironmental condition inside biofilm such as pH and hypoxia, and so on (Gebreyohannes et al., 2019; Jolivet-Gougeon et al., 2014).

Studies have reported that the bacteria obtained from aquatic sources in Northern Malaysia have the potential to form biofilm and have shown AR patterns. The predominant microorganisms identified are *Aeromonas* sp. and *Vibrio* sp. from the coastal sea water and sediments. Moreover, both organisms are pathogens to humans and aquatic animals and have been isolated from seafoods from different parts of Northern Malaysia. Thus, the capacity to form biofilm and AR affects the humans residing in the coastal areas (Odeyemi and Ahmad, 2017).

The continuous acquaintance of biofilm with antibiotics (at low concentration) results in alteration of taxonomical community structure and metabolism and could lead to cell death. Additionally, the abundance of antibiotic-resistant bacteria increases, leading to evolution and affecting trophic levels of the food web and ultimately disrupting the functioning of the ecosystem (Matviichuk et al., 2022).

12.4 IMPACTS OF BIOFILM AND AR IN AQUATIC NICHES

The continuous discharge of wastewater effluents from anthropogenic activities from hospital, household, and industrial waste into the aquatic environment makes them highly polluted with antibiotic resistome and toxic pollutants. Antibiotics from clinical settings and agricultural surface runoff end up in aquatic ecosystems. This leads to the accumulation of recalcitrant pollutants called trace organic compounds (TOrCs). Indigenous microorganisms present in aquatic ecosystems are continuously exposed to recalcitrant compounds such as antibiotics at varying concentrations (Penesyan et al., 2020).

Antibiotics are extensively used in aquaculture industries to enhance the health and growth of fish. Studies estimate that the continuous increased usage of antibiotics may lead to AR in the coming decades. Admistration of antibiotics to fish through food and water and also excretion of antibiotic residues by fish has become a significant route for surface water contamination. Selective pressure of antibiotics results in the advent of ARB and also increases the transmission of antibiotic-resistant genes in marine ecosystems. Moreover, the selection pressure may cause loss of microbial communities involved in nutrient cycling, decomposition, and primary productivity. Studies have reported that antibiotics in aquatic niches cause increased incidence of toxic cyanobacteria, resulting in eutrophication in freshwater systems and posing serious health problems to humans. Additionally, the presence of antibiotic-resistant organisms affects fish by altering the gut microbiota and consumption of such fish poses serious health issues to consumers (Monahan et al., 2021; Adenaya et al., 2023). Bhattacharyya et al. 2023 reported the richness of ARGs in the gut of freshwater macroinvertebrates.

Some studies have detected the existence of pathogens containing ARGs in well water, which may likely spread to humans and animals. Moreover, some reports have suggested their presence in ground water and also in crop fields through irrigation water (Lin et al., 2021).

Triclosan, a bactericidal compound, is widely used as a preservative or as an antiseptic agent in household products such as soaps, toothpastes, cleaning lotions, etc. Triclosan is a recalcitrant compound that is not removed through wastewater treatment methods and eventually reach rivers. Several studies have reported the toxicological effects of triclosan. At high concentrations (EC_{50}= 900 µg/ml) it inhibits the photosynthetic ability of autotrophs. Moreover, triclosan at 10 µg/L concentration affects microbial communities, algal biomass, biofilm structure, and functions (Ricart et al., 2010).

Environmental biofilms are crucial for the functioning of ecosystems, but the presence of contaminants disturbs their health and normal functioning. The anaerobic ammonium oxidation (Anammox) process, an important process in wastewater treatment by the biofilm community, oxidizes ammonia into nitrogen gas under anaerobic condition. Anaerobic ammonia-oxidizing bacteria are highly sensitive to some inhibitory compounds that affect the Annamox process. Discharge of effluents containing antibiotics, especially norfloxacin (in trace amounts), can significantly retard the oxidation process (Zhang. X et al., 2019).

It was reported that disposable facial mask serves as a substrate for biofilm formation and also enriched the ARGs and pathogens in the aquatic system. Researchers reported the presence of various opportunistic human pathogens in disposable face masks obtained from urban river including *Acinetobacter, Escherichia-Shigella, Bacillus,* and *Klebsiella.* These pathogens serve as ARG carriers and pose great hazards to the environment and human health. Furthermore, the wastewater effluent discharge into urban rivers also aggravates the load of ARGs and pathogens in the biofilm developed on the discarded masks (Liu et al., 2023).

In the past few decades, aquatic ecosystems have been greatly disturbed due to discharge of effluents without proper treatment. One of the important pollutants that negatively impact ecosystems is heavy metals. These pollutants modulate the structure and function of microbial populations and also acquire resistance mechanisms in order to thrive under environmental stress. Due to increased environmental selective pressure, heavy metal resistance may lead to the progression of AR in microbial communities. Some studies have reported that biofilm formation can be induced at subinhibitory concentration of heavy metals such as lead, nickel, and cadmium. In addition, these metals can promote microbial adhesion, biofilm formation, and finally resistance (de Araújo and de Oliveira, 2020).

River biofilms contain almost 90% of whole microbial flora such as algae, bacteria, and fungi and represent the primary production source. Contamination of river biofilm causes the bioaccumulation and biomagnification of pollutants in the respective trophic level (Aubertheau et al., 2017).

Bioaccumulation of pharmaceutical drugs and their residues in fish have been reported in urban rivers. According to Ondarza et al. (2019) there is an abundance of pharmaceutical products, antibiotics, and their residues in the gills, liver, and flesh of fish, including erythromycin (0.7–5.6 μg/kg), caffeine (13 μg/kg), sulfamethoxazole, trimethoprim, sulfathiazole (0.9–5.5 μg/kg), and norfluoxetine, a byproduct of the antidepressant fluoxetine (1.1–9.1 μg/kg). The adverse impacts of bioaccumulation include imbalance in the gut microorganisms, growth, immune system, reproduction and digestion of aquatic fauna in the urban river system. This leads to the propagation and increase of antimicrobial resistance in the niches and gradual spread into humans (Ondarza et al., 2019: Reddy et al., 2022).

Other studies have reported the presence of excessive quantities of antibiotics from the effluents of the pharmaceutical industry in Hyderabad, India. In 2006, the amount of ciprofloxacin in the effluent was high at around 31,000 μg/ml, equivalent to 45 kg/day. Similarly, a study conducted in 2016 reported that along with ciprofloxacin, other antibiotics such as levofloxacin, linezolid, moxifloxacin, ampicillin, doxycycline, etc., were found to be predominant in the effluent. Improper discharge of antibiotic- containing effluents from industries causes contamination of the Musi River of Hyderabad, which affects aquatic environments, humans, and animals (Gandra et al., 2017).

12.5 ACQUISITION AND SPREAD OF ARGS IN BIOFILM

ARGs have become a worldwide threat to human and animal well-being in the past few years. As a result of increased and overuse of antibiotics, under selection pressure of antibiotics, the advent of ARB and ARGs are evolving rapidly. The prevalence of antibiotics in ecosystems, in particular aquatic systems, has received international interest due to the possible side effects to humans, animals, and aquatic ecosystems (Wang et al., 2024).

The location of ARGs in the bacteria may be on chromosomes, mobile genetic elements (MGEs) such as plasmids, transposons, or integrons, or found as intracellular DNA (iDNA) or extracellular DNA (eDNA) in the milieu. In mobile genetic elements, the ARGs are mainly located at the accessory regions, which are not essential for maintenance and mobilization. In addition to HGT mechanisms, membrane vesicles have become an important reservoir of environmental DNA in aquatic environments, since they are capable of gene transfer between bacteria (Abe et al., 2020). Plasmid-containing ARGs are highly mobile and more abundant in the extracellular environment than the chromosomal ones. The perseverance of chromosomal DNA in the milieu is higher than for the plasmid DNA. Thus, lower frequency of ARGs in the plasmid is expected in environmental samples such as soil and sediments. Insertion sequences (IS), a simple mobile genetic element, converts the non-transmissible elements into transmissible ones (Zarei-Baygi and Smith, 2021).

Bacteria gain resistance towards antibiotics through mutation and horizontal gene transfer (HGT). In the latter case, the susceptible cells attain resistance via horizontal transfer of ARGs from resistant cells or by means of extracellular DNA. The major HGT mechanisms developed by bacteria are conjugation, transformation, and transduction. In intracellular DNA, the ARGs are disseminated via conjugation or transduction; whereas in extracellular DNA, ARGs are spread through the transformation process. Mao and co-workers reported that the persistence of eDNA was higher compared to iDNA in river sediments than water, where eDNA may function as a pool for the spread of ARGs. eDNA is highly stable for months in soil or sediments and found to be 3–4 orders higher in magnitude in sediments than in water (Guo et al., 2017; Mao et al., 2014)

Class I integron (intI1) are reported as a crucial medium for the transmission of ARGs in urban lakes. Studies have shown that selection pressure of antibiotics along with heavy metals act as foremost triggers for the dissemination of ARGs in urban lakes. Moreover, a plethora of ARGs was reported in urban lakes with high anthropogenic impacts (Yang et al., 2017).

Bacteriophages can mediate HGT via transduction mechanism where the phage transfers genetic material from one bacterium to other. They are group of viruses (primarily scavenges bacteria) and also a reservoir of DNA in the environment. Phages are highly resistant to treatment and cannot be completely killed or removed by water disinfection methods including chlorination and UV. Several studies have reported the incidence of phage-carrying ARGs of diverse antibiotics such as aminoglycoside, macrolides, tetracyclines, and so on from rivers, WWTPs, seas, and rivers. For instance, *Acinetobacter baumannii*, a commensal in water bodies and a multidrug-resistant pathogen, transfers bla_{NDM-1}-resistant gene through phage transduction (Abe et al., 2020).

Researchers have reported that tetracycline resistance gene *tet(M)* detected in marine bacteria was transferred to human-enteric bacteria using *in vitro* studies. Donors used for the study were fish pathogens such as *Lactococcus garvieae* and *Vibrio* spp. and were capable of transmission of resistance genes to recipient human-enteric bacteria, *Enterococcus fecalis,* and *Escherichia coli,* respectively. These pathogens colonize the gut of fish and form biofilm and hence lead to the distribution of ARGs (Neela et al., 2009).

Due to the high absorption capacity of biofilm, several pollutants from water bodies such as metals, antibiotic residues, and resistant genes can be absorbed. Biofilm also serves as a binding site for ARB. Studies have shown that ARGs are more abundant in water biofilm than in water and sediments. The existence of antibiotic residues and heavy metals can induce the selection of resistance genes by biofilm and trigger HGT between resistant and susceptible bacteria (Lin et al., 2021).

Irrigation is an important process in agriculture. Some studies have reported the presence of ARGs in irrigation water, where it can spread through crops and soil to humans and animals. Farmers who are in contact with such crops and fields could be at high risk of infection due to ARB (Lin et al., 2021; Bougnom et al.,

2020). In a study, the distribution of veterinary antibiotics and resistant genes on farmland soil suggested that ARGs and ARB can penetrate into the deeper layers of soil and perhaps reach the ground water system by means of vertical transport (Huang et al., 2013). Dissemination of ARGs is higher between cells in the biofilm and was achieved through HGT and genetic exchange mechanisms. It was confirmed that the gene transfer frequency was greater in adjacent cells attached to glass beads when compared to the cells in aqueous phase. Therefore, it is clear that biofilms are the pools in which ARGs and ARB flourish. Detachment of ARB from the biofilm may disseminate into the environment and can challenge human well-being (Abe et al., 2020).

12.6 AQUATIC ECOSYSTEM – A RESERVOIR OF ARGS

Aquatic ecosystems are the utmost plausible sites of ARG pools, in which bacteria from diverse environments including clinical, terrestrial, and human commensals flourish. Humans and animals are highly subjected to the aquatic environment in day-to-day activities (Abe et al., 2020). Effluents containing antibiotics from hospitals, households, pharmaceutical industries, and agricultural field runoff are discharged water bodies. The interface of antimicrobials with the indigenous microbial population induces the proliferation of ARB and ARGs in the environment, thereby generating resistant novel pathogens, which affects human health (Lin et al., 2021). Other anthropogenic activities such as livestock farming aquaculture, etc., are sources of antibiotics in the aquatic ecosystem. Human-intervened surface waters are susceptible to huge amounts of ARGs and ARB due to the release of effluents from treatment plants and feces from the environment. Hence, surface water acts as source of ARGs, ARB, and MGEs (Zhang et al., 2021) (Figure 12.1).

One of the major hotspots for antibiotic resistomes is wastewater treatment plants (WWTPs). Sewage from multiple sources are received in WWTPs. Hospital effluents contain high concentration of drug remains, infectious materials, and pathogens. Around 20–97% of the antibiotics ingested by humans and animals are excreted as active molecules and reach the sewage treatment plants (Lorenzo et al., 2018). During treatment, the gut microbiota of humans is exposed to antibiotic dosages, and may be released into the sewage through human feces. Several studies have reported the high frequency of antibiotics, ARGs, and ARB in WWTPs (Lorenzo et al., 2018). WWTPs are considered as crucial systems for the transmission of drug resistance to aquatic niches because of three reasons: (i) release of antibiotics and resistomes from municipal sewage; (ii) optimal conditions for the genetic exchange mechanisms among bacterial community during treatment process; and (iii) higher levels of AR in effluents than the unprocessed effluents (Lorenzo et al., 2018).

Freshwater ecosystems are prone to antibiotic contamination from different sources including sewage effluents, surface and field runoff, aquaculture, etc. Some researchers have detected high levels of ARB in river systems (~ 98% of total identified bacteria) and lakes (~ 77%). The presence of highly epidemic antibiotic-restistant pathogens such as vancomycin-restistant *Enterococcus,* extended spectrum β-lactamase producing Enterobacteriaceae, and methicillin-resistant *Staphylococcus*

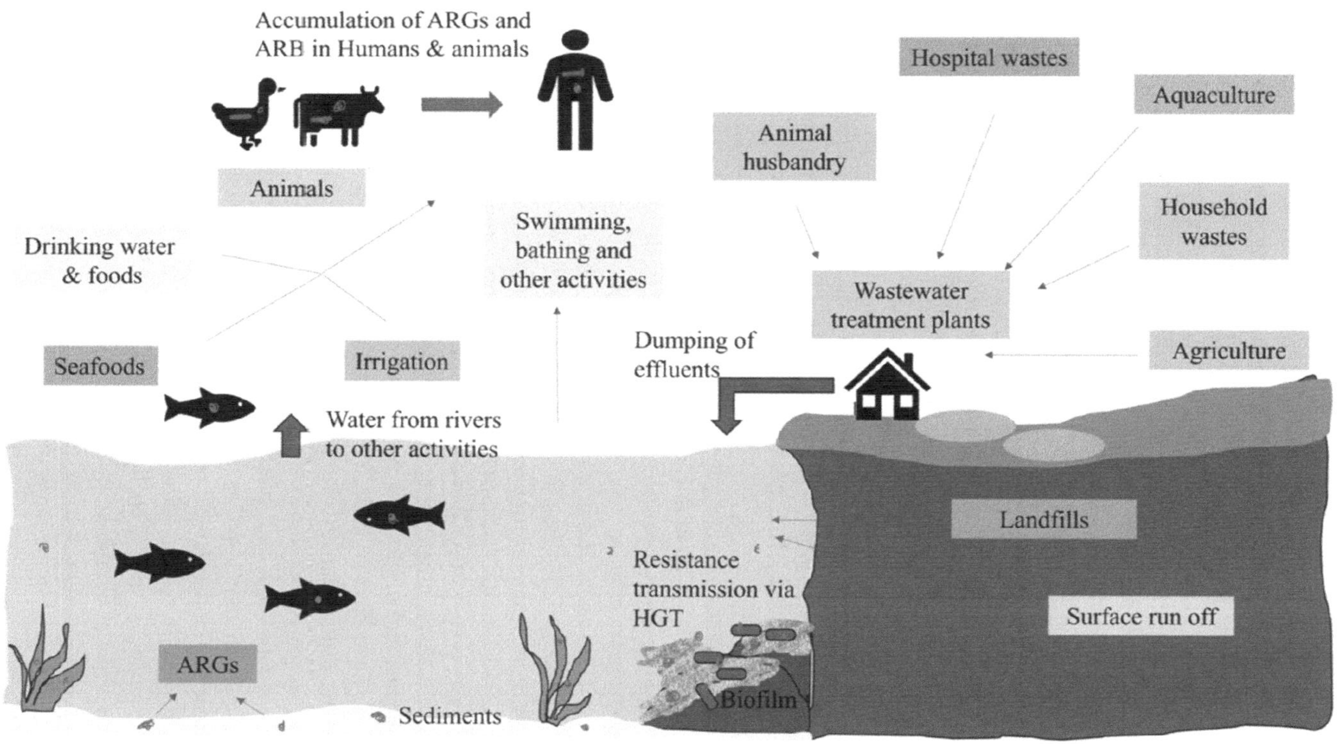

FIGURE 12.1 Aquatic ecosystem acts as a pool of ARGs and ARB. The acquisition and distribution of ARG and ARB into humans and animals from the aquatic environment includes drinking water, irrigation, swimming, bathing, animals, fishing, and other recreational activities.

aureus was detected. In fact, dilution and degradation of drugs do not completely remove the distribution of ARB and ARGs in freshwater systems (Nnadozie and Odume, 2019). In another study, antibiotic resistome was analyzed in a constructed wetland reservoir, where around 342 ARG subtypes belonging to 18 ARG types were detected from water and sediments. They reported the plentitude of bacitracin and multidrug resistance genes in both water and sediments. A metagenomic study revealed several mobile genetic elements and also human pathogens such as *S. aureus*, *Escherichia coli*, *Salmonella enterica*, and *Klebsiella pneumonia* (Li et al., 2023).

Dang et al. (2020) explored the antibiotic resistome of water and sediments of the Danjiangkou reservoir using metagenomic analysis. They analyzed around 436 ARG subtypes belonging to 20 ARG types from water and sediments. The richness of ARG was higher in the sediments than the water and the prevalent resistances gene identified were bacitracin and vancomycin resistance genes in water and sediments, respectively. The predominance of ARGs was higher throughout dry season than the wet climate. The frequency of gene transfer was higher in water than the sediments and the ARB pathogens identified were *E. coli*, *K. pneumoniae*, and *Pseudomonas aeruginosa*.

ARGs encodes for tetracycline resistance (*tet(M)* and *tet(S)*) and sulphonamide resistance (*sul1-sul3*) were detected in aquatic niches like sea water, sea sediments, aquaculture, and fish. These ARGs are reported to diffuse and accumulate in water biofilms (Abe et al., 2020).

Cyanobacterial blooms diminish the water quality and negatively impact the aquatic environment. Such blooms can disrupt the bacterial community and also human health. Studies suggest that ARGs enrichment in freshwater is strongly linked with the cyanobacterial blooms. The relative abundance of ARGs and pathogenic microorganisms is increased during a *Planktothrix* bloom as observed in this study (Zhang et al., 2020).

Due to urbanization and industrialization, urban water systems are highly impacted with microbial contamination as well as environmental pollution. Several researchers have detected the existence of ARB and ARGs in various urban lakes, rivers, streams, and WWTPs. In a study, urban water ecosystems including ponds, drinking water resources, influents, and effluents of wastewater were studied for the detection of resistomes. They identified around 237 ARGs having resistance towards multiple drugs and the most abundant resistance genes observed were aminoglycoside and β-lactamase resistance genes (Hou et al., 2021).

Furthermore, studies reported that wild birds, especially those in contact with human or agricultural fields, act as a vector for transmission of AR in aquatic ecosystems. Recently, studies have reported the presence of ARB in both resident and migratory gulls globally, which mediates the transmission of resistance from human activities to the aquatic system (Zeballos-Gross et al., 2021).

Jia et al. investigated the presence of various antibiotics in the Ba River, China. High incidence of antibiotics such as sulfonamides, macrolides, tetracyclines, and quinolones were detected in the range of 85.7–100%. Tetracycline was the predominant antibiotic in both water ($\sim 8.54 \times 10^2$ ng/L) and sediments ($\sim 2.08 \times 10^3$ μg/kg). In addition, they reported the abundance of ARGs such as *sul*1, *tet*A, *tet*Z, bla_{TEM}, *gyr*A, *erm*F, and *cml*A in the riverine system (Jia et al., 2018).

In India, the Ganges is considered as the holy river, and has been heavily contaminated with antibiotic resistomes and metal ions due to the release of untreated/treated domestic wastes into the river system. During the monsoon season, river floods enter nearby cities and may cause potential risks to the public health. High-throughput metagenomic sequencing has shown abundance of ARGs such as *tuf, par*Y, *mfd,* and *ile*S in water and sediments and the predominant antibiotics reported are β-lactam, multidrug/efflux, and elfamycin. Taxonomical studies discovered the abundance of phyla such as *Proteobacteria* and *Actinobacteria* and fecal contamination indicator genera, *Bacteroides,* in the river samples (Reddy and Dubey, 2019).

12.7 ASSESSMENT OF AR IN AQUATIC BIOFILMS

Advances in genomic studies provide new horizons to comprehend AR mechanisms and their transmission among microorganisms. Previously, the data associated with resistance was limited due to the traditional culturable techniques. With advancements in genomics and metagenomics, multiple environmental samples can be sequenced and analyzed rapidly. Thus far, several environmental samples have been metagenomically sequenced and are available publically via databases such as MG-RAST, CAMERA, and IMG/M, and offer information about microbial communities. Other databases provide precise information regarding the ecology, transmission, and progression of AR in the environment including the ARG Database (ARGD), the Integron Database (INTEGRALL), Palzkill, USEARCH, the Comprehensive Antibiotic Resistance Database (CARD), the Bush, DIAMOND, and so on (Balcázar et al., 2015).

In the past few years, metagenomics has been often used for the detection of AR in aquatic systems. This method also provides information about metal ion resistance and the ARG transmission routes through the detection of key genes responsible for genetic mobilization in the microorganisms such as mobile genetic elements or phage genomes. In aquaculture, metagenomics can be used to evaluate the impact of AR in fish farming. The release of antibiotics from aquaculture is a major issue as it can transmit and accumulate in the niches and remains in the sediments for prolonged period (Niegowska et al., 2021). In a study, the abundance of ARGs was detected in Chinese mitten crab (*Eriocheir sinensis*) aquaculture environment through metagenomics. Bacterial phyla such as *Proteobacteria, Actinobacteria,* and *Bacteroidetes* were plentiful in water samples, whereas in sediment samples the dominance of *Proteobacteria, Verrucomicrobia, Bacteroidetes,* and *Chloroflexi* were detected. ARGs of Bacitracin and multidrug resistance were profound in the water and sediments, respectively. The authors concluded that opportunistic pathogens may gain ARGs or genetic elements through HGT, which may cause potential health hazards to humans (Fang et al., 2019).

The use of antibiotics as a prophylactic measure in aquaculture is a common practice in order to prevent and treat fish diseases, leading to evolution of ARGs and ARB and their entrance into the food chain. Prophylactic usage of antibiotic sulfamethoprim (25% sulfadiazine; 5% trimethoprim) in aquaculture pond of Silver Carp increased the presence of resistance genes. qPCR studies have detected the abundance of ARGs such as *sul1, sul2, dfrA1, tetA,* bla_{TEM} and class 1 integron, *int1*. Shotgun metagenomics was applied to infer the resistance profile, which showed abundance of sulfonamide and tetracycline ARGs. The prophylactic use of sulfonamides triggers the progression of resistance in aquaculture and should be constrained (Patil et al., 2020). Further, some researchers have aimed to eradicate antibiotics (e.g., norfloxacin, oxytetracycline, rifampicin) in mariculture wastewater by using moving bed biofilm reactors (MBBRs). After treatment, the plethora of ARGs and ARB increases in the MBBR due to the selection pressure of antibiotics and thereby alters the diversity, structure, etc., of the biofilm (Li et al., 2017).

Humans and animals are closely associated with rivers and streams. Anthropogenic intervention and urbanization leads to the discharge of untreated wastewater effluents into waterbodies. This affects the health of receiving waterbodies, their composition, microbial structure, fish, and finally humans. With wastewater treatment, most of the bacteria including coliforms can be cleared from the effluents but the removal of persistent pathogenic bacteria and ARGs is still an alarming concern. In a study, researchers investigated the bacterial community structure and ARGs in a WWTP and the effluent receiving water. Using pyrosequencing, the microbial structure of both WWTP and receiving water body was investigated and illumina high-throughput sequencing was employed for the identification of ARGs. Where WWTP removed most of the pathogens from the wastewater, but the richness of ARGs and ARB was slightly higher in downstream samples than the upstream ones (Tang et al., 2016).

Recently, Tang et al. studied the distribution of ARB and ARGs in various industrial wastewaters and receiving river systems. They found around 166 pathogens from 6 water samples, where *Salmonella enterica* and *P. aeruginosa* were the most prevalent pathogens followed by *Fusarium graminearum* and *Magnaporthe oryzae*. Bacitracin was found to be the abundant antibiotic succeeded by chloramphenicol, sulfonamide, tetracycline, and aminoglycoside. Effluents from a pharmaceutical company manufacturing Chinese medicine contained several ARGs including *bacA, mexW, mexF, sul1,* and *oprn*. The frequency of virulence factors in wastewater and their co-occurrence with resistant pathogens has potential microbiological risks (Tang et al., 2021). The distribution of antimicrobial resistance genes in several environments was studied extensively, but there are fewer studies related to salt lake environment. Liang et al. studied AR profiles, microbial community structure, and their associations between antibiotic resistomes and microbiota in four salt lakes through shotgun metagenomic studies. Around 175 ARGs as subtypes of 19 ARGs were identified that showed resistance towards multidrugs such as bacitracin and macrolide-lincosamide-streptogramin (MLS) with an abundance of 71.2%. Additionally, with increased concentration of salinity, the abundance of ARGs also declined in the salt lakes. Microbial genera including *Azoarcus*, *Kocuria, Aeromonas,* and *Pseudomonas* co-occurred with ARGs, implying that these organisms are probable hosts carrying ARGs into the salt lakes (Liang et al., 2021).

In drinking water systems, biofilm acts as a hotspot for the transmission of ARGs and resistance bacteria, a source of contamination. In biofilm, the non-pathogenic, antibiotic-resistant bacteria can transmit the ARGs to pathogens through HGT, causing health risks especially via drinking water consumption. In a study, the

TABLE 12.1
Metagenomic Studies on Different Aquatic Biofilms, the Abundance of ARGs and Microorganisms

Sampling site	Biofilm	ARGs	Microorganisms	References
River sediments	River biofilm	*Bla*FOX, *cph*A, *bla*PER, *bla*IMP, *tet, ant, erm*	Microorganisms belongs to orders such as Aeromonadales, Chromatiales,	(Kneis et al., 2022)
Mesocosmic wetland	Epiphytic biofilm	*mfd, gyrA, gyrB, parC,*	Phyla Proteobacteria, Actinobacteria and Planctomycetes	(Ohore et al., 2021)
Annular bioreactor with South Saskatchewan River water	Microcosm river biofilm	*aph6-Id, aph3"-Ib, tetC, tetG, sul1, sul2, int1, blaOXA-2, blaOXA-21, rrsA*	Proteobacteria – *Bradyrhizobium, Methylibium, Methylorubrum, Novosphingobium, Sphingobium, Variovax. Mycobacterium and Mycobacteroides*	Flores-Vargas et al., 2023
Sediment and water samples of *Eriocheir sinensis* freshwater aquaculture	-	*bacA, acrB, mexF, macB, smeE,*	*Aeromonas* spp., *Bacillus cereus, Klebsiella oxytoca, Mycobacterium fortuitum, Pseudomonas aeruginosa*	Fang et al., 2019
Seawater, metal surface, rocks	Marine biofilms	AcrB/AcrD/AcrF family encodes for efflux transporters such as enterobactin confers resistance	Acidobacteria, α-, β-, γ-, δ-, and ε-proteobacteria, cyanobacteria, Bacteroidetes	Zhang, W et al., 2019
Lake and sea water	-	*bacA*, bla_{TEM} (hub: $bla_{TEM\text{-}175}$)	-	Yang et al., 2019
River	Epilithic biofilm	*aac(6')-IIa, mecA, sul1*, bla_{VIM}, *acrA, dfrA27*, $bla_{CTX\text{-}M}$, *mefA*, bla_{TEM}, bla_{DHA}, *sulA, aph(3')-III, aph(3')-Ib, cat, catB3, ermC, ermY, ermX, aadB, arnA, vat(A), etc*	*Nitrospira, α-, β-, γ-* and *δ-proteobacteria, Bacteroidetes, Aeromonas, Clostridium, Pseudomonas* spp.	Matviichuk et al., 2022

(*continued*)

TABLE 12.1 (Continued)
Metagenomic Studies on Different Aquatic Biofilms, the Abundance of ARGs and Microorganisms

Sampling site	Biofilm	ARGs	Microorganisms	References
Drinking water distribution system (DWDS)	DWDS biofilm	*aph(3)-IIb, smeD,*	*Pseudomonas, Stenotrophomonas maltophilia*	Ke et al., 2023
Reclaimed water	Reclaimed water distribution system biofilm	*qnrA,* $bla_{TEM,}$ *vanA, sul1,*	Proteobacteria, Fibrobacteres, Synergistetes, α-proteobacteria, Actinobacteria, Cyanobacteria, Firmicutes	Garner et al., 2018
Aerobic biofilm reactors (tigecycline concentration)	Experimental aerobic biofilm	*tetA, tetC, tetG, tetX, bl2d_oxa1, ant3ia, ant6ia, catb3, cml_e1, sul2, aph6id*	*Flavobacterium*, *Cytophaga*, *Leadbetterella*, *Flexibacter*, *Chitinophaga*, *Enterobacteriaceae*, *Pseudomonadaceae*, *Sphingobacterium*	Tian et al., 2019
Small rivers	River biofilm	$bla_{CTX\text{-}M}$, *qnrS, sul1, ermB*	-	Proia et al., 2016

incidence of ARGs and pathogens in water and biofilm samples of a drinking water system was studied. qPCR studies showed decreases in ARG abundance after drinking water treatment process, but the relative concentrations were not significantly varied between the treated as well as untreated samples. The incidence of ARGs is higher in biofilms when compared to bulk and tap water samples. qPCR analysis suggested the abundance of opportunistic pathogens in biofilm than the water samples (Chen et al., 2020).

Su and co-workers confirmed the abundance of ARGs and virulence factors in South China Sea through metagenomic analysis. Overall, around 140 ARG subtypes and 155 virulence factors were identified, where the predominant ARG was multidrug resistance followed by bacitracin. The main virulence factor detected was flagella. They found a significant association between ARGs and virulence factors through a Pearson correlation study. Oceans serve as reservoirs of various ARGs and virulence factors, risking the safety of seafood and human health (Su et al., 2022). Some of the studies related to ARGs in aquatic biofilm are listed in Table 12.1.

12.8 CONCLUSION

Antibiotic resistance is a significant ecological concern that affects the health of animals and humans. The aquatic environment is a dumpsite for the discharge of wastes intentionally or unintentionally from different sources. Aquatic environment pollution is a worldwide concern due to the frequency of AR detected in water, sediments, and microorganisms, which threatens public health. Without prior treatment, wastewater from hospitals, municipal waste, household, agricultural, aquaculture, surface runoff, poultry, and other industrial effluents are often dumped into waterbodies directly. Several studies have reported the existence of ARGs and ARB in sediments, surface water, and biofilm communities. Due to progression of AR, microbial communities in biofilm have been altered in structure and function and lead to imbalance in aquatic ecosystems. Similarly, the presence of antibiotic resistomes have been reported in drinking water, crops, and seafoods, which threatens human health. Advancements in sequencing techniques and metagenomics are promising approaches for the detection of ARGs and their risk prediction in the environment. Constant monitoring of WWTP effluents should be required to reduce the transmission of ARGs in aquatic environments.

REFERENCES

Abe, K., N. Nomura, and S. Suzuki. 2020. Biofilms: Hot spots of horizontal gene transfer (HGT) in aquatic environments, with a focus on a new HGT mechanism. *FEMS Microbiol Ecol* 96(5): fiaa031.

Adenaya, A., M. Berger, T. Brinkhoff, M. Ribas-Ribas, and O. Wurl. 2023. Usage of antibiotics in aquaculture and the impact on coastal waters. *Mar Pollut Bull* 188:114645.

Aubertheau, E., T. Stalder, L. Mondamert, M. C. Ploy, C. Dagot, and J. Labanowski. 2017. Impact of wastewater treatment plant discharge on the contamination of river biofilms by pharmaceuticals and antibiotic resistance. *Sci Total Environ* 579: 1387–1398.

Balcázar, J. L., J. Subirats, and C. M. Borrego. 2015. The role of biofilms as environmental reservoirs of antibiotic resistance. *Front Microbiol* 6: 1216.

Besemer, K. 2015. Biodiversity, community structure and function of biofilms in stream ecosystems. *Res Microbiol* 166(10): 774–781.

Bhattacharyya, S., A. C. Eagar, J. Engohang-Ndong, and L. G. Leff. 2023. Antibiotic resistance gene abundance and bacterial community composition in macroinvertebrates of an urban stream. *Freshw Biol* 68(7): 1107–1121.

Bougnom, B. P., S. Thiele-Bruhn, V. Ricci, C. Zongo, and L. J. V. Piddock. 2020. Raw wastewater irrigation for urban agriculture in three African cities increases the abundance of transferable antibiotic resistance genes in soil, including those encoding extended spectrum β-lactamases (ESBLs). *Sci Total Environ* 698: 134201.

Burns, A. and D. S. Ryder. 2001. Potential for biofilms as biological indicators in Australian riverine systems. *Ecol Manag Restor* 2(1): 53–64.

Chao, Y., Y. Mao, Z. Wang, and T. Zhang. 2015. Diversity and functions of bacterial community in drinking water biofilms revealed by high-throughput sequencing. *Sci Rep* 5(1): 10044.

Chen, J., W. Li, J. Zhang, W. Qi, Y. Li, S. Chen, and W. Zhou. 2020. Prevalence of antibiotic resistance genes in drinking water and biofilms: The correlation with the microbial community and opportunistic pathogens. *Chemosphere* 259:127483.

Dang, C., Y. Xia, M. Zheng, T. Liu, W. Liu, Q. Chen, and J. Ni. 2020. Metagenomic insights into the profile of antibiotic resistomes in a large drinking water reservoir. *Environ Int* 136:105449.

Das, M. K., S. Das, and P. K. Srivastava. 2023. An overview on the prevalence and potential impact of antimicrobials and antimicrobial resistance in the aquatic environment of India. *Environ Monit Assess* 195(9):1015.

Davey, M. E. and G. A. O'toole. 2000. Microbial biofilms: From ecology to molecular genetics. *Microbiol Mol Biol Rev* 64(4):847–867.

de Araújo, L.C.A. and M. B. M. de Oliveira. 2020. Effect of heavy metals on the biofilm formed by microorganisms from impacted aquatic environments. In *Bacterial biofilms,* ed. S. Dincer and M. Sümengen19-32. London, UK: IntechOpen.

Fang, H., K. Huang., J. Yu., C. Ding., Z. Wang., C. Zhao., H. Yuan., Z. Wang., S. Wang., J. Hu., and Y. Cui. 2019. Metagenomic analysis of bacterial communities and antibiotic resistance genes in the Eriocheir sinensis freshwater aquaculture environment. *Chemosphere* 224:202–211.

Fernandes, G., M. C. Bastos, J. P. R. de Vargas, T. Le Guet, B. Clasen, and D. R. Dos Santos. 2020. The use of epilithic biofilms as bioaccumulators of pesticides and pharmaceuticals in aquatic environments. *Ecotoxicology* 29:1293–1305.

Flemming, H. C. and J. Wingender. 2010. The biofilm matrix. *Nat Rev Microbiol* 8:623–633.

Flemming, H. C., E. D. van Hullebusch, T. R. Neu, P. H. Neilsen, T. Seviour, P. Stoodley, J. Wingender, and S. Wuertz. 2023. The biofilm matrix: Multitasking in a shared space. *Nat Rev Microbiol* 21(2):70–86.

Flores-Vargas, G., D. R. Korber, and J. Bergsveinson. 2023. Sub-MIC antibiotics influence the microbiome, resistome and structure of riverine biofilm communities. *Front Microbiol* 14: 1194952.

Fu, J., Y. Zhang., S. Lin., W. Zhang., G. Shu., J. Lin., H. Li., F. Xu., H. Tang., G. Peng., L. Zhao. 2021. Strategies for interfering with bacterial early-stage biofilms. *Front Microbiol* 12:675843.

Gandra, S., J. Joshi., A. Trett., A. S. Lamkang, and R. Laxminarayan. 2017. *Scoping Report on Antimicrobial Resistance in India.* Washington: Center for Disease Dynamics, Economics & Policy.

Garner, E., C. Chen., K. Xia., J. Bowers., D. M. Engelthaler., J. McLain., M. A. Edwards, and A. Pruden. 2018. Metagenomic characterization of antibiotic resistance genes in full-scale reclaimed water distribution systems and corresponding potable systems. *Environ Sci Technol* 52(11):6113–6125.

Gebreyohannes, G., A. Nyerere., C. Bii., and D. B. Sbhatu. 2019. Challenges of intervention, treatment, and antibiotic resistance of biofilm-forming microorganisms. *Heliyon* 5(8):e02192.

Guo, X. P., Y. Yang., D. P. Lu., Z. S. Niu., J. N. Feng., Y. R. Chen., F. Y. Tou., E. Garner., J. Xu., M. Liu., and M. F. Hochella Jr. 2017. Biofilms as a sink for antibiotic resistance genes (ARGs) in the Yangtze Estuary, *Water Res* 129:277–286.

Hou, L., L. Zhang., F. Li., S. Huang., J. Yang., C. Ma., D. Zhang., C. P. Yu., and A. Hu. 2021. Urban ponds as hotspots of antibiotic resistome in the urban environment. *J Hazard Mater* 403:124008.

Huang, X., C. Liu., K. Li., F. Liu., D. Liao., L. Liu., G. Zhu., and J. Liao. 2013. Occurrence and distribution of veterinary antibiotics and tetracycline resistance genes in farmland soils around swine feedlots in Fujian Province, China. *Environ Sci Pollut Res* 20:9066–9074.

Jia, J., Y. Guan., M. Cheng., H. Chen., J. He., S. Wang., and Z. Wang. 2018. Occurrence and distribution of antibiotics and antibiotic resistance genes in Ba River, China. *Sci Total Environ* 642:1136–1144.

Jolivet-Gougeon, A. and Bonnaure-Mallet, M. 2014. Biofilms as a mechanism of bacterial resistance. *Drug Discov Today Technol* 11:49–56.

Ke, Y., W. Sun., X. Chen., Y. Zhu., X. Guo., W. Yan, and S. Xie. 2023. Seasonality determines the variations of biofilm microbiome and antibiotic resistome in a pilot-scale chlorinated drinking water distribution system deciphered by metagenome assembly. *Environ Sci Technol* 57(31):11430–11441.

Kneis, D., T. U. Berendonk., S. K. Forslund, and S. Hess. 2022. Antibiotic resistance genes in river biofilms: A metagenomic approach toward the identification of sources and candidate hosts. *Environ Sci Technol* 56(21):14913–14922.

Li, S., S. Zhang., C. Ye., W. Lin., M. Zhang., L. Chen., J. Li. and X. Yu. 2017. Biofilm processes in treating mariculture wastewater may be a reservoir of antibiotic resistance genes. *Mar Pollut Bull* 118(1–2):289–296.

Li, X., J. Xie., C. Ding., H. Du., S. Gao., W. Ma., F. Liang., H. Zhang. and A. Wang. 2023. Occurrence, fate and potential health risks of antibiotic resistomes in a constructed wetlands-reservoir ecosystem for drinking water source improvement. *Sci Total Environ* 901:166055.

Liang, H., F. Wang., R. Mu., J. Huang., R. Zhao., X. Li., K. Yu, and B. Li. 2021. Metagenomics analysis revealing the occurrence of antibiotic resistome in salt lakes. *Sci Total Environ* 790:148262.

Lin, Z., T. Yuan., L. Zhou., S. Cheng., X. Qu., P. Lu, and Q. Feng. 2021. Impact factors of the accumulation, migration and spread of antibiotic resistance in the environment. *Environ Geochem Health* 43:1741–1758.

Liu, Y. J., Z. H. Li., Y. T. He., L. Yuan, and G. P. Sheng. 2023. Antibiotic resistomes in face-mask biofilm along an urban river: Multiple drivers and co-occurrence with human opportunistic pathogens. *J Hazard Mater* 455:131587.

Lorenzo, P., A. Adriana., S. Jessica., B. Carles., F. Marinella., L. Marta., B. J. Luis, and S. Pierre. 2018. Antibiotic resistance in urban and hospital wastewaters and their impact on a receiving freshwater ecosystem. *Chemosphere* 206:70–82.

Lundström, S. V., M. Östman., J. Bengtsson-Palme., C. Rutgersson., M. Thoudal., T. Sircar., H. Blanck., K. M. Eriksson., M. Tysklind., C. F. Flach, and D. J. Larsson. 2016. Minimal selective concentrations of tetracycline in complex aquatic bacterial biofilms. *Sci Total Environ* 553:587–595.

Mao, D., Y. Luo., J. Mathieu., Q. Wang., L. Feng., Q. Mu., C. Feng, and P. J. J. Alvarez. 2014. Persistence of extracellular DNA in river sediment facilitates antibiotic resistance gene propagation. *Environ Sci Technol* 48(1):71–78.

Matviichuk, O., L. Mondamert., C. Geffroy., M. Gaschet., C. Dagot, and J. Labanowski. 2022. River biofilms microbiome and resistome responses to wastewater treatment plant effluents containing antibiotics. *Front Microbiol* 13:795206.

Mishra, S., Y. Huang., J. Li., X. Wu., Z. Zhou., Q. Lei., P. Bhatt, and S. Chen. 2022. Biofilm-mediated bioremediation is a powerful tool for the removal of environmental pollutants. *Chemosphere* 294:133609.

Monahan, C., Nag, R., Morris, D. and Cummins, E. 2021. Antibiotic residues in the aquatic environment–current perspective and risk considerations. *J Environ Sci Health*, Part A, 56:733–751.

Neela, F. A., L. Nonaka., M. H. Rahman, and S. Suzuki. 2009. Transfer of the chromosomally encoded tetracycline resistance gene *tet (M)* from marine bacteria to *Escherichia coli* and *Enterococcus faecalis*. *World J Microbiol Biotechnol* 25:1095–1101.

Niegowska, M., I. Sanseverino., A. Navarro, and T. Lettieri. 2021. Knowledge gaps in the assessment of antimicrobial resistance in surface waters. *FEMS Microbiol Ecol* 97(11): fiab140.

Nnadozie, C. F. and O. N. Odume. 2019. Freshwater environments as reservoirs of antibiotic resistant bacteria and their role in the dissemination of antibiotic resistance genes. *Environ Pollut* 254:113067.

Odeyemi, O. A. and A. Ahmad. 2017. Population dynamics, antibiotics resistance and biofilm formation of *Aeromonas* and *Vibrio* species isolated from aquatic sources in Northern Malaysia. *Microb Pathog* 103:178–185.

Ohore, O. E., S. Zhang., S. Guo., F. G. Addo., B. Manirakiza, and W. Zhang. 2021. Ciprofloxacin increased abundance of antibiotic resistance genes and shaped microbial community in epiphytic biofilm on Vallisneria spiralis in mesocosmic wetland. *Biores Technol* 323:124574.

Ondarza, P. M., S. P. Haddad., E. Avigliano., K. S. Miglioranza, and B. W. Brooks. 2019. Pharmaceuticals, illicit drugs and their metabolites in fish from Argentina: implications for protected areas influenced by urbanization. *Sci total Environ* 649:1029–1037.

Patil, H. J., J. Gatica., A. Zolti., A. Benet-Perelberg., A. Naor., B. Dror., A. Al Ashhab., S. Marman., N. A. Hasan., R. R. Colwell, and D. Sher. 2020. Temporal resistome and microbial community dynamics in an intensive aquaculture facility with prophylactic antimicrobial treatment. *Microorganisms* 8(12):1984.

Penesyan, A., I. T. Paulsen., M. R. Gillings., S. Kjelleberg, and M. J. Manefield. 2020. Secondary effects of antibiotics on microbial biofilms. *Front Microbiol* 11:2109.

Proia, L., D. von Schiller., A. Sànchez-Melsió., S. Sabater., C. M. Borrego., S. Rodríguez-Mozaz, and J. L. Balcázar. 2016. Occurrence and persistence of antibiotic resistance genes in river biofilms after wastewater inputs in small rivers. *Environ Pollut* 210:121–128.

Rather, M. A., K. Gupta, and M. Mandal. 2021. Microbial biofilm: formation, architecture, antibiotic resistance, and control strategies. *Braz J Microbiol* 52:1701–1718.

Reddy, B. and S. K. Dubey. 2019. River Ganges water as reservoir of microbes with antibiotic and metal ion resistance genes: High throughput metagenomic approach. *Environ Pollut* 246:443–451.

Reddy, S., K. Kaur., P. Barathe., V. Shriram., M. Govarthanan, and V. Kumar. 2022. Antimicrobial resistance in urban river ecosystems. *Microbiol Res* 263:127135.

Ricart, M., H. Guasch., M. Alberch., D. Barceló., C. Bonnineau., A. Geiszinger., J. Ferrer., F. Ricciardi., A. M. Romaní., S. Morin, and L. Proia. 2010. Triclosan persistence through wastewater treatment plants and its potential toxic effects on river biofilms. *Aquat Toxicol* 100(4):346–353.

Rivera, S. F., V. Vasselon., N. Mary., O. Monnier., F. Rimet, and A. Bouchez. 2021. Exploring the capacity of aquatic biofilms to act as environmental DNA samplers: Test on macroinvertebrate communities in rivers. *Sci Total Environ* 763:144208.

Sentenac, H., A. Loyau., J. Leflaive, and D. S. Schmeller. 2022. The significance of biofilms to human, animal, plant and ecosystem health. *Funct Ecol* 36(2):294–313.

Su, H., X. Hu., W. Xu., Y. Xu., G. Wen, and Y. Cao. 2022. Diversity, abundances and distribution of antibiotic resistance genes and virulence factors in the South China Sea revealed by metagenomic sequencing. *Sci Total Environ* 814:152803.

Tang, J., Y. Bu., X. X. Zhang., K. Huang., X. He., L. Ye., Z. Shan, and H. Ren. 2016. Metagenomic analysis of bacterial community composition and antibiotic resistance genes in a wastewater treatment plant and its receiving surface water. *Ecotoxicol Environ Saf* 132:260–269.

Tang, Y., Z. Liang., G. Li., H. Zhao, and T. An. 2021. Metagenomic profiles and health risks of pathogens and antibiotic resistance genes in various industrial wastewaters and the associated receiving surface water. *Chemosphere* 283:131224.

Tian, Z., R. Liu., H. Zhang., M. Yang, and Y. Zhang. 2019. Developmental dynamics of antibiotic resistome in aerobic biofilm microbiota treating wastewater under stepwise increasing tigecycline concentrations. *Environ Int* 131:105008.

Toyofuku, M., T. Inaba., T. Kiyokawa., N. Obana., Y. Yawata, and N. Nomura. 2016. Environmental factors that shape biofilm formation. *Biosci Biotechnol Biochem* 80(1):7–12.

Vishwakarma, V. 2020. Impact of environmental biofilms: Industrial components and its remediation. *J Basic Microbiol* 60(3):198–206.

Wang, M., A. Masoudi., C. Wang., C. Wu., Z. Zhang., X. Zhao., Y. Liu., Z. Yu, and J. Liu. 2024. Impacts of net cages on pollutant accumulation and its consequence on antibiotic resistance genes (ARGs) dissemination in freshwater ecosystems: Insights for sustainable urban water management. *Environ Int* 183:108357.

Yang, Y., C. Xu., X. Cao., H. Lin, and J. Wang. 2017. Antibiotic resistance genes in surface water of eutrophic urban lakes are related to heavy metals, antibiotics, lake morphology and anthropic impact. *Ecotoxicol* 26:831–840.

Yang, Y., Z. Li., W. Song., L. Du., C. Ye., B. Zhao., W. Liu., D. Deng., Y. Pan., H. Lin, and X. Cao. 2019. Metagenomic insights into the abundance and composition of resistance genes in aquatic environments: Influence of stratification and geography. *Environ Int* 127:371–380.

Zarei-Baygi, A. and A. L. Smith. 2021. Intracellular versus extracellular antibiotic resistance genes in the environment: Prevalence, horizontal transfer, and mitigation strategies. *Biores technol* 319:124181.

Zeballos-Gross, D., Z. Rojas-Sereno., M. Salgado-Caxito., P. Poeta., C. Torres, and J. A. Benavides. 2021. The role of gulls as reservoirs of antibiotic resistance in aquatic environments: A scoping review. *Front Microbiol* 12:703886.

Zhang, Q., Z. Zhang., T. Lu., W. J. G. M. Peijnenburg., M. Gillings., X. Yang., J. Chen., J. Penuelas., Y. G. Zhu., N. Y. Zhou, and J. Su. 2020. Cyanobacterial blooms contribute to the diversity of antibiotic-resistance genes in aquatic ecosystems. *Commun Biol* 3(1):737.

Zhang, W., W. Ding., Y. X. Li., C. Tam., S. Bougouffa., R. Wang., B. Pei., H. Chiang., P. Leung., Y. Lu, and J. Sun. 2019. Marine biofilms constitute a bank of hidden microbial diversity and functional potential. *Nature Commun* 10(1):517.

Zhang, X., Z. Chen., Y. Ma., N. Zhang., Q. Pang., X. Xie., Y. Li, and J. Jia. 2019. Response of Anammox biofilm to antibiotics in trace concentration: Microbial activity, diversity and antibiotic resistance genes. *J Hazard mater* 367:182–187.

Zhang, Y., G. Shen., S. Hu., Y. He., P. Li, and B. Zhang. 2021. Deciphering of antibiotic resistance genes (ARGs) and potential abiotic indicators for the emergence of ARGs in an interconnected lake-river-reservoir system. *J Hazard Mater* 410:124552.

13 Role of Biofilms in the Remediation of Aquatic Pollution

Ganga G.

13.1 INTRODUCTION

Biofilms play a significant role in the remediation of aquatic pollution, as they are a common and effective mechanism for managing and degrading contaminants in aquatic environments. A biofilm is a structured community of microorganisms encased in a self-produced matrix of extracellular polymeric substance (EPS), which can be composed of polysaccharides, proteins, and DNA. These biofilms can form on various surfaces, including rocks, sediments, and submerged structures in aquatic systems (Dar and Bhat, 2020). Mishra et al. (2022) reported that because of its high capacity to absorb, immobilize, or degrade toxins, biomass, and their broad adaptability, biofilm-mediated bioremediation is a promising method for the removal of environmental pollutants (Figure 13.1).

Biofilms contribute to the remediation of aquatic pollution in the following ways (Sonawane et al., 2022):

Biodegradation of Pollutants: Biofilms are composed of diverse microbial populations, including bacteria, fungi, and algae, which are capable of degrading a wide range of organic and inorganic pollutants. These microorganisms can break down contaminants into less harmful substances through metabolic processes, including aerobic and anaerobic degradation (Doukani et al., 2023).

Nutrient Uptake: Biofilms are highly efficient in taking up and retaining nutrients from the surrounding water. By sequestering nutrients, they can help mitigate nutrient pollution in aquatic ecosystems, such as excess nitrogen and phosphorus that lead to eutrophication (Lisle John, 2020).

Metal Removal and Immobilization: Some biofilm-forming microorganisms are capable of adsorbing and immobilizing heavy metals through chemical reactions or ion exchange. This can be crucial in reducing the toxicity of metal pollutants in aquatic systems (Jasu and Ray, 2021).

Sediment Stabilization: Biofilms help stabilize sediments by binding particles and preventing erosion. This is particularly important in situations where sediment resuspension can release previously buried pollutants back into the water column (Kendall Valentine and Giulio Mariotti, 2020).

DOI: 10.1201/9781003487203-13

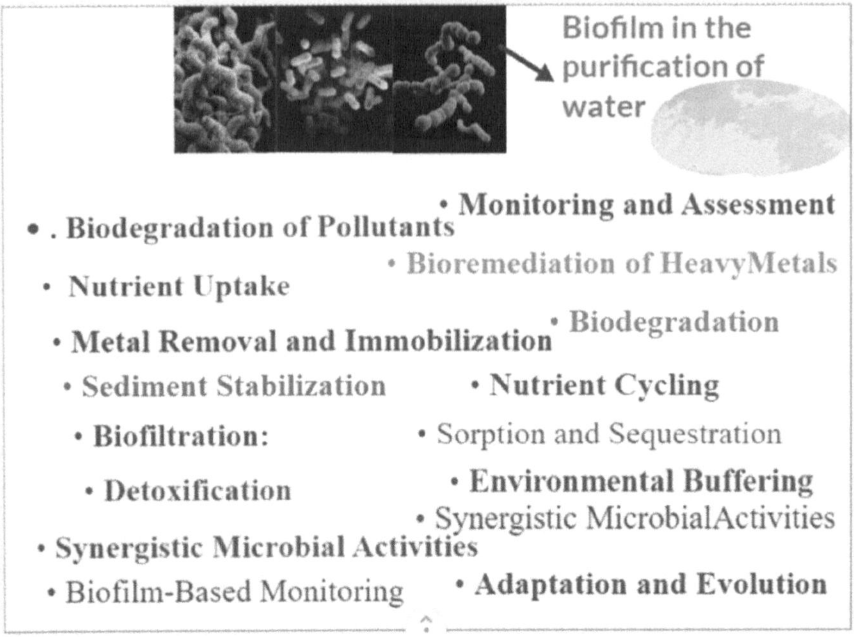

FIGURE 13.1 Roles of biofilm in purification of water.

Biofiltration: Biofilms can be used in constructed wetlands, biofilters, and other engineered systems to treat contaminated water. They act as a natural filter, providing a habitat for pollutant-degrading microorganisms to thrive and remove contaminants from the water (Cercado et al., 2012).

Detoxification: In some cases, biofilms can facilitate the detoxification of certain pollutants. Microorganisms in biofilms can transform toxic compounds into less harmful forms, reducing the impact on aquatic life (Md. Manjurul Haque et al., 2022).

Monitoring and Assessment: The presence and composition of biofilms can be used as bioindicators for water quality assessment. Changes in biofilm communities can indicate shifts in the health of the aquatic ecosystem and the presence of pollutants (Wilson et al., 2017).

Biodegradation: Biofilms are composed of diverse microbial communities with specialized metabolic capabilities. They excel in breaking down organic pollutants, such as hydrocarbons, pesticides, and industrial chemicals, through enzymatic reactions, turning them into less harmful or non-toxic compounds. These microorganisms create a detoxifying environment for polluted areas (Morohoshi et al., 2018).

Bioremediation of Heavy Metals: Some microorganisms within biofilms have the capacity to immobilize heavy metals by adsorbing them to EPS or through the formation of less soluble metal compounds. This minimizes the mobility and bioavailability of toxic metals, reducing their impact on ecosystems (Tayang and Songachan, 2021).

Nutrient Cycling: Biofilms are essential for nutrient cycling in aquatic ecosystems. They uptake, transform, and recycle essential nutrients like nitrogen and phosphorus. This not only reduces excess nutrient levels in the water but also maintains ecological balance (Paul et al., 1991).

Sorption and Sequestration: The EPS matrix in biofilms provides functional groups that adsorb and sequester pollutants, such as organic compounds and heavy metals. The biofilm acts as a physical barrier, reducing the mobility and exposure of contaminants to the surrounding environment (Hao Cheng et al., 2022).

Environmental Buffering: Biofilms protect microorganisms from harsh environmental conditions, including UV radiation, wave action, and desiccation. The EPS matrix shields these microorganisms and supports their persistence in challenging environments (Sharma, 2022).

Synergistic Microbial Activities: Within biofilms, different microbial species communicate, participating in complementary metabolic activities. Some microbes may break down complex pollutants into simpler forms, while others can further metabolize these degradation products (Saraiva et al., 2021).

Biofilm-Based Monitoring: The composition and structure of biofilms are sensitive indicators of changing water quality and pollutant levels. Monitoring shifts in the biofilm community can serve as an early warning system for pollution events (Kumar et al., 2023).

Adaptation and Evolution: Biofilm communities are capable of adapting and evolving over time in response to changes in pollution levels. This adaptability allows them to maintain their pollutant-degrading capabilities even as the composition of pollutants changes (Parrilli et al., 2022). It is essential to note that the effectiveness of biofilms in remediating aquatic pollution depends on various factors, including the specific pollutants present, environmental conditions, and the composition of the biofilm community. Furthermore, researchers continue to study and develop biofilm-based approaches for more efficient and targeted pollutant remediation in aquatic ecosystems. Biofilms are composed of a complex matrix of EPS and a diverse community of microorganisms. This matrix, along with the microbial inhabitants, plays a critical role in the biodegradation of pollutants within biofilms.

13.2 COMPONENTS OF BIOFILMS IN BIODEGRADATION

The key components and their functions in the biodegradation of pollutants in biofilms are as follows (Morohoshi et al., 2018):

Microorganisms: The microorganisms within biofilms are the primary agents responsible for the biodegradation of pollutants. They include bacteria, fungi, algae, and protozoa. Different microorganisms have varying metabolic capabilities and can break down a wide range of organic and inorganic pollutants. For example, some bacteria are well-known for their ability to degrade hydrocarbons, while others can metabolize organic matter, nutrients, or toxic substances.

Extracellular Polymeric Substance (EPS): The EPS makes up the matrix that holds the biofilm together. This matrix consists of various components, including polysaccharides, such as exopolysaccharides, which serve as a structural framework and provide protection for the microbial cells. They also play a role in adhesion to surfaces and in the retention of water and nutrients. The EPS also contains proteins that can serve as enzymes and transporters. Enzymes are crucial for catalyzing the biodegradation reactions of pollutants. For example, enzymes like dehydrogenases, oxygenases, and hydrolases are involved in the breakdown of various organic contaminants. eDNA is involved in the structural stability of the biofilm and acts as a binding agent. It also plays a role in horizontal gene transfer, enabling the sharing of genetic material and metabolic capabilities among microorganisms in the biofilm.

Water Channels and microenvironment: Biofilms contain channels and pores within the EPS matrix that facilitate the movement of water and nutrients. These channels create microenvironments where different microorganisms may thrive, allowing for synergistic interactions among the community members. In these microenvironments, metabolic processes can occur, promoting the degradation of pollutants. Quorum

Sensing and signaling molecules: Some microorganisms within biofilms use quorum sensing mechanisms and signaling molecules to communicate with each other. This enables coordinated responses to changes in the environment and the efficient allocation of metabolic resources for pollutant degradation. Nutrient Availability: Biofilms can accumulate and retain nutrients from the surrounding water, providing a continuous source of essential elements required for microbial metabolism. This ensures that microorganisms have the resources they need to carry out biodegradation processes. Redox Gradient: Biofilms often exhibit complex redox gradients, creating zones with varying oxygen concentrations. These gradients can be important for the biodegradation of contaminants that require specific conditions, such as aerobic or anaerobic metabolism.

The combination of microorganisms with diverse metabolic capabilities and the protective, structurally supportive EPS matrix makes biofilms highly effective in biodegrading a wide range of pollutants. The interactions and cooperation among the microorganisms in biofilms allow them to target and break down pollutants more efficiently than individual free-floating microorganisms in the water column.

13.3 NUTRIENT UPTAKE IN BIOFILMS

Nutrient uptake in biofilms is a complex process that involves the interaction of microorganisms with their surrounding environment. Biofilms are communities of microorganisms, such as bacteria, fungi, and algae, that adhere to surfaces and are encased in a matrix of EPS. The EPS matrix is essential for biofilm formation and provides a protective environment for the microorganisms within it. Nutrient uptake in biofilms can be influenced by various factors, including the type of microorganisms present, the availability of nutrients, the physical and chemical properties of the environment, and the structure of the biofilm itself. Here are some key points to consider regarding nutrient uptake in biofilms:

(i) Nutrient Gradients: Biofilms often develop nutrient gradients, where the availability of nutrients varies across the biofilm. Nutrients are typically more abundant near the outer layers of the biofilm, where they can diffuse from the surrounding environment, and become progressively scarcer toward the interior of the biofilm.
(ii) Mass Transfer Limitations: Nutrient uptake in the deeper layers of a biofilm can be limited by mass transfer. As biofilms grow and thicken, it becomes more challenging for nutrients to diffuse through the EPS matrix to reach the microorganisms deeper within the biofilm. This can lead to nutrient limitation for these cells.
(iii) Microbial Diversity: Biofilms often consist of a diverse community of microorganisms with varying metabolic capabilities. Different species within the biofilm may compete for or cooperate in utilizing different nutrients. Some microorganisms may produce exoenzymes to break down complex organic compounds in the EPS matrix or surrounding environment to release nutrients for the community.
(iv) EPS Role: The EPS matrix in biofilms plays a crucial role in nutrient uptake. It can serve as a reservoir for nutrients, protecting them from diffusion or washout. The EPS can also act as a substrate for some microorganisms to metabolize. Additionally, the EPS can help retain water, facilitating the transport of dissolved nutrients.
(v) Nutrient Preferences: Different microorganisms within a biofilm may have preferences for specific nutrients. Some may be facultative anaerobes, relying on the availability of oxygen, while others may thrive in anoxic regions of the biofilm. Nutrient preferences can further shape the microbial community structure within the biofilm.
(vi) Dynamic Conditions: Nutrient availability can change over time due to factors like environmental fluctuations, nutrient input, and metabolic activities of the microorganisms. Biofilms can adapt to these changing conditions by adjusting their structure and metabolic activities.

In summary, nutrient uptake in biofilms is a dynamic and complex process influenced by a combination of physical, chemical, and biological factors. Understanding the nutrient dynamics within biofilms is crucial for various applications, including wastewater treatment, bioremediation, and medical context, as biofilms can have both positive and negative effects on human and environmental health.

13.4 METAL REMOVAL AND IMMOBILIZATION PROCESS BY BIOFILM

Biofilms can play a significant role in the removal and immobilization of metals in various environmental and industrial settings. Microorganisms within biofilms possess unique metabolic capabilities that allow them to interact with metal ions and facilitate their removal and immobilization. The processes involved in metal removal and immobilization by biofilms include the following (Yonghui et al., 2022):

(a) Adsorption and Ion Exchange: Biofilms contain EPS that can adsorb metal ions through physical and chemical interactions. The EPS matrix, which includes proteins, polysaccharides, and DNA, provides binding sites for metal ions. Metal cations can be attracted to negatively charged functional groups on EPS components, allowing for adsorption.
(b) Biological Reduction: Some microorganisms within biofilms, such as certain types of bacteria, are capable of reducing metal ions to less toxic or less soluble forms. This is often seen with metals like iron and manganese. For example, in the presence of suitable electron donors, iron-reducing bacteria can convert soluble ferric iron (Fe^{3+}) into less soluble ferrous iron (Fe^{2+}).
(c) Biomineralization: Biofilms can facilitate the precipitation of metal ions as metal sulfides or metal oxides through biomineralization. Certain bacteria, like sulfate-reducing bacteria, can produce hydrogen sulfide (H_2S) as a metabolic byproduct, which can react with metal ions to form insoluble metal sulfides. This process can effectively remove metals from the aqueous phase.
(d) Metal Accumulation: Some biofilm-forming microorganisms have the ability to accumulate and concentrate metals within their cells. This process is known as metal bioaccumulation or metal sequestration. Once metals are accumulated by these microorganisms, they become less available in the surrounding environment.
(e) Microbial Consortia: Biofilms often consist of diverse microbial consortia with complementary metabolic capabilities. Different microorganisms can work together to detoxify and immobilize metals. For instance, one species may produce an organic ligand that binds to metal ions, making them less bioavailable, while another may reduce the metal ions to less toxic forms.
(f) Sorption to Biofilm Matrix: The EPS matrix of biofilms can act as a sorbent for metal ions. The EPS can trap metal ions, reducing their mobility and making them less available for release into the environment.
(g) Redox Cycling: Some bacteria in biofilms can engage in redox cycling, where they undergo repeated oxidation and reduction reactions with metals. This can lead to the transformation of metals and their immobilization in less soluble forms.
(h) Biofilm Structural Changes: Biofilms can adapt their structure in response to changes in metal concentrations. For example, increased metal concentrations can lead to a denser biofilm matrix, reducing metal diffusion and enhancing metal retention. The effectiveness of metal removal and immobilization by biofilms depends on the specific biofilm community, the types and concentrations of metals, and the environmental conditions. Biofilms can be used in various applications, including bioremediation of metal-contaminated sites, wastewater treatment, and in biotechnological processes for recovering valuable metals from industrial effluents.

13.5 SEDIMENT STABILIZATION BY BIOFILM

Biofilm formation can play a crucial role in the stabilization of sediments in aquatic environments. Sediment stabilization by biofilms helps prevent erosion, maintain water quality, and provide a habitat for a variety of organisms. Biofilms contribute to sediment stabilization in the following ways (Chen et al., 2017; Gerbersdorf et al., 2020):

(1) EPS Production: Biofilms produce EPS, a gel-like matrix of polysaccharides, proteins, and other biomolecules. The EPS acts as a cohesive agent that binds sediment particles together, creating a stable structure. It also increases the sediment's resistance to hydraulic forces, such as water currents or waves.
(2) Biofilm Adhesion: Microorganisms within the biofilm adhere to sediment particles and form a network of cells. The attachment of microbial cells to sediment grains enhances sediment stability by increasing its resistance to detachment.
(3) Biostabilization: Microbial activity in the biofilm can physically trap sediment particles, preventing their transport by water flow. Bacterial filaments and extracellular structures extend into the sediment, effectively anchoring the sediments in place. Secretion of Mucilage: Some biofilm-forming microorganisms secrete mucilage, a slimy substance that aids in the adhesion of sediment particles. Mucilage can also act as a binding agent, contributing to sediment stability.
(4) Macroorganism Colonization: Biofilms provide a substrate for the attachment and growth of macroorganisms such as algae, aquatic plants, and invertebrates. These organisms, in turn, contribute to sediment stabilization by their root systems or by creating physical structures that hold sediment in place.
(5) Reduction of Water Flow: Biofilms slow down water flow near the sediment surface, reducing erosive forces. This decreased flow velocity results from the friction created by the biofilm matrix and the biofilm's influence on flow patterns.
(5) Nutrient Uptake: Microorganisms within the biofilm can take up and store excess nutrients like phosphorus and nitrogen. By reducing nutrient concentrations in the water column, biofilms can indirectly affect sediment stabilization. Elevated nutrient levels can lead to increased algal growth on the sediment surface, potentially destabilizing sediments.
(6) Microbial Activities: Microbial metabolic processes can chemically bind sediment particles together. For instance, the precipitation of metal oxides or sulfides can cement sediment particles and improve stability.
(7) Diversity of Microbial Communities: Different microbial species within biofilms contribute to sediment stabilization through their varied metabolic activities, nutrient uptake, and physical interactions with sediment particles. Sediment stabilization by biofilms is essential for maintaining the ecological health of aquatic ecosystems. It helps to prevent sediment erosion, which can lead to increased turbidity, reduced light penetration, and altered nutrient cycling. Furthermore, it provides habitats for benthic organisms and supports the overall biodiversity of aquatic systems. In situations where sediment destabilization is a concern, promoting the growth and health of biofilms can be an effective strategy for sediment control and water quality management.

13.6 BIOFILTRATION BY BIOFILMS

Biofiltration by biofilms is a wastewater and air treatment process that relies on the activities of microorganisms growing in a biofilm matrix to remove contaminants from the water or air. Biofiltration is an eco-friendly and cost-effective method for treating a wide range of organic and inorganic pollutants. Biofilms are essential in this process because they provide a suitable environment for diverse microbial communities to thrive and carry out the desired biodegradation and removal processes. Here's how biofiltration by biofilms works.

13.6.1 Water Biofiltration

(i) Biofilm Formation: In water biofiltration, a support medium (often made of materials like activated carbon, peat, or sand) is used to create a surface for biofilm formation. This medium provides a substrate for the attachment and growth of microorganisms.
(ii) Contaminant Uptake: Wastewater containing organic and sometimes inorganic contaminants is passed through the biofilter. The biofilm that forms on the support medium contains a wide variety of microorganisms, including bacteria, fungi, and protozoa, each with specific metabolic capabilities. These microorganisms break down and consume the contaminants as a food source, removing them from the water.
(iii) Biodegradation: The microorganisms within the biofilm metabolize the contaminants through various biochemical processes, converting them into simpler, less harmful substances such as carbon dioxide, water, and biomass. This biodegradation is often enhanced by the presence of oxygen, which is provided through aeration or oxygen diffusion.
(iv) Maintenance: To ensure efficient biofiltration, the biofilm requires maintenance. Regular monitoring and control of environmental parameters, such as temperature, pH, and nutrient levels, can optimize the microbial activity within the biofilm.

13.6.2 Air Biofiltration

Air biofiltration operates on a similar principle but is designed to treat volatile organic compounds (VOCs) and odorous gases in the air. The process involves the following steps:

(i) Biofilm Formation: A support medium, often composed of organic materials like compost, wood chips, or synthetic materials, provides a substrate for the biofilm to grow. Microorganisms form a biofilm on the surface of this medium.
(ii) Airflow through the Biofilter: Contaminated air is forced through the biofilter, allowing the biofilm to come into contact with the VOCs or odorous compounds.
(iii) Biodegradation: The microorganisms within the biofilm metabolize the VOCs and odorous compounds, converting them into less harmful substances, such

as carbon dioxide, water, and microbial biomass. This process may be aerobic or anaerobic, depending on the specific microorganisms involved and the availability of oxygen.

(iv) Maintenance: Regular maintenance and monitoring of environmental conditions are critical for optimizing the biofiltration process. Parameters like temperature, humidity, and nutrient levels are adjusted to support the microbial community.

Biofiltration by biofilms is a sustainable and environmentally friendly approach to wastewater and air treatment. It offers several advantages, including high treatment efficiency, low operational costs, and minimal production of secondary waste. This technology is widely used in wastewater treatment plants, industrial processes, and air pollution control to remove a wide range of contaminants while minimizing the environmental impact.

13.7 DETOXIFICATION PROCESS BY BIOFILMS

Feng et al. (2022) reported that biofilms can play a significant role in the detoxification of various pollutants in different environmental and industrial settings (Feng et al., 2022). Microorganisms within biofilms have the ability to biodegrade, immobilize, or otherwise mitigate the toxic effects of contaminants. Here's an overview of the detoxification processes by biofilms:

(i) Biodegradation: Biofilms often contain diverse microbial communities with various metabolic capabilities. Some microorganisms within biofilms can degrade organic pollutants, such as hydrocarbons, pesticides, and industrial chemicals. These microorganisms can produce enzymes that break down complex organic compounds into simpler, less toxic substances. For example, hydrocarbon-degrading bacteria can break down oil and other hydrocarbons, reducing their toxicity.

(ii) Metabolic Transformation: Microorganisms in biofilms can transform toxic compounds into less harmful or non-toxic forms through metabolic pathways. This may involve the addition of functional groups, such as hydroxyl or carboxyl groups, to chemical structures, making them more amenable to biodegradation.

(iii) Sorption and Sequestration: The EPS produced by biofilms can adsorb and immobilize contaminants. These substances can act as a physical barrier, reducing the mobility of pollutants and preventing them from interacting with the surrounding environment or organisms.

(iv) Chemical Precipitation: Some microorganisms in biofilms can induce the precipitation of metal ions as metal sulfides or metal oxides, which are often less soluble and, therefore, less toxic. This can be particularly relevant in detoxifying heavy metals from contaminated sites.

(v) Ion Exchange: Biofilms can facilitate the exchange of ions between the contaminated environment and the microbial community. This can help to reduce the concentration of toxic ions in the environment while allowing the biofilm community to take up less harmful ions.

(vi) Detoxification of Organic Pollutants: Certain microorganisms can enzymatically transform or detoxify organic pollutants. For example, enzymes like cytochrome P450 can catalyze the degradation of aromatic compounds and other organic toxins.

(vii) Microbial Consortia: Biofilms often consist of complex microbial consortia in which different species collaborate in the detoxification process. This cooperative action allows for more efficient removal of toxic compounds from the environment.

(viii) Adaptation to Toxicants: Microbial communities within biofilms can adapt to the presence of toxicants over time. This can include developing resistance mechanisms, changes in gene expression, or metabolic shifts that enable them to cope with and detoxify the contaminants.

(ix) Environmental Buffering: The biofilm matrix can provide a buffering effect, reducing the exposure of microbial cells to toxicants. This can protect the microorganisms within the biofilm from direct contact with harmful compounds.

(x) Reduced Environmental Impact: Biofilm-based detoxification processes often produce fewer harmful byproducts and waste, making them environmentally friendly compared to some conventional treatment methods. Biofilm-mediated detoxification processes are valuable in applications such as bioremediation of contaminated sites, wastewater treatment, and the treatment of industrial effluents. They offer a natural and sustainable approach to reducing the presence of toxic pollutants in the environment while minimizing the impact on ecosystems and human health.

Monitoring and assessing water quality using biofilms can provide valuable insights into the health and condition of aquatic ecosystems. Biofilms are bioindicators that respond to changes in water quality by altering their composition and structure.

Biofilms can be used for monitoring and assessing water quality as follows:

(i) Biofilm Composition and Diversity: The composition and diversity of microorganisms within a biofilm are influenced by water quality parameters such as temperature, pH, nutrient levels, and the presence of pollutants. A shift in the microbial community structure within a biofilm can serve as an early indicator of changes in water quality.

(ii) Metabolic Activity: Microorganisms within biofilms exhibit specific metabolic activities that can be linked to water quality parameters. For example, the rate of oxygen consumption or production within a biofilm can provide insights into the level of organic pollution or the availability of oxygen in the water.

(iii) Nutrient Cycling: Biofilms play a key role in nutrient cycling in aquatic ecosystems. Monitoring the rates of nutrient uptake, transformation, and release by biofilms can provide information about nutrient availability and cycling in the water.

(iv) Bioaccumulation and Biomagnification: Some biofilm-dwelling microorganisms can accumulate pollutants or heavy metals. By analyzing the metal content of biofilms, it is possible to assess the extent of pollution in the water and understand the potential for biomagnification in the food web.

(v) Biofilm Thickness and Structure: Changes in the physical characteristics of biofilms, such as their thickness, density, and structural integrity, can indicate alterations in water quality. Disruption of biofilm structure may result from changes in flow, sedimentation, or the presence of pollutants.

(vi) Biofilm Response to Contaminants: Biofilms respond to the presence of specific contaminants by altering their gene expression and metabolic activities. This responsiveness can be used to detect the presence of contaminants and assess their impact on the ecosystem.

(vii) Biofilm Growth Rate: The growth rate of biofilms can be indicative of water quality conditions. Faster biofilm growth may be associated with nutrient-rich waters, while slower growth could be linked to nutrient limitation.

(viii) Biofilm Biomass: Monitoring the biomass of biofilms can provide insights into the overall health of an aquatic ecosystem. Changes in biofilm biomass may be linked to variations in nutrient availability and the presence of pollutants.

(ix) Genetic and Metagenomic Analysis: Modern molecular techniques, such as DNA sequencing, can be used to analyze the genetic diversity of microorganisms within biofilms. This can provide detailed information about the identity and potential functions of the microorganisms.

(x) Biofilm Bioassays: Biofilm-based bioassays involve exposing biofilms to water samples and assessing changes in their health and vitality. The response of biofilms to different water quality parameters can be quantified and used to evaluate water quality.

(xi) Biofilm-Based Sensors: Biofilm-based sensors can be deployed in aquatic environments to continuously monitor water quality. These sensors can be designed to measure parameters such as pH, dissolved oxygen, and specific pollutants based on the response of the biofilm.

13.8 MONITORING AND ASSESSMENT OF WATER QUALITY BY BIOFILM

Douterelo et al. (2016) reported that biofilm monitoring and assessment can be valuable tools for water quality management, environmental protection, and early detection of pollution incidents. By studying the responses of biofilms to changing water quality conditions, researchers and environmental professionals can gain insights into the health and resilience of aquatic ecosystems. Bioremediation by biofilms is a natural and sustainable approach to cleaning up contaminated environments using microbial communities embedded within biofilms. Biofilms are aggregates of microorganisms encased in a matrix of EPS. They have several advantages for bioremediation due to their ability to adapt to different environmental conditions, protect microorganisms from toxic compounds, and efficiently degrade or immobilize contaminants.

Bioremediation by biofilms works as follows:

(i) Contaminant Uptake: Biofilms can form on various surfaces, including those contaminated with pollutants. Microorganisms within biofilms attach to these surfaces and begin to colonize. The biofilm matrix acts as a protective shield for microorganisms against harsh environmental conditions and toxic compounds.
(ii) Metabolic Diversity: Biofilms often contain diverse microbial communities with different metabolic capabilities. Some microorganisms are specialized in degrading specific contaminants, while others contribute to the overall degradation process through various metabolic pathways.
(iii) Biodegradation: Microorganisms within biofilms metabolize contaminants, breaking them down into less toxic or non-toxic forms. This process can involve the production of enzymes that catalyze the breakdown of complex organic compounds, such as hydrocarbons, chlorinated solvents, or pesticides.
(iv) Immobilization: In addition to biodegradation, biofilms can immobilize contaminants by adsorption or precipitation. The EPS matrix can capture and sequester heavy metals or other pollutants, reducing their mobility and bioavailability.
(v) Redox Reactions: Biofilms can facilitate redox reactions that change the chemical state of contaminants. For example, in anaerobic biofilms, some microorganisms can reduce toxic metals, making them less soluble and less harmful.
(vi) Biofilm Consortia: Microbial communities within biofilms often consist of consortia of microorganisms with complementary roles. These consortia can work together to efficiently detoxify and degrade contaminants.
(v) Sorption and Sequestration: The EPS matrix of biofilms contains functional groups that can bind to contaminants. This sorption process can effectively reduce the concentration of pollutants in the surrounding environment.
(vi) Nutrient Management: Biofilms can efficiently manage and recycle nutrients, which are often essential for the growth of contaminant-degrading microorganisms. This nutrient recycling can enhance the bioremediation process. Long-Term Remediation Biofilms can persist in contaminated environments over an extended period, allowing for long-term remediation efforts. As long as suitable conditions are maintained, biofilms can continue to contribute to the detoxification of contaminants. (vii) Monitoring and Control: Bioremediation by biofilms can be monitored and controlled by adjusting environmental parameters such as temperature, pH, and nutrient availability. This allows for optimization of the bioremediation process.

13.9 BIOREMEDIATION BY BIOFILM

Mishra et al. (2022) observed that bioremediation by biofilms is used in various contexts, including the cleanup of contaminated soil, groundwater, and sediments, as well as the treatment of industrial wastewater and the management of oil spills.

It offers a sustainable and environmentally friendly solution to mitigate the impact of pollution and restore contaminated environments. The success of bioremediation by biofilms depends on factors such as the type of contaminants, environmental conditions, and the specific microbial communities involved.

Biofilms can play a significant role in the bioremediation of oil spills. Bioremediation is a process that uses microorganisms to break down and degrade pollutants, such as hydrocarbons found in crude oil. Biofilms, composed of diverse microbial communities, provide an ideal environment for the growth of oil-degrading microorganisms and can enhance the efficiency of oil spill cleanup.

Biofilms contribute to the bioremediation of oil spills as follows:

(i) Biofilm Formation: When oil spills occur in aquatic environments, microorganisms capable of degrading hydrocarbons quickly colonize the oil-contaminated areas. These microorganisms form biofilms on the oil-water interface, solid surfaces, or on floating debris.

(ii) Diverse Microbial Communities: Biofilms can contain a diverse array of hydrocarbon-degrading microorganisms, including hydrocarbonoclastic bacteria and fungi. These microorganisms utilize hydrocarbons as a carbon and energy source for their growth and metabolism.

(iii) Synergistic Activities: Within biofilms, different microbial species can collaborate and engage in complementary metabolic activities. Some bacteria may specialize in breaking down complex hydrocarbons into simpler compounds, while others may metabolize the breakdown products further.

(iv) Extracellular Enzymes: Microorganisms within biofilms produce extracellular enzymes that help solubilize and degrade hydrocarbons. These enzymes can act on the hydrophobic nature of oil, making it more accessible for microbial degradation.

(v) Nutrient Recycling: Biofilm communities can recycle and retain essential nutrients such as nitrogen and phosphorus. This nutrient recycling helps support the growth of oil-degrading microorganisms within the biofilm.

(vi) Reduced Mass Transfer Limitations: Biofilms improve mass transfer of hydrocarbons by providing a concentrated and localized population of oil-degrading microorganisms in close proximity to the oil. This reduces mass transfer limitations and enhances the rate of oil degradation.

(vii) Protection from Environmental Factors: Biofilms offer protection to oil-degrading microorganisms from harsh environmental conditions, such as UV radiation, wave action, and desiccation. The EPS matrix of biofilms acts as a physical barrier, shielding the microorganisms.

(viii) Enhanced Biodegradation Rates: Biofilms can significantly enhance the rates of biodegradation of oil compared to free-floating microorganisms in the water column. This efficiency is due to the concentrated and synergistic activities of the microbial community within the biofilm.

(ix) Long-Term Remediation: Biofilms have the potential for long-term remediation. Once established, they can persist on the oil-contaminated surfaces, gradually degrading the oil over an extended period.

(x) Adaptation and Evolution: Biofilm communities can adapt and evolve over time in response to changes in the composition and concentration of oil in the environment. This adaptability allows biofilms to maintain their oil-degrading capabilities even as the oil composition changes. Bioremediation of oil spills using biofilms is a natural and sustainable approach to mitigating the environmental impact of oil contamination. It can be applied in a variety of settings, from marine environments affected by large oil spills to industrial contexts with localized oil contamination. Properly managed bioremediation efforts using biofilms can significantly reduce the ecological damage caused by oil spills.

13.10 CONCLUSION

In conclusion, biofilms represent a pivotal frontier in the fight against water pollution. Their innate ability to absorb, detoxify, and biodegrade a variety of contaminants makes them a promising tool for sustainable water purification. As research continues to uncover their complex interactions and potential applications, harnessing biofilms effectively could lead to significant advancements in environmental remediation strategies, paving the way towards cleaner and safer water resources for future generations.

REFERENCES

Amrita, J., Rani Ray, R. Biofilm mediated strategies to mitigate heavy metal pollution: A critical review in metal bioremediation. *Biocatalysis Agri Biotechnol.* 2021;37:102183, ISSN 1878-8181. https://doi.org/10.1016/j.bcab.2021.102183.

Cercado, B., Auria, R., Cardenas, B., Revah, S. Characterization of artificially dried biofilms for air biofiltration studies. *J Environ Sci Health A Tox Hazard Subst Environ Eng.* 2012;47(7):940–948. https://doi.org/10.1080/10934529.2012.667292. PMID: 22486663.

Chen X. D., Zhang C. K., Zhou Z., Gong Z., Zhou J. J., Tao J. F., Paterson D. M., Feng Q. Stabilizing effects of bacterial biofilms: EPS penetration and redistribution of bed stability down the sediment profile. *J Geophys Res: Biogeosci,* 2017;122(12):3113–3125 https://doi.org/10.1002/2017JG004050

Dar, S. A., Bhat, R. A. (2020). Aquatic Pollution Stress and Role of Biofilms as Environment Cleanup Technology. In: Qadri, H., Bhat, R., Mehmood, M., Dar, G. (eds) *Fresh Water Pollution Dynamics and Remediation.* Springer, Singapore. https://doi.org/10.1007/978-981-13-8277-2_16

Doukani, K., Boukirat, D., Boumezrag, A., Bouhenni, H., Bounouira, Y. (2023). Biodegradation of Pollutants. In: Ali, G.A.M., Makhlouf, A.S.H. (eds) *Handbook of Biodegradable Materials*. Springer, Cham. https://doi.org/10.1007/978-3-031-09710-2_10

Douterelo, I., Jackson, M., Solomon, C., et al. Microbial analysis of in situ biofilm formation in drinking water distribution systems: Implications for monitoring and control of drinking water quality. *Appl Microbiol Biotechnol.* 2016;100:3301–3311. https://doi.org/10.1007/s00253-015-7155-3

Feng, H., Xu, L., Chen, R., Ma, X., Qiao, H., Zhao, N., Ding, Y., Wu, D. Detoxification mechanisms of electroactive microorganisms under toxicity stress: A review. *Front Microbiol.* 2022 Nov 29;13:1084530. https://doi.org/10.3389/fmicb.2022.1084530. PMID: 36523836; PMCID: PMC9744814.

Hao, C., Zhimao, M., Youshao, W., Liu, D., Sun, Y. Role of extracellular polymeric substances in metal sequestration during mangrove restoration. *Chemosphere.* 2022;306:135550, ISSN 0045-6535. https://doi.org/10.1016/j.chemosphere.2022.135550.

Haque, M. M., Haque, M. A., Mosharaf, M. K., Islam, M. S., Md. Mynul, I., Hasan, M., Molla, H. A., Haque, M. A. Biofilm-mediated decolorization, degradation and detoxification of synthetic effluent by novel biofilm-producing bacteria isolated from textile dyeing effluent. *Environ Pollut.* 2022;314:120237, ISSN 0269-7491. https://doi.org/10.1016/j.envpol.2022.12023

Kumar, S., Nguyen, A. T., Goswami, S., Ferracane, J., Koley, D. Real-time monitoring of biofilm formation using a noninvasive impedance-based method. *Sensors and Actuators B: Chemical.* 2023;376(Part A):133034, ISSN 0925-4005. https://doi.org/10.1016/j.snb.2022.133034.

Lisle, J. T. nutrient removal and uptake by native planktonic and biofilm bacterial communities in an anaerobic aquifer. *Front Microbiol.* 2020;11, ISSN=1664-302X. https://doi.org/10.3389/fmicb.2020.01765

Mishra, S., Huang, Y., Li, J., Wu, X., Zhou, Z., Lei, Q., Bhatt, P., Chen, S. Biofilm-mediated bioremediation is a powerful tool for the removal of environmental pollutants. *Chemosphere.* 2022 May;294:133609. https://doi.org/10.1016/j.chemosphere..133609. Epub 2022 Jan 17. PMID: 35051518.

Morohoshi, T., Oi, T., Aiso, H., Suzuki, T., Okura, T., Sato, S. Biofilm formation and degradation of commercially available biodegradable plastic films by bacterial consortiums in freshwater environments. *Microbes Environ.* 2018 29;33(3):332–335. https://doi.org/10.1264/jsme2.ME18033. Epub 2018 Aug 28. PMID: 30158390; PMCID:

Parrilli, E., Tutino, M. L., Marino, G. Biofilm as an adaptation strategy to extreme conditions. *Rend Fis Acc Lincei.* 2022;33:527–536. https://doi.org/10.1007/s12210-022-01083-8

Paul, B. J., Duthie, H. C., Taylor, W. D. Nutrient cycling by biofilms in running waters of differing nutrient status. *J N Am Benthol Soc.* 1991;10(1):31–41. https://doi.org/10.2307/1467762

Sabine U. G., Koca K., de Beer, D., Chennu, A., Christian, N., Risse-Buhl, U., Weitere, M., Eiff, O., Wagner, M., Aberle, J., Schweikert, M., Terheiden, K. Exploring flow-biofilm-sediment interactions: Assessment of current status and future challenges. *Water Res.* 2020;185:116182, ISSN 0043-1354. https://doi.org/10.1016/j.watres.2020.116182.

Saraiva, J. P., Worrich, A., Karakoç, C., Kallies, R., Chatzinotas, A., Centler, F., Nunes da Rocha, U. Mining synergistic microbial interactions: A roadmap on how to integrate multi-omics data. *Microorganisms.* 2021 Apr 14;9(4):840. https://doi.org/10.3390/microorganisms9040840. PMID: 33920040; PMCID: PMC8070991.

Sharma, P. Role and significance of biofilm-forming microbes in phytoremediation -A review. *Environ Technol Innovat.* 2022;25:102182, ISSN 2352–1864. https://doi.org/10.1016/j.eti.2021.102182.

Sonawane, J. M., Rai, A. K., Sharma, M., Tripathi, M., Prasad, R. Microbial biofilms: Recent advances and progress in environmental bioremediation. *Sci Total Environ.* 2022;824:153843. doi: 10.1016/j.scitotenv.2022.153843. Epub 2022 Feb 14. PMID: 35176385.

Tayang A. and L. S. Songachan. Microbial bioremediation of heavy metals. *Curr Sci.* 2021;120(6):25

Valentine, K., Mariotti, G. Does eutrophication affect the ability of biofilms to stabilize muddy sediments? Estuarine. *Coastal Shelf Sci.* 2020;232:106490, ISSN 0272-7714, https://doi.org/10.1016/j.ecss.2019.106490.

Xing, Y., Tan, S., Liu, S., Xu, S., Wan, W., Huang, Q., Chen, W. Effective immobilization of heavy metals via reactive barrier by rhizosphere bacteria and their biofilms. *Environ Res*. 2022;207:112080, ISSN 0013-9351. https://doi.org/10.1016/j.envres.2021.112080.

14 Marine Organisms and Epibiotic Biofilms

Santhilatha Pandrangi, Kaveti Sahithi, Kerodawala Zainab, and Suseela Lanka

14.1 INTRODUCTION

Biofilms are microbial colonies or communities attached to a substratum and enveloped by a self-produced extracellular polymeric substance (EPS) (Percival et al., 2011; Lazar and Chifiriuc, 2010; Donlan, 2002; Jain et al., 2007). Biofilm-associated cells are sessile (Bhinu, 2005) and are manifested across diverse habitats, ranging from aquatic ecosystems and living tissues to medical devices (Donlan and Costerton, 2002) and water piping systems (Percival et al., 2011; Donlan and Costerton, 2002; Flemming, 2002).

Flourishing in both natural and man-made settings, these resilient structures often conceal harmful byproducts and toxins within their matrix. Whether benign or pathogenic, biofilms thrive in aqueous environments, clinging to moist surfaces, plant roots, and various animal hosts. Some biofilms serve beneficial roles in rivers or wastewater treatment, while others pose a threat, fostering infections in humans and animals. The ubiquity of biofilms in nature underscores their dual nature, acting as symbiotic communities or triggering infections in the intricate tapestry of living organisms (Percival et al., 2011).

Biofilm communities exhibit a fascinating diversity, often comprising a myriad of microorganisms ranging from bacteria and fungi to algae, yeasts, protozoa, and more. Nature orchestrates these intricate assemblies, incorporating non-living debris and corrosion products into the mix. A remarkable example of this microbial symphony is found in dental plaque biofilms (Marsh, 2004), where over 500 bacterial species coexist (Percival et al., 2011).

The size of a biofilm is a dynamic spectrum. On one end, these communities can be incredibly subtle, escaping the notice of the naked eye, forming just a few cell layers thick. For instance, the biofilms stealthily residing on our kitchen counter might elude visual detection, especially if diligent cleaning practices are in place. On the other end of the spectrum, biofilms can grow considerably thicker, expanding to several inches in certain environments (Percival et al., 2011). The scale of biofilm formations spans from the inconspicuous to the substantial, reflecting the adaptability and versatility as they live in complex microbial ecosystems (Costerton et al., 1995).

Tiny, free-floating bacteria (called Planktonic bacteria) encounter a submerged surface where in just a few minutes, these bacteria transform into builders, creating

DOI: 10.1201/9781003487203-14

a slimy layer called extracellular polymeric substance (EPS) while sticking to the surface. This slimy layer starts the formation of a biofilm community like a tiny, bustling city. It grows quickly, creating a complex structure in three dimensions by the influence of the environment (Percival et al., 2011). Surprisingly, these biofilm communities can pop up within a few hours. Biofilms have a unique way of spreading. They can break off into little groups or even release individual bacteria into their surroundings, a bit like spreading seeds. This helps them attach to new surfaces or join other biofilms downstream. So, in this microscopic world, the bacteria are not just hanging out; they are creating intricate communities as they go.

14.2 MARINE BIOFILMS

Within the vast expanse of the marine realm, a hidden world thrives in the form of ubiquitous marine biofilms. These intricate microbial communities act as agile responders to the ever-shifting rhythms of the marine environment, showcasing a breathtaking diversity in structures, functions, and metabolisms. Despite their significance, our understanding of the intricate microbial tapestry and the profound impact of natural marine biofilms on global geochemical cycles (Qian et al., 2022) remains somewhat limited.

Microbial cues, akin to the subtle language of the microscopic world, govern the interaction between microorganisms, their surroundings and fellow benthic organisms (Qian et al., 2022). This orchestration influences the succession and metamorphosis of these communities, revealing a captivating narrative of adaptation and interaction within the marine ecosystem.

Moreover, marine biofilms play a pivotal role as mediators of marine biofouling, a phenomenon in which accumulation of living matter takes place on different surfaces that significantly reverberates across various marine industries, shaping the challenges faced by ocean-related endeavours (Dürr & Thomason, 2010; Yebra et al., 2004). In unravelling the secrets of marine biofilms, we delve into a world where microbial communities hold the keys to understanding not only the delicate balance of life beneath the waves but also the broader implications on the intricate web of marine industries.

14.3 EPIBIOTIC BIOFILMS

Epibiotic biofilms form intricate communities of microorganisms on surfaces in aquatic environments. Comprising bacteria, algae, and other microorganisms, they adhere to each other and the substrate, influencing nutrient cycling, microbial interactions, and the overall health of marine ecosystems. Understanding epibiotic biofilms is crucial due to their significant ecological impacts. They contribute to fouling marine structures, influence the settlement of larvae on marine organisms, and participate in biogeochemical processes. Researchers from diverse fields like microbiology, ecology, and environmental science aim to unravel the composition, dynamics, and ecological functions of these biofilms. Epibionts, organisms living on the surface of another organism (Karatan and Watnick, 2009) (basibiont), form harmless interactions (Wahl et al., 2012; Wahl, 2008). Unlike parasitism, where one

TABLE 14.1
Overview of Biofilm Types and Characteristics in Diverse Environments

Biofilm type	Characteristics
General biofilms	Microbial colonies with self-produced EPS, found in diverse habitats
Composition	Diverse communities of bacteria, algae, fungi, etc. (e.g., dental plaque with 500+ bacterial species)
Size dynamics	Ranges from subtle to substantial structures with versatility
Interaction	Formation of biofilms is closely related to quorum sensing (QS) (Donlan, 2002) (cell-cell communication that allows bacteria for interaction between two) (Waters and Bassler, 2005) (Irie and Parsek, 2008; Steinberg et al., 2011) signals orchestrated by signalling molecules (Dobretsov et al., 2009), influencing lifecycle of marine organisms and interaction with marine ecosystem.
Marine & Epibiotic biofilms	Ubiquitous and play a major role in global biogeochemical cycles and maintaining host and marine ecosystem health (Armstrong et al., 2001)
Epibionts in marine life	Bacterial species, barnacles found on larger marine organisms, etc.

benefits at the others expense, epibiosis is neutralistic or commensalistic. Examples include bacteria, barnacles, remoras, and algae, often found on larger marine organisms like whales and sea turtles (Lescinsky, 2001). While there is no direct harm to the host, indirect effects impact surface qualities, influencing crucial ecological functions in marine organisms and aquatic ecosystems (Table 14.1).

14.4 EPIBIOTIC BIOFILMS AND MARINE ORGANISMS

In aquatic settings, the presence of biofilms on solid surfaces is widespread, especially on the outer surfaces of marine organisms. These biofilms, composed of bacteria, algae, and various microorganisms, serve as dynamic interfaces connecting the host with its surrounding environment. Inhabiting the host surfaces, these play a crucial role in influencing exchange of information, energy, and substances, thereby modulating both azoic and vital interactions of the host (Figure 14.1).

The impact of marine epibiotic biofilms on the ecology of their hosts can be substantial. They have the potential to limit the host's access to light, gases, and nutrients, consequently influencing interactions with other organisms such as foulers or pathogens. The effects of these biofilms on hosts can range from harmful to favorable, contingent upon the nature of the interaction and environmental conditions (Harder et al., 2012).

The outcomes of such biofilms may manifest in various ways, from alterations in color and contour to hindrances in mating and shifts in interactions among species. Ongoing research in this field underscores the importance of comprehending the reciprocal relationship between hosts and epibiotic biofilms, providing a more thorough ecological perspective.

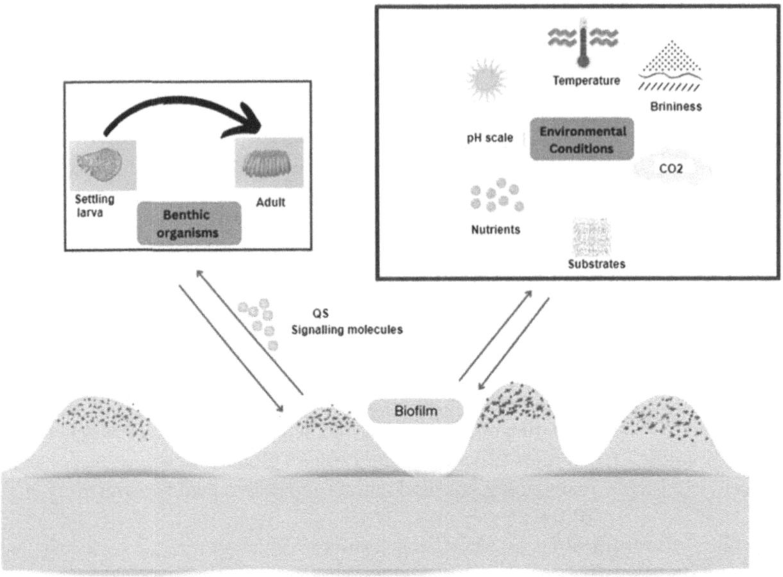

FIGURE 14.1 This figure depicts the relationship between Benthic organisms and biofilms which is mutually beneficial. Signaling molecules released by biofilms play a major role in their interactions serving as chemical communication, helps in Benthos growth and activities while benthos influence the structure of biofilms. The relationship between marine environment and biofilms is multifaceted. Biofilms contribute to nutrient cycling and biogeochemical processes in marine ecosystems and in return changes in marine environment influence stability of biofilms.

14.5 COMPOSITION OF MARINE EPIBIOTIC BIOFILMS

Marine epibiotic biofilms are complex communities formed on surfaces submerged in the aquatic environment. Comprising mainly bacteria and diatoms, these biofilms play a crucial role in the ecology of marine ecosystems and are particularly relevant in the context of developing environmentally benign antifouling (AF) coatings (Maria Salta et al., 2013).

A detailed examination of the composition of these biofilms reveals key components and their interactions as follows:

14.5.1 Bacterial Component

Diversity: Marine biofilms exhibit a diverse array of bacterial species, ranging from Gram-negative to Gram-positive (Pompilio et al., 2021), and encompassing both aerobic and anaerobic bacteria.

Functionality: Bacteria within these biofilms contribute to various ecological processes, including nutrient cycling, organic matter degradation, and nitrogen fixation.

14.5.2 Diatom Component

Abundance: Diatoms, unicellular photosynthetic algae with intricate silica cell walls, are a prominent component of marine biofilms. Their abundance is influenced by factors such as light availability and nutrient concentrations.

Role in Biofilm Structure: Diatoms contribute significantly to the biofilm structure and can create microenvironments that influence the settlement of other organisms.

14.5.3 Macrofouling Interactions

Settlement Induction: The underlying biofilm is found to induce larval and spore settlement of macrofoulers (Antunes et al., 2019; Dobretsov et al., 2006; Gölzhäuser and Wöll, 2010; Wahl, 1989). This process is critical in understanding the dynamics of biofilm interactions.

Species Specificity: The composition of the biofilm can influence the settlement preferences of macrofoulers, with variations observed based on exposure sites and biofilm composition.

14.5.4 Variability

Spatial and Temporal Dynamics: Studies indicate a variation in the species composition of marine biofilms is based on geographical locations and environmental changes. Understanding these dynamics is essential for predicting biofilm behavior (Thakur et al., 2004).

Natural Multispecies vs. Monospecific Species: The choice between a natural multispecies biofilm and a monospecific species biofilm can impact the outcomes of biofilm/macrofouling interactions, highlighting the importance of considering biofilm composition in research. Most of the biofilms in marine ecosystem are multispecies (Dobretsov, 2010).

14.5.5 Challenges and Considerations

Reproducibility: Challenges related to reproducibility in biofilm studies emphasize the need for standardized methodologies and experimental conditions.

Influence of Exposure Sites: Differences in exposure sites can lead to variations in biofilm composition and subsequent interactions with macrofoulers.

In summary, marine epibiotic biofilms are intricate ecosystems shaped by the interplay of bacteria, diatoms, and their interactions with macrofoulers. Understanding the composition, dynamics, and influencing factors of these biofilms is crucial for advancing research on environmentally friendly antifouling coatings and comprehending the complexities of marine ecological systems.

14.6 ECOLOGICAL CONSEQUENCES OF EPIBIOTIC FILMS ON MARINE ORGANISMS

Epibiotic biofilms on marine organisms exert profound ecological consequences by serving as dynamic interfaces between the host and its environment. These biofilms are omnipresent on solid surfaces and play a crucial role in shaping the ecological landscape of aquatic systems. They impact the fluxes of information, energy, and matter across the host outer surfaces, influencing key ecological processes (Harder, 2009; Costerton et al., 1987).

Epibiotic biofilms modulate the host's access to vital resources. Epibiotic biofilms can restrict or enhance the availability of light, gases, and nutrients, thereby shaping the ecological niche of the host organism. This, in turn, influences the organism interactions with its environment and other marine life. Furthermore, the biofilms have a significant impact on the interactions between the host and external entities such as foulers, consumers, and pathogens. By altering the properties of the host outer surfaces, epibiotic biofilms can either promote or hinder the attachment of other organisms, thus influencing community dynamics and species relationships. The ecological impact of epibiotic biofilms is not uniform; it varies based on different conditions. In some instances, these biofilms may confer benefits to the host, while in others, they may pose challenges or even detrimentally affect the organism. In essence, the presence of epibiotic biofilms on marine organisms contributes to the intricate web of ecological relationships in aquatic ecosystems. Understanding these consequences is essential for a comprehensive grasp of the functioning and resilience of marine communities in the face of environmental variability.

14.7 HOST DEFENSE TOWARDS EPIBIOTIC BIOFILMS

The defense mechanisms of marine organisms against epibiotic biofilms involve a robust innate immune response. When confronted with infection, the host initiates immediate defense mechanisms using immune cells, receptors, and humoral factors. Key components such as mannose binding lectins, collectins, and antibodies play a pivotal role by binding to invading microbes, facilitating recognition, and promoting phagocytosis by innate immune cells like polymorphonuclear neutrophils (PMNs) (Alhede et al., 2014), macrophages, and dendritic cells (Figure 14.2).

Crucial in recognizing pathogen-associated molecular structures, Toll-like receptors (TLRs) initiate signaling pathways that prompt phagocytosis (Kotsougiani et al., 2010) and the secretion of antimicrobial products. Dendritic cells (DCs) act as a bridge between innate and adaptive immunity (Banchereau et al., 2000), presenting microbial fragments to T cells, leading to long-lasting protective immunity. Specific factors, like lactoferrin and the host defense peptide LL-37 (Britigan et al., 1993), actively inhibit the formation of epibiotic biofilms (Overhage et al., 2008; Hell et al., 2010). Nevertheless, bacteria can counteract these defenses by secreting proteases capable of degrading these antimicrobial factors.

Polymorphonuclear neutrophils (PMNs) (Gunther et al., 2009) play a vital role in countering biofilms (Guenther et al., 2009), causing an increased oxidative burst,

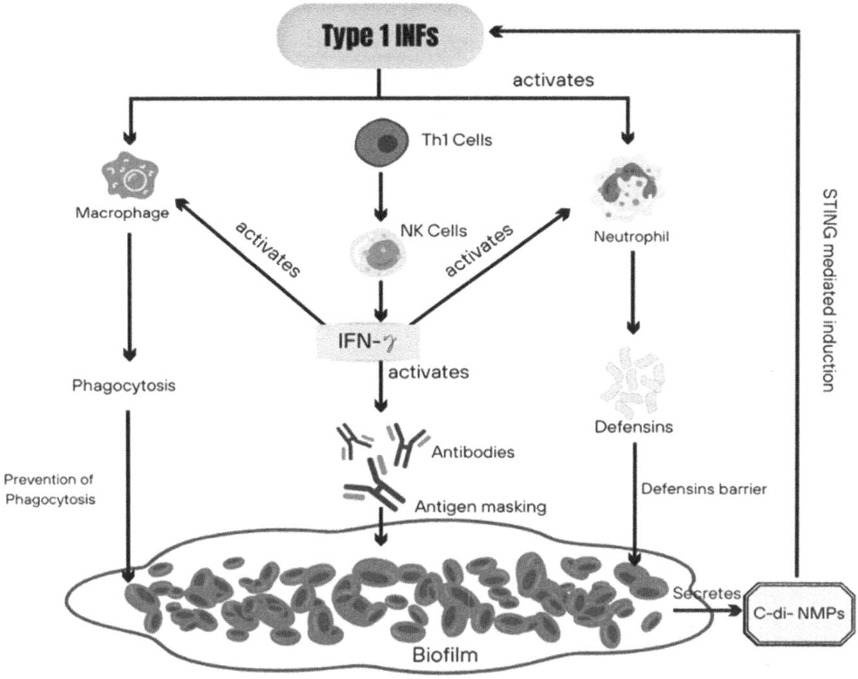

FIGURE 14.2 Immune cells such as macrophages and neutrophils recognize and engulf microbes associated with biofilms. Antibodies produced by the host recognize specific antigens on the surface of biofilm by a tagging mechanism, which promotes disruption of biofilm adhesion.

enhanced glucose uptake, and a high L-lactate concentration in chronic infections. In the case of *S. aureus* (a major pathogen responsible for diseases) (Fournier and Philpott, 2005) biofilms, PMNs contribute to biofilm mass reduction through phagocytosis, degranulation of lactoferrin and DNA release (Meyle et al., 2010; Meyle et al., 2012). The efficacy of this defense varies with biofilm maturation.

Interestingly, the attachment of PMNs to *S. epidermidis* biofilms and subsequent phagocytosis appears opsonization-independent, which suggests the involvement of biofilm-specific PMN signaling molecules (Gunther et al., 2009). In summary, host defenses against epibiotic biofilms involve a coordinated innate immune response with various immune cells like neutrophils (Gertrud Maria Hänsch, 2012) and antimicrobial factors. The effectiveness of these defenses depends on factors such as the biofilm maturity, the specific microorganisms involved, and the interplay between the host immune system and biofilm-specific mechanisms.

Moving on to the *Cryptococcus neoformans* capsule (Martinez and Casadevall, 2005), it is made up of glucuronoxylomannan, a critical component in extracellular biofilm matrix. This polysaccharide is essential for the formation of biofilm, where microorganisms adhere to surfaces. Antibodies against *C. neoformans* provide protection in cryptococcosis (Martinez and Casadevall, 2005) by enhancing survival, promoting phagocytosis, aiding antigen presentation, activating complement (Brown,1991), and modulating immune protein expression.

In a study conducted by Martinez and Casadevall, monoclonal antibodies against glucuronoxylomannan were found to inhibit biofilm formation. This discovery suggests that such antibodies play a crucial role in preventing *C. neoformans* biofilm establishment, offering new insights into their protective functions (Martinez and Casadevall, 2005).

14.8 DISCUSSION

The content provides insights into marine epibiotic biofilms, emphasizing their diverse composition (extracellular matrices are as diverse as biofilms) (Branda et al., 2005), ecological impact, and host defense mechanisms. These biofilms, composed mainly of bacteria and diatoms, play a crucial role in marine ecosystems and are pertinent to the development of environmentally friendly antifouling coatings. Their ecological consequences vary, influencing resource availability and interactions within aquatic ecosystems. Marine organisms deploy robust innate immune responses to counter these biofilms, with factors like biofilm maturity affecting the effectiveness of these defenses. Additionally, it highlights the role of glucuronoxylomannan in *Cryptococcus neoformans* biofilm formation and the inhibitory effects of antibodies, providing potential avenues for protective research. This summary captures the key aspects of the complex dynamics involving marine epibiotic biofilms.

14.9 CONCLUSION

In conclusion, marine epibiotic biofilms, comprising diverse microorganisms like bacteria and diatoms, play a pivotal role in marine ecosystems. Understanding their composition is essential for developing eco-friendly antifouling coatings. These biofilms have significant ecological consequences, influencing resource availability and shaping interactions within aquatic ecosystems. Marine organisms deploy robust innate immune responses to counter biofilms, with defense effectiveness influenced by factors like biofilm maturity. The inhibitory effects of antibodies on *Cryptococcus neoformans* biofilm formation suggest potential avenues for protective research. Overall, grasping the dynamics of marine epibiotic biofilms is crucial for environmental sustainability and marine industry advancements.

REFERENCES

Alhede M, Bjarnsholt T, Givskov M, Alhede M. Pseudomonas aeruginosa biofilms: mechanisms of immune evasion. Adv Appl Microbiol. 2014;86:1–40. doi: 10.1016/B978-0-12-800262-9.00001-9. PMID: 24377853.

Antunes J, Leão P, Vasconcelos V. Marine biofilms: diversity of communities and of chemical cues. Environ Microbiol Rep. 2019 Jun;11(3):287–305. doi: 10.1111/1758-2229.12694. Epub 2018 Oct 15. PMID: 30246474.

Armstrong, E., Yan, L., Boyd, K.G. et al. The symbiotic role of marine microbes on living surfaces. Hydrobiologia 461, 37–40 (2001).

Banchereau J, Briere F, Caux C, Davoust J, Lebecque S, Liu YJ, Pulendran B, Palucka K. Immunobiology of dendritic cells. Annu Rev Immunol. 2000;18:767–811. doi: 10.1146/annurev.immunol.18.1.767. PMID: 10837075.

Bhinu VS. Insight into biofilm-associated microbial life. J Mol Microbiol Biotechnol. 2005;10(1):15–21. doi: 10.1159/000090344. PMID: 16491022.

Branda SS, Vik S, Friedman L, Kolter R. Biofilms: the matrix revisited. Trends Microbiol. 2005 Jan;13(1):20–6. doi: 10.1016/j.tim.2004.11.006. PMID: 15639628.

Britigan BE, Hayek MB, Doebbeling BN, Fick RB Jr. Transferrin and lactoferrin undergo proteolytic cleavage in the Pseudomonas aeruginosa-infected lungs of patients with cystic fibrosis. Infect Immun. 1993 Dec;61(12):5049–55. doi: 10.1128/iai.61.12.5049-5055.1993. PMID: 8225581; PMCID: PMC281282.

Brown EJ. 1991.Complement receptors and phagocytosis. *Curr Opinion Immunol* 3:76–82.

Costerton JW. 1999. Introduction to biofilm. *Int J Antimicrob Agents*. May;11(3-4):217–221; discussion 237-9. doi: 10.1016/s0924-8579(99)00018-7. PMID: 10394973.

Costerton JW, Cheng KJ, Geesey GG, Ladd TI, Nickel JC, Dasgupta M, Marrie TJ. 1987. Bacterial biofilms in nature and disease. *Annu Rev Microbiol* 41:435-464. doi: 10.1146/annurev.mi.41.100187.002251

Costerton JW, Lewandowski Z, Caldwell DE, Korber DR, Lappin-Scott HM. 1995. Microbial biofilms. *Annu Rev Microbiol* 49:711–745. doi: 10.1146/annurev.mi.49.100195.003431. PMID: 8561477.

Dobretsov S. 2010. Marine biofilms. in Dürr S, Thomason JT (eds.) *Biofouling, Chap*. 9. Wiley-Blackwell, Oxford.

Dobretsov S, Dahms HU, Qian PY. 2006. Inhibition of biofouling by marine microorganisms and their metabolites. *Biofouling* 22(1): 43–54. doi: 10.1080/08927010500504784.

Dobretsov S, Teplitski M, Paul V. 2009. Mini-review: quorum sensing in the marine environment and its relationship to biofouling. *Biofouling* 25(5):413-427. doi:10.1080/08927010902853516. PMID: 19306145.

Donlan RM, Costerton JW. 2002. Biofilms: survival mechanisms of clinically relevant microorganisms. *Clin Microbiol Rev* 15(2):167–193. doi: 10.1128/CMR.15.2.167-193.2002. PMID: 11932229; PMCID: PMC118068.

Donlan RM. 2002. Biofilms: microbial life on surfaces. *Emerg Infect Dis* 8(9):881-890. doi: 10.3201/eid0809.020063. PMID: 12194761; PMCID: PMC2732559.

Dürr S, Thomason JC. 2010. *Biofouling*. doi: 10.1002/9781444315462.

Flemming HC. 2002. Biofouling in water systems-cases, causes and countermeasures. *Appl Microbiol Biotechnol* 59(6):629–640.

Fournier B, Philpott DJ,". 2005."Recognition of Staphylococcus aureus by the innate immune system Rev 18(3):521–540.

Gölzhäuser A, Wöll C. 2010. Interfacial systems chemistry: out of the vacuum-through the liquid- into the cell. *ChemPhysChem*. 11(15):3201-3213. doi: 10.1002/cphc.201000488

Guenther F, Stroh P, Wagner C, Obst U, Hänch GM. 2009.Phagocytosis of staphylococci biofilms by polymorphonuclear neutrophils: S. aureus and S. epidermidis differ with regard to their susceptibility towards the host defense. *Int J Artif Org* 32(9):565-573.

Gunther F, Wabnitz GH, Stroh P, Prior B, Obst U, Samstag Y, Wagner C, Hansch GM. 2009. Host defense against Staphylococcus aureus biofilms infection: phagocytosis of biofilms by polymorphonuclear neutrophils (PMN). *Mol Immunol* 46:1805-1813.

Hänsch GM. 2012. Host defence against bacterial biofilms: Mission Impossible? *Int Scholar Res Notices* 2012:17, Article ID 853123. https://doi.org/10.5402/2012/853123

Harder T. 2009 Marine epibiosis: concepts, ecological consequences and host defence. In Flemming HC, Murthy PS, Venkatesan R, Cooksey K (eds.) *Marine and Industrial*

Biofouling. Springer Series on Biofilms, vol 4. Springer, Berlin, Heidelberg. https://doi.org/10.1007/978-3-540-69796-1_12.

Harder T, Campbell AH, Egan S, Steinberg PD. 2012. Chemical mediation of ternary interactions between marine holobionts and their environment as exemplified by the red alga Delisea pulchra. *J Chem Ecol* 38(5):442–450. doi: 10.1007/s10886-012-0119-5. Epub 2012 Apr 25. PMID: 22527059.

Hell E, Giske CG, Nelson A, Römling U, Marchini G. 2010. Human cathelicidin peptide LL37 inhibits both attachment capability and biofilm formation of Staphylococcus epidermidis. *Lett Appl Microbiol* 50(2):211–215.

Irie Y, Parsek MR. 2008. Quorum sensing and microbial biofilms. Curr Top Microbiol Immunol 322:67–84. doi: 10.1007/978-3-540-75418-3_4. PMID: 18453272.

Jain A, Gupta Y, Agrawal R, Khare P, Jain SK. 2007. Biofilms--a microbial life perspective: a critical review. *Crit Rev Ther Drug Carrier Syst* 24(5):393-443. doi: 10.1615/critrevtherdrugcarriersyst.v24.i5.10. PMID: 18197780.

Karatan E, Watnick P. 2009. Signals, regulatory networks, and materials that build and break bacterial biofilms. *Microbiol Mol Biol Rev* 73(2):310-347.

Kotsougiani D, Pioch M, Prior B, Heppert V, Hansch GM, Wagner C. 2010. Activation of T lymphocytes in response to persistent bacterial infection: induction of CD11b and of toll-like receptors on T cells. *Int J Inflam* 2010:10, Article ID 526740.

Lazăr V, Chifiriuc MC. 2010. Architecture and physiology of microbial biofilms. *Roum Arch Microbiol Immunol* 69(2):95-107. PMID: 21235137.

Lescinsky HL. 2001. Epibionts. in Briggs D, Crowther P (eds.). *Palaeobiology II.* Wiley-Blackwell. 460–63. doi:10.1002/9780470999295.ch112. ISBN 978-0-632-05149-6.

Marsh PD. 2004. Dental plaque as a microbial biofilm. *Caries Res* 38(3):204-211. doi: 10.1159/000077756. PMID: 15153690.

Martinez LR, Casadevall A. 2005. Specific antibody can prevent fungal biofilm formation and this effect correlates with protective efficacy. *Infect Immun* 73:6350–6362.

Meyle E, Stroh P, Günther F, Hoppy-Tichy T, Wagner C, Hänsch GM. 2010. Destruction of bacterial biofilms by polymorphonuclear neutrophils: relative contribution of phagocytosis, DNA release, and degranulation. *Int J Artif Organs*. Sep;33(9):608–20. doi: 10.1177/039139881003300906. PMID: 20890882.

Meyle E, Brenner-Weiss G, Obst U, Prior B, Hänsch GM. 2012. Immune defense against S. epidermidis biofilms: components of the extracellular polymeric substance activate distinct bactericidal mechanisms of phagocytic cells. *Int J Artif Organs*. Oct;35(10):700–12. doi: 10.5301/ijao.5000151. PMID: 23065886.

Overhage J, Campisano A, Bains M, Torfs ECW, Rehm BHA, Hancock REW. 2008. Human host defense peptide LL-37 prevents bacterial biofilm formation. *Infection and Immunity* 76 (9): 4176-4182.

Percival S.L, Malic S, Cruz H, Williams DW. 2011. Biofilms and veterinary medicine. In *Springer Series on Biofilms,* vol. 6. Springer-Verlag, Berlin Heidelberg. doi: 10.1007/978-3-642-21289-5_2

Pompilio A, Scribano D, Sarshar M, Di Bonaventura G, Palamara AT, Ambrosi C. 2021. Gram-negative bacteria holding together in a biofilm: the acinetobacter baumannii way. *Microorganisms* 9(7):1353. doi: 10.3390/microorganisms9071353. PMID: 34206680; PMCID: PMC8304980.

Qian PY, Cheng A, Wang R, Zhang R. 2022. Marine biofilms: diversity, interactions and biofouling. *Nat Rev Microbiol* 20(11):671–684. doi: 10.1038/s41579-022-00744-7. Epub 2022 May 25. PMID: 35614346.

Salta M, Wharton JA, Blache Y, Stokes KR, Briand JF. Marine biofilms on artificial surfaces: structure and dynamics. Environ Microbiol. 2013 Nov;15(11):2879–93. doi: 10.1111/1462-2920.12186. Epub 2013 Jul 19. PMID: 23869714.

Steinberg PD, Rice SA, Campbell AH, McDougald D, Harder T. 2011. Interfaces between bacterial and eukaryotic neuroecology. Integr Comp Biol. 51(5):794–806. doi: 10.1093/icb/icr115. Epub 2011 Sep 5. PMID: 21893590.

Thakur, Narsinh & Anil, Arga & Muller, W.E.G.. (2004). Culturable epibacteria of the marine sponge Ircinia fusca: Temporal variations and their possible role in the epibacterial defense of the host. Aquat. Microb. Ecol. 37. 10.3354/ame037295.

Wahl M. 1989. Marine epibiosis. I. Fouling and antifouling: some basic aspects. *Mar Ecol Prog Ser* 58:175–189.

Wahl M. 2008. Ecological lever and interface ecology: epibiosis modulates the interactions between host and environment. Biofouling 24:427–438. 10.1080/08927010802339772.

Wahl M, Goecke F, Labes A, Dobretsov S, Weinberger F. 2012. The second skin: ecological role of epibiotic biofilms on marine organisms. Front Microbiol. 3:292. doi: 10.3389/fmicb.2012.00292. PMID: 22936927; PMCID: PMC3425911.

Waters CM, Bassler BL. 2005. Quorum sensing: cell-to-cell communication in bacteria. Annu Rev Cell Dev Biol 21:319–346. doi: 10.1146/annurev.cellbio.21.012704.131001. PMID: 16212498.

Yebra DM, Kiil S, Dam-Johansen K. 2004. Antifouling technology-past, present and future steps towards efficient and environmentally friendly antifouling coatings. *Prog Org Coat* 50(2):75–104. doi:10.1016/j.porgcoat.2003.06.001

15 Biofilm in Aquaculture Microorganisms

Naji Naseef Pathoor, Natarajan Chandrakala, Jayaraman Eswari, Sellamuthu Shanmugapriya, R.V. Geetha, Rajesh Kanna Gopal, Saravanan Shanmugam, Gulshan Wadhwa, and Pitchaipillai Sankar Ganesh

15.1 INTRODUCTION

Biofilms are complex communities of microorganisms, mostly bacteria, that stick to surfaces and constitute a protective matrix of extracellular polymeric substance (EPS). Biofilms in aquaculture have the ability to grow on a variety of underwater structures including tank surfaces, pipes, nets, and aquatic organism's gills and skin (Zhao et al., 2023a). Biofilm formation is typically thought to occur as follows: Bacteria is ability to colonize on the surfaces,and further it form microcolonies, finally to form biofilms (Crouzet et al., 2014). The development of biofilm is influenced by the organism's constituents, surface properties, and aqueous environmental conditions. Detachment from biofilms can occur due to internal or external causes. However, numerous organisms use dispersal to colonize new areas (Sauer et al., 2002).

Aquaculture is now recognized as a vital part of food production, with its primary objective of increasing productivity and output on a continuous basis. The broadening of pond areas and the implementation of extensive farming practices in existing ponds hold great promise for aquacultures future global growth. However, biofilms in aquatic ecosystems are intricate and ever-changing communities characterized by a high diversity of microbes in which the microorganisms interact with advantages from increased utilization of nutrients, waste elimination, protection from toxins and antibiotics, refuge from enemies, and storage of enzyme activities (Neagu et al., 2017) .

A significant priority for water retailers is to ensure the preservation of drinking water quality throughout the entire journey from the treatment plant to consumer's taps. The water distribution systems face continual exposure to fluctuating water flows containing biodegradable organic wastes and microorganisms. These elements may stem from natural biofilms within the wastewater treatment facilities or incidents that occur within the water supply system (Aneesh M.R., 2021; September et al., 2007). Additionally, biofilm provides supplemental natural sustenance and living conditions, reducing territorial impacts. Additionally, it enhances the facilitation of organic matter breakdown, and encourages nitrification through microbial activity. Moreover, it encourages heterotrophic production of single-cell proteins

DOI: 10.1201/9781003487203-15

(Avnimelech, 2007; Crab et al., 2007; Hari et al., 2004). Altogether this chapter aims to outline the functions of biofilms in aquaculture systems by elucidating their roles as well as remedial measures for managing biofilm formation.

15.2 FACTORS INFLUENCING BIOFILM FORMATION IN AQUACULTURE

Biofilm formation is an unpredictable and intricate procedure that includes the initial adherence of bacterial cells to the substrate, physiological modifications within the microbes, multiplication of adhered cells to form microcolonies, and, finally, biofilm maturation (Puttamreddy et al., 2010). Microorganisms in biofilms have distinct characteristics from their free-living planktonic counterparts. These differences include altered physiological processes and increased resistance to immune responses and antimicrobial agents, making biofilms a possible source of chronic and persistent illnesses (Wei & Ma, 2013).

Biofilm formation in aquaculture environments is determined from chemical, biological, and physical factors. Understanding these factors is crucial for tackling biofilm growth and its impact on aquaculture systems. Outlined are a few major factors affecting the development of biofilm in aquaculture environments.

15.2.1 The Status of the Water

15.2.1.1 Nutritional Supply

The presence of essential nutrients like nitrogen and phosphorus compounds influences the growth of microorganisms, potentially contributing to the formation of biofilms. Biofilm development can be stimulated by an abundance of nutrients derived from sources such as waste, uneaten feed, and organic matters (Rashid et al., 2016). The crc locus, which encodes the catabolite suppressor protein, is also related to biofilm formation (Davey & O'toole, 2000).

15.2.1.2 Oxygen Concentration Levels

Oxygen constitutes one of among the most essential water-quality indicators. It can impact the growth and health of aquatic creatures, as well as the overall efficiency of the aquaculture sector system (Abu Shmeis, 2018). Low oxygen levels may encourage the growth of anaerobic bacteria and affect biofilm composition (Kulakowski et al., 2023; Yang et al., 2021).

15.2.1.3 pH Levels

The pH of water can influence the growth and metabolic processes of microorganisms. Some algal and bacterial species prefer specific pH ranges, which may impact biofilm structure (Gatamaneni et al., 2018; Jin & Kirk, 2018).

15.2.1.4 Temperature

Elevated temperatures promote bacterial growth and can accelerate the formation of biofilms. Nevertheless, higher temperatures may alter the composition of the biofilm community. Temperature fluctuations play a significant role in influencing the size and physical characteristics of biofilms (Villanueva et al., 2011; Yang et al., 2021).

15.2.1.5 Water Circulation and Hydrodynamics

Fluid flow has a substantial effect on biofilm attributes. The motion of bulk flow impacts a variety of parameters, including the rate of growth (Boyle & Lappin-Scott, 2006; Conrad & Poling-Skutvik, 2018; Picioreanu et al., 2009; Thomen et al., 2017) structure (Conrad & Poling-Skutvik, 2018; Stoodley et al., 1998); shape (Biswas et al., 2018; Rusconi et al., 2011; Taherzadeh et al., 2010); cell concentration (Garrett et al., 2008; Guimerà et al., 2016; Moreira et al., 2015); and detachment (Radu et al., 2012). Optimal flow rates can help combat excess development of biofilm and reduce the risk of biofouling (Krsmanovic et al., 2021).

15.2.1.6 Organic Matter

High levels of biological material, such as uneaten feed and feces, can provide nutrients for biofilm-forming microorganisms. Proper disposal actions are essential for controlling the amount of organic matter in the aquaculture sector (Dauda et al., 2019; Horvathova et al., 2016).

15.2.1.7 Competition and Interaction

The development of biofilms is influenced by connections among microbial species, including competition and interaction. Advantageous microbes may compete with potential pathogens, influencing the overall health of the aquaculture system (Fang et al., 2020; Rendueles & Ghigo, 2015a).

15.2.1.8 Aquatic Chemistry

In aquatic environments, physicochemical properties of water, such as ionic strength and hardness, can have a significant impact on biofilm adhesion. Generally, higher ionic strength can help microbes adhere to surfaces more effectively. It happens since ions in water can neutralize the charges on both microbial cells and surfaces, allowing for more robust interactions and attachment (Zheng et al., 2021).

15.3 EMERGENCE OF BIOFILM IN AQUACULTURE ENVIRONMENT

The formation of biofilms in aquaculture settings can have a beneficial as well as detrimental impact on the overall health and productivity of the system. In aquaculture, biofilms occur on a variety of surfaces, including pipes, nets, and tank surfaces. The formation of these biofilms is influenced by environmental factors such as water quality and organic matter.

15.3.1 Beneficial Elements of Biofilms in Aquaculture

15.3.1.1 Nutrient Cycling

Biofilms contribute to the metabolic process by breaking down organic matter and recycling nutrients in water. The nutrient cycling is aided by enzymes generated by microorganisms in the biofilm, as well as the chemical and physical relationships between cell membranes and the matrix. The process of nutrient cycling may take place throughout an individual isogenic group as well as between organisms. This may formulate an equilibrium and harmonious ecosystem (Bamford et al., 2023a).

15.3.1.2 Aquatic Quality Improvement

Biofilms serve as natural water cleaners by removing excessive nutrients and particulates from water, enabling higher water quality. Several investigations have shown that the existence of nitrifying microbes toward the biofilm bring on a decrease in ammonium quantity in the generated water (Hossain et al., 2022). The microbes that induce nitrification are ammonia-oxidizing bacteria (AOB), ammonia-oxidizing archaea (AOA), and nitrite-oxidizing bacteria (NOB) (Yin et al., 2018; Zhang et al., 2009). Indeed, the study's noticed decline in ammonium levels, as well as a rise in nitrite and nitrate concentrations, shows that nitrifying bacteria within the biofilm play a crucial part in managing water quality (Lu et al., 2021). Additionally, microalgae absorb ammonium and use it to produce new biomass (Thompson et al., 2002). Even though uncontrollable nitrification may affect the ecological equilibrium in the distribution system, leading to water-quality issues (Lytle et al., 2007). As an example, rise in bacterial activity during nitrification may result in a drop in system pH, which leads to rust of transportation materials (Bremer et al., 2001; Lytle et al., 2007).

15.3.1.3 Advantageous Microorganisms (AMs)

Various microbes capable of forming biofilms can be beneficial in aquaculture setups, including algae, bacteria, fungi, and viruses. Microalgae utilize photosynthesis to absorb CO_2 and release oxygen, contributing to the well-being of aquatic animals. Bacteria and fungi play a key role in decomposing organic matter in sediments, aiding in maintaining water cleanliness. Application of AMs may enhance the disease resistance of common aquatic species in aquaculture systems (C. Wang et al., 2022). They are typically classified into three types: such as (i) single strains; (ii) multiple strains; and (iii) compound strains with synergy. Their functions revolve around two fundamental aspects: 1) Their objectives include regulating microdysbiosis, sustaining microeubiosis, promoting the host's health, and encouraging the proliferation of AMs and their products of metabolism in ecology. 2) These microbes are essential for the decomposition of organic matter such as waste and feed parts or leftovers. Their vital function lies in maintaining a dynamic ecological balance among organisms from diverse natural kingdoms and establishing microbial equilibrium in both water and sediments. Additionally, they contribute to creating a favorable environment for the thriving of various fish species (Zhou et al., 2009).

15.3.2 Negative Elements of Biofilms in Aquaculture

Pathogenic Microorganisms

Harmful microorganisms, such as pathogenic bacteria and fungi, can be found in biofilms, endangering aquatic organisms and causing diseases and decreased productivity. *Tenacibaculum* spp. were found in mariculture-associated biofilm in ponds and aquaculture equipment in China. These bacteria can cause an ulcerative disease called *Tenacibaculosis* in fishes. This disease is predominantly observed in various marine species (Deng et al., 2021; Dussud et al., 2018; Hou et al., 2021; McCormick et al., 2014). Yet, globally, *V. cholerae*, the causative agent of numerous historical epidemics, has been responsible for diseases in humans throughout various periods. Moreover, the species *V. mimicus* produces cholera toxins, leading to gastroenteritis and is transmitted through raw oysters, fish, turtle eggs, and prawns. *V. vulnificus*, a hazardous food-borne pathogen linked to seafood, is believed to have a 50% mortality rate in severe infections (Jones & Oliver, 2009; Ezhil Nilavan, S., & Mothadaka, M. P., 2023; M. M. Silva et al., 2019; Strom & Paranjpye, 2000).

Clogging of Equipment

An overgrowth of biofilms can clog pipes, filters, and other aquaculture equipment, reducing water flow and operational efficiency. This influences the circulation of nutrients, oxygen, and other essential elements, resulting in additional strain loss and requiring more energy to pump water through the system. This can lead to increased operating expenses and decreased energy effectiveness (Mahapatra, 2015). Biofilm overgrowth can disrupt water treatment processes such as ultraviolet (UV) sterilization and chemical treatments by reducing their efficacy for preserving water quality (Chen et al., 2022; Luo et al., 2022).

This can be summarized as the unneeded development and aggregates of living organisms on surfaces. This biofouling can have a negative impact on aquaculture infrastructure, including nets and cages (Flemming, 2011). Biofouling happens when microbes such as bacteria and diatoms attach to the ship's hull. As microbes mature and reproduce, the biofilm matrix binds the cells to the hull. As a result, the surface becomes slimy with biofilm, allowing larger organisms to attach more easily. Biofouling may adversely affect ships by increasing fuel consumption, reducing speed, and increasing maintenance costs. Ships can introduce species that are invasive into new environments, causing essential sustainable and economic impacts (Bamford et al., 2023b; Samrot et al., 2021).

Competition for Resources

Biofilms may compete for nutrients with aquatic organisms, limiting the growth and development of cultured species (Rendueles & Ghigo, 2015b). The following points outline the impacts of nutrient imbalance caused by biofilms:

The competition for nutrients can cause a disparity in the nutrient composition of water, potentially affecting the physiological processes of aquatic organisms. This lack of balance can impact osmoregulation, metabolism, and reproductive abilities (Nadell et al., 2016).

Physiological Stress

It may result from increased competition for resources among cultured species. This stress compromises the immune system, rendering aquatic organisms more vulnerable to diseases and environmental stressors (Lories et al., 2020).

In conclusion, while biofilms play important roles in nutrient cycling and ecosystem functioning in aquaculture, their excessive growth might prove problematic. Setting up proactive management strategies is critical for maximizing biofilms' positive contributions while minimizing potential negative impacts on the overall health of the aquaculture ecosystem.

15.4 BARRIERS POSED BY BIOFILMS IN AQUACULTURE

Biofilms in aquaculture may cause a number of barriers and challenges that can have an impact on human health. Depositing organic waste and microplastics in aquatic water results in pathogenic bacteria forming more biofilms, increasing the rate of infection in aquatic animals, and eventually affecting humans (Figure 15.1). Many pathogens have the potential to infect humans via aquaculture. Certainly, *Vibrio cholerae*, among the human pathogenic bacteria mentioned earlier, is widely recognized as the causative agent of numerous global epidemics throughout generations, as confirmed by scientific studies (Ojeda Rodriguez and Kahwaji, 2023). *Vibrio fluvialis*, found in marine environments, is a food-borne pathogen that results in diarrhea and occasional extraintestinal cases, regardless of the state of the economy (Ramamurthy et al., 2014). *V. alginolyticus* has been linked to skin ulcers, otitis, and ocular inflammation

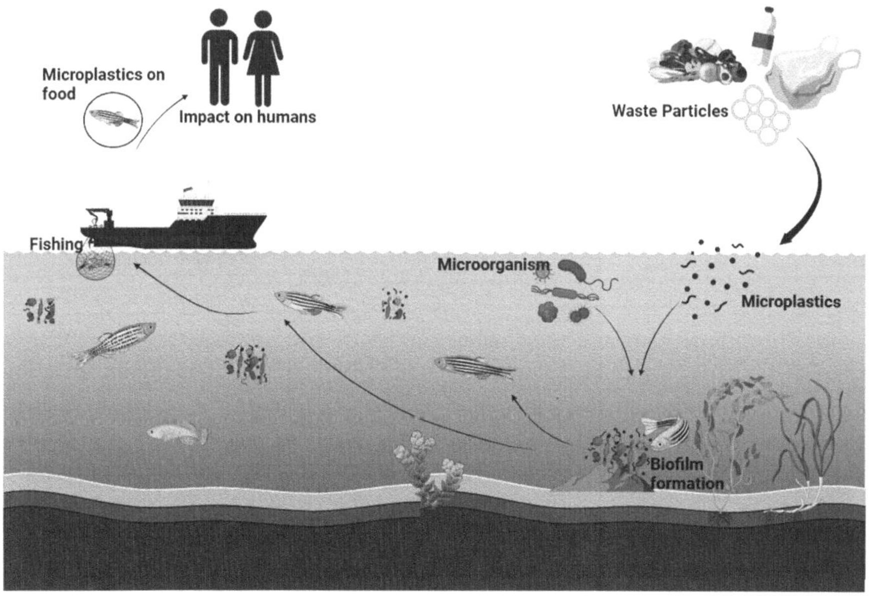

FIGURE 15.1 Impact of organic waste and microplastics on biofilm formation in aquatic environments. This figure illustrates the consequential effects of depositing organic waste and microplastics in aquatic water, highlighting the subsequent increase in biofilm formation by pathogenic bacteria. The elevated biofilm presence intensifies the risk of infection in aquatic animals, ultimately posing potential threats to human health.

(Schmidt et al., 2014). Similarly, *Vibrio* species are the most common pathogens among farmed animals exposed to microplastics in the aquatic environment. *Vibrio* spp. and *Escherichia coli* have been found in microplastics from both the beach and the Forth Estuary in Scotland, which raises public health concerns (Quilliam et al., 2014; Rodrigues et al., 2019). Plastics coated with biofilms release scent signals that stimulate consumption, linking to the circulation and buildup of plastics in food chains. This leads to the prevalence of these materials in diverse seafood products and raises questions about the impact on the animal's intestinal microbiome (Barboza et al., 2018; X. Wang et al., 2016).

15.5 FOULING OF AQUACULTURE INFRASTRUCTURE CAN IMPACT FISH AND SHELLFISH PRODUCTION

The fouling of aquaculture structures, characterized by the unwanted accumulation of substances such as biofilms, algae, barnacles, and other organisms on surfaces within aquaculture facilities, can significantly impact fish and shellfish production. This phenomenon affects various components of aquaculture infrastructure, including nets, cages, pipes, and equipment, leading to negative consequences for both the health and growth of cultured organisms and the operational efficiency of aquaculture systems (Bannister et al., 2019).

In shellfish farming, biofouling poses a threat to species like mussels and oysters, causing physical damage to shells and compromising their structural integrity. This damage, attributed to activities of endolithic species or calcareous tube-dwelling polychaetes, makes shells vulnerable to predators and reduces overall product quality. Biofouling also interferes with feeding mechanisms, fosters competition for resources, and alters environmental conditions, contributing to decreased growth and yield (Fitridge et al., 2012).

In fish farming, particularly on cage nets, the fouling of aquaculture infrastructure has diverse impacts. It can restrict water exchange, leading to poor water quality, reduced dissolved oxygen levels, and compromised removal of excess feed and waste (Archana et al., 2019). Fouling organisms on cage netting may act as reservoirs for pathogenic microorganisms, heightening the risk of disease transmission to cultured fish. Additionally, biofouling accumulations add weight to the infrastructure, resulting in cage deformation and structural fatigue. This degradation not only affects aquaculture efficiency but also escalates production costs due to the need for additional maintenance and repairs (Comas et al., 2021). In summary, the fouling of aquaculture infrastructure reverberates throughout the industry, influencing the health, growth, and overall productivity of both fish and shellfish in aquaculture systems.

15.6 MONITORING AND EVALUATION OF BIOFILMS IN AQUACULTURE

Aquaculture is a vital contributor to global food production, addressing the rising demand for seafood while mitigating pressure on wild fisheries (Verdegem et al., 2023). However, the intricate systems of aquaculture face challenges posed by biofilms. These biofilms, encompassing bacteria, algae, and other microorganisms, can risk aquatic health and operational efficiency (Muhammad et al., 2020). Effective monitoring and evaluation of biofilms are crucial for responsible aquaculture management. Biofilms, if unchecked, can compromise water quality by altering oxygen levels and nutrient cycling, creating an unfavourable environment for cultured species (Zhao et al., 2023b). Accumulation on infrastructure components, such as pipes and tanks, can impede water flow and reduce system efficiency. Proactive monitoring is essential for sustaining a productive aquaculture environment.

Various techniques are employed for the detection and quantification of biofilms in aquaculture settings (Table 15.1). Microscopic analysis, encompassing light and electron microscopy, stands as a vital tool for visualizing biofilm structures. Staining methods, such as crystal violet or fluorescent dyes, enhance visualization, facilitating the quantification of biofilm biomass and providing valuable insights into their structure and composition (Relucenti et al., 2021; Wilson et al., 2017). Molecular methods, such as Polymerase Chain Reaction (PCR) and quantitative PCR (qPCR), offer a molecular perspective on biofilm detection by targeting specific genes associated with biofilm-forming microorganisms, enabling precise identification and quantification (N. B. S. Silva et al., 2021). Next-generation sequencing further contributes to understanding biofilm diversity. Biochemical analysis involves scrutinizing the metabolic activities of biofilm communities, where enzyme assays and metabolic profiling serve as indicators for monitoring biofilm growth and activity (Yeor-Davidi et al., 2020). Finally, advanced imaging technologies like Confocal Laser Scanning Microscopy (CLSM) and Scanning Electron Microscopy (SEM) provide detailed three-dimensional visualizations, offering insights into biofilm structures and interactions within aquaculture systems (Wilson et al., 2017).

Assessing the impact of biofilms in aquaculture involves monitoring key aspects. Biofilms can alter water-quality parameters, necessitating regular assessment for dissolved oxygen, pH, and nutrient levels to gauge their influence on overall water quality. Their presence also affects fish health, demanding frequent health assessments, including observation for stress signs, diseases, skin lesions, and abnormal behavior (Nesse et al., 2023). Moreover, biofilms on aquaculture infrastructure can diminish system efficiency, requiring routine inspection and cleaning of components like pipes, tanks, and nets to prevent issues such as clogging and reduced water flow. Additionally, understanding biofilms' role in nutrient cycling is crucial, prompting assessments of nutrient concentrations to optimize management strategies and prevent imbalances within aquaculture systems (Pandey et al., 2014).

The monitoring and evaluation of biofilms in aquaculture require a combination of techniques for detection, quantification, and assessing their impact on the overall system. Integrating traditional methods with advanced technologies enables a comprehensive understanding of biofilm dynamics. By employing these techniques, aquaculture practitioners can make informed decisions to mitigate the negative effects of biofilms and maintain a healthy and sustainable aquatic environment. Continued research and technological advancements in this field are essential for enhancing our ability to manage biofilm-related challenges in aquaculture systems.

TABLE 15.1
Techniques for the Detection and Quantification of Biofilms in Aquaculture Settings

Techniques		Application	Reference
Microscopic analysis	Light microscopy	Useful for obtaining initial insights into the composition and organization of biofilms	(Wilson et al. (2017); Relucenti et al. (2021)
	Scanning Electron Microscopy (SEM)	Captures the fine details and spatial relationships within biofilms at a nanoscale level	
	Staining methods (crystal violet or fluorescent dyes)	Highlights biofilm biomass, makes it more visible, and facilitates quantitative analysis	
Molecular methods	Polymerase Chain Reaction (PCR)	Help to gain insight into the identity and prevalence of microorganisms contributing to biofilm development	Silva et al. (2021)
	Quantitative PCR (qPCR)	Offers a quantitative snapshot of the microbial composition within the biofilm matrix	
Biochemical analysis	Enzyme assays	Reveals the metabolic potential and degradation capacity of the biofilm	(Yeor-Davidi et al., 2020)
	Metabolic profiling	Unveils the functions of biofilm	
Advanced imaging technologies	Confocal Laser Scanning Microscopy (CLSM)	Provides a detailed 3D reconstruction for a thorough understanding of biofilm complexity	(Wilson et al., 2017)

15.7 BIOFILM ECOLOGY IN VARIOUS AQUACULTURE SYSTEMS

Biofilms play a significant role in the ecology of diverse aquaculture systems and understanding their dynamics in different settings is crucial for maintaining water quality, supporting fish health, and optimizing overall system efficiency.

In freshwater aquaculture, biofilm dynamics are influenced by factors such as nutrient availability, water flow, and temperature. Surfaces within tanks, raceways, or ponds serve as substrates for biofilm formation (Moreno Osorio et al., 2021). Nutrient-rich environments, coupled with adequate water flow, can lead to the development of robust biofilms. The composition and structure of freshwater biofilms impact water-quality parameters and nutrient cycling within the system. Monitoring and managing biofilm dynamics in freshwater settings are essential to ensure optimal conditions for aquatic organisms (Moyal et al., 2023).

Marine aquaculture environments present unique challenges and opportunities for biofilm ecology. Substrates in marine systems, such as nets, cages, and infrastructure, can support the growth of biofilms. The dynamics are influenced by salinity, nutrient availability, and the diversity of microorganisms present. Marine biofilms contribute to nutrient cycling, affecting water quality and providing a substrate for filter-feeding organisms (De Carvalho, 2018). Understanding the specificities of biofilm development in marine aquaculture is crucial for maintaining a balanced and sustainable aquatic environment.

Comparing biofilm ecology across different aquaculture settings reveals subtle differences and shared characteristics. While freshwater systems may exhibit biofilm growth on tank surfaces and substrates, marine environments may feature biofilms on floating cages and nets. The microbial composition of biofilms varies between these settings, influenced by the distinct characteristics of freshwater and marine ecosystems (Schoina et al., 2022) .Despite these differences, the overarching role of biofilms in nutrient cycling, water-quality maintenance, and providing a habitat for diverse microorganisms remains consistent. Efforts to manage and optimize biofilm dynamics in aquaculture require a tailored approach based on the specific conditions and requirements of each system (Balcázar et al., 2015).

In summary, biofilm ecology in various aquaculture systems is a dynamic and intricate field that necessitates a thorough understanding of the factors influencing biofilm dynamics. Tailoring management strategies to the unique characteristics of freshwater and marine environments is essential for promoting a healthy and sustainable aquaculture ecosystem.

15.8 STRATEGIES FOR CONTROLLING BIOFILMS

Marine biofilms present challenges in various industries, and effective control measures are crucial for maintaining structures and equipment in marine environments. Strategies for controlling marine biofilms encompass physical methods, chemical approaches, and biological control measures (Figure 15.2). Physical methods involve mechanical interventions to remove or disrupt marine biofilms from surfaces. Techniques such as scrubbing, brushing, pressure washing, and the use of technologies like ultrasonication and high-pressure water jets are effective in physically

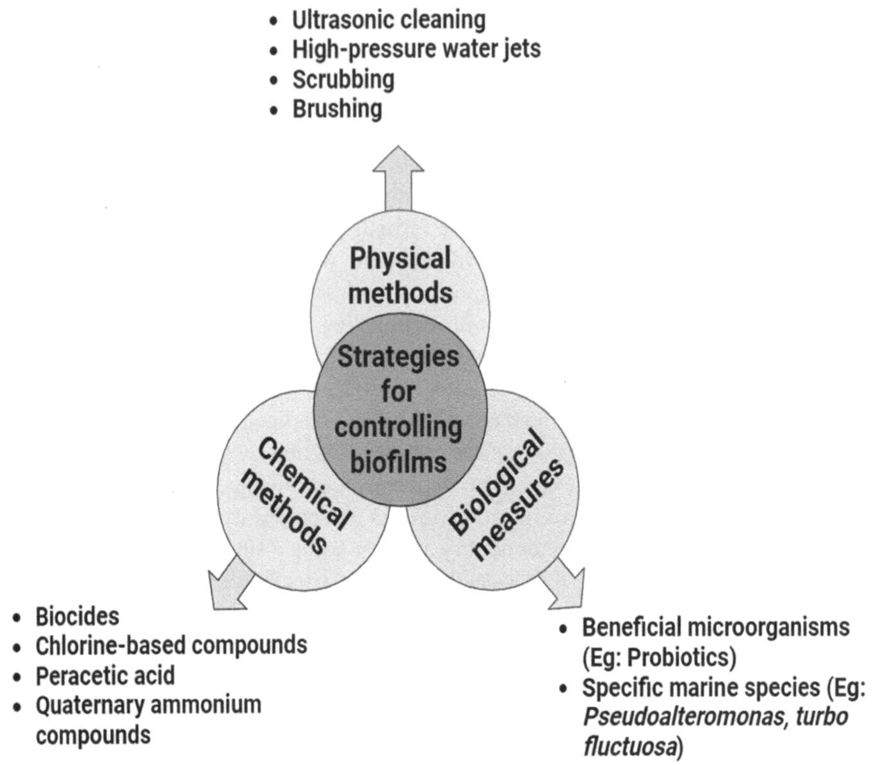

FIGURE 15.2 Strategies to control aquatic biofilms. Physical, chemical, and biological control measures.

detaching and preventing biofilm formation on marine structures (Li et al., 2021; Song & Cui, 2020). Regular application of physical methods helps prevent the accumulation of persistent marine biofilms. Chemical approaches utilize antimicrobial agents to inhibit marine biofilm formation or disperse existing biofilms. Biocides, chlorine-based compounds, peracetic acid, and quaternary ammonium compounds are common chemical options. These substances target the microorganisms within the marine biofilm matrix, disrupting their growth and attachment (Bajpai, 2015; Li et al., 2021). Careful consideration of the chemical's impact on marine life and the environment is crucial for responsible application in marine settings. Biological control measures leverage natural processes and organisms to manage marine biofilms. This includes the introduction of beneficial microorganisms like probiotics, which outcompete harmful marine biofilm-forming bacteria. Additionally, the use of grazing organisms, such as specific marine species like *Pseudoalteromonas* and *Turbo fluctuosa*, contributes to biofilm control by consuming or disrupting their structure ("Biofouling and Mitigation Methods: A Review," 2022; Fang et al., 2020). Biological control measures aim to establish a balanced microbial community, fostering sustainable, long-term marine biofilm management.

Implementing a combination of these strategies is often the most effective approach to controlling biofilms in marine environments. Physical methods offer immediate relief, chemical approaches provide targeted control, and biological control measures contribute to sustainable, long-term management. Regular monitoring and assessment are crucial to tailoring these strategies to the specific conditions of the marine structures, ensuring a comprehensive and proactive approach to marine biofilm control.

15.9 CONCLUSION

In conclusion, understanding the factors that influence biofilm formation, recognizing both the positive and negative aspects of biofilms in aquaculture, and implementing proactive monitoring and management strategies are critical steps toward ensuring a healthy and sustainable aquatic environment in the face of biofilm-related challenges. Continued research and technological advancements will help us better address and overcome these challenges in aquaculture systems.

REFERENCES

Abu Shmeis, R. M. 2018. Water Chemistry and Microbiology *Comprehensive Analytical Chemistry.* 81, 1–56 https://doi.org/10.1016/bs.coac.2018.02.001

Aneesh M. R. 2021. Quality of drinking water and sanitation in India. *Indian Journal of Human Development*, *15*(1), 138–152. https://doi.org/10.1177/09737030211003658

Archana, S., Sundaramoorthy, B., & Mohamed Faizullah, M. 2019. Review on impact of biofouling in aquafarm infrastructures. *International Journal of Current Microbiology and Applied Sciences*, *8*(7), 2942–2953. https://doi.org/10.20546/ijcmas.2019.807.365

Avnimelech, Y. 2007. Feeding with microbial flocs by tilapia in minimal discharge bio-flocs technology ponds. *Aquaculture*, *264*(1–4), 140–147. https://doi.org/10.1016/j.aquaculture.2006.11.025

Bajpai, P. 2015. The control of microbiological problems**Some excerpts taken from Bajpai P (2012). Biotechnology for pulp and paper processing with kind permission from Springer Science1Business Media. In *Pulp and Paper Industry* (pp. 103–195). Elsevier. https://doi.org/10.1016/B978-0-12-803409-5.00008-2

Balcázar, J. L., Subirats, J., & Borrego, C. M. 2015. The role of biofilms as environmental reservoirs of antibiotic resistance. *Frontiers in Microbiology*, *6*, 1216 https://doi.org/10.3389/fmicb.2015.01216

Bamford, N. C., MacPhee, C. E., & Stanley-Wall, N. R. 2023a. Microbial Primer: An introduction to biofilms – what they are, why they form and their impact on built and natural environments. *Microbiology*, *169*(8), 001338 https://doi.org/10.1099/mic.0.001338

Bamford, N. C., MacPhee, C. E., & Stanley-Wall, N. R. 2023b. Microbial Primer: An introduction to biofilms – what they are, why they form and their impact on built and natural environments. *Microbiology*, *169*(8). https://doi.org/10.1099/mic.0.001338

Bannister, J., Sievers, M., Bush, F., & Bloecher, N. 2019. Biofouling in marine aquaculture: A review of recent research and developments. *Biofouling*, *35*(6), 631–648. https://doi.org/10.1080/08927014.2019.1640214

Barboza, L. G. A., Dick Vethaak, A., Lavorante, B. R. B. O., Lundebye, A.-K., & Guilhermino, L. 2018. Marine microplastic debris: An emerging issue for food security, food safety and human health. *Marine Pollution Bulletin*, *133*, 336–348. https://doi.org/10.1016/j.marpolbul.2018.05.047

Biswas, I., Ghosh, R., Sadrzadeh, M., & Kumar, A. 2018. Near wall void growth leads to disintegration of colloidal bacterial streamer. *Journal of Colloid and Interface Science*, *522*, 249–255. https://doi.org/10.1016/j.jcis.2018.03.074

Boyle, J. D., & Lappin-Scott, H. 2006. Quantification of the effect of flowrate on the rates of arrival and attachment to glass of *Pseudomonas aeruginosa*. *Biofouling*, *22*(2), 117–123. https://doi.org/10.1080/08927010600579157

Bremer, P. J., Webster, B. J., & Wells, D. B. 2001. Biocorrosion of copper in potable water. *Journal AWWA*, *93*(8), 82–91. https://doi.org/10.1002/j.1551-8833.2001.tb09269.x

Chen, Y., Li, Y., Yang, S., Chiang, T. Y., Zhu, X., & Hu, J. 2022. Controlling biofilm growth and its antibiotic resistance in drinking water by combined UV and chlorination processes. *Water*, *14*(22), 3643. https://doi.org/10.3390/w14223643

Comas, J., Parra, D., Balasch, J. C., & Tort, L. 2021. Effects of fouling management and net coating strategies on reared gilthead sea bream juveniles. *Animals*, *11*(3), 734. https://doi.org/10.3390/ani11030734

Conrad, J. C., & Poling-Skutvik, R. 2018. Confined flow: Consequences and implications for bacteria and biofilms. *Annual Review of Chemical and Biomolecular Engineering*, *9*(1), 175–200. https://doi.org/10.1146/annurev-chembioeng-060817-084006

Crab, R., Avnimelech, Y., Defoirdt, T., Bossier, P., & Verstraete, W. 2007. Nitrogen removal techniques in aquaculture for a sustainable production. *Aquaculture*, *270*(1–4), 1–14. https://doi.org/10.1016/j.aquaculture.2007.05.006

Crouzet, M., Le Senechal, C., Brözel, V. S., Costaglioli, P., Barthe, C., Bonneu, M., Garbay, B., & Vilain, S. 2014. Exploring early steps in biofilm formation: Set-up of an experimental system for molecular studies. *BMC Microbiology*, *14*(1), 253. https://doi.org/10.1186/s12866-014-0253-z

Dauda, A. B., Ajadi, A., Tola-Fabunmi, A. S., & Akinwole, A. O. 2019. Waste production in aquaculture: Sources, components and managements in different culture systems. *Aquaculture and Fisheries*, *4*(3), 81–88. https://doi.org/10.1016/j.aaf.2018.10.002

Davey, M. E., & O'toole, G. A. 2000. Microbial biofilms: From ecology to molecular genetics. *Microbiology and Molecular Biology Reviews*, *64*(4), 847–867. https://doi.org/10.1128/MMBR.64.4.847-867.2000

De Carvalho, C. C. C. R. 2018. Marine biofilms: A successful microbial strategy with economic implications. *Frontiers in Marine Science*, *5*. https://doi.org/10.3389/fmars.2018.00126

Deng, H., Fu, Q., Li, D., Zhang, Y., He, J., Feng, D., Zhao, Y., Du, G., Yu, H., & Ge, C. 2021. Microplastic-associated biofilm in an intensive mariculture pond: Temporal dynamics of microbial communities, extracellular polymeric substances and impacts on microplastics properties. *Journal of Cleaner Production*, *319*, 128774. https://doi.org/10.1016/j.jclepro.2021.128774

Dussud, C., Meistertzheim, A. L., Conan, P., Pujo-Pay, M., George, M., Fabre, P., Coudane, J., Higgs, P., Elineau, A., Pedrotti, M. L., Gorsky, G., & Ghiglione, J. F. 2018. Evidence of niche partitioning among bacteria living on plastics, organic particles and surrounding seawaters. *Environmental Pollution*, *236*, 807–816. https://doi.org/10.1016/j.envpol.2017.12.027

Ezhil Nilavan, S., & Mothadaka, M. P. (2023). Vibrio mimicus and its antimicrobial resistance in fish and aquatic environments. In *Handbook on Antimicrobial Resistance* (pp. 1–21). Springer Nature Singapore. https://doi.org/10.1007/978-981-16-9723-4_19-1

Fang, K., Park, O.-J., & Hong, S. H. 2020. Controlling biofilms using synthetic biology approaches. *Biotechnology Advances*, *40*, 107518. https://doi.org/10.1016/j.biotechadv.2020.107518

Fitridge, I., Dempster, T., Guenther, J., & de Nys, R. 2012. The impact and control of biofouling in marine aquaculture: A review. *Biofouling*, *28*(7), 649–669. https://doi.org/10.1080/08927014.2012.700478

Flemming, H.-C. 2011. *Microbial Biofouling: Unsolved Problems, Insufficient Approaches, and Possible Solutions* 5:81–109. https://doi.org/10.1007/978-3-642-19940-0_5

Garrett, T. R., Bhakoo, M., & Zhang, Z. 2008. Bacterial adhesion and biofilms on surfaces. *Progress in Natural Science*, *18*(9), 1049–1056. https://doi.org/10.1016/j.pnsc.2008.04.001

Gatamaneni, B. L., Orsat, V., & Lefsrud, M. 2018. Factors affecting growth of various microalgal species. *Environmental Engineering Science*, 35(10), 1037–1048. https://doi.org/10.1089/ees.2017.0521

Gizer G., Önal U., Ram M., & Şahiner N. 2022. Biofouling and mitigation methods: A review. *Biointerface Research in Applied Chemistry*, *13*(2), 185. https://doi.org/10.33263/BRIAC132.185

Guimerà, X., Dorado, A. D., Bonsfills, A., Gabriel, G., Gabriel, D., & Gamisans, X. 2016. Dynamic characterization of external and internal mass transport in heterotrophic biofilms from microsensors measurements. *Water Research*, *102*, 551–560. https://doi.org/10.1016/j.watres.2016.07.009

Hari, B., Madhusoodana Kurup, B., Varghese, J. T., Schrama, J. W., & Verdegem, M. C. J. 2004. Effects of carbohydrate addition on production in extensive shrimp culture systems. *Aquaculture*, *241*(1–4), 179–194. https://doi.org/10.1016/j.aquaculture.2004.07.002

Horvathova, T., Babik, W., & Bauchinger, U. 2016. Biofilm feeding: Microbial colonization of food promotes the growth of a detritivorous arthropod. *ZooKeys*, *577*, 25–41. https://doi.org/10.3897/zookeys.577.6149

Hossain, S., Chow, C. W. K., Cook, D., Sawade, E., & Hewa, G. A. 2022. Review of nitrification monitoring and control strategies in drinking water system. *International Journal of Environmental Research and Public Health*, *19*(7), 4003. https://doi.org/10.3390/ijerph19074003

Hou, D., Hong, M., Wang, Y., Dong, P., Cheng, H., Yan, H., Yao, Z., Li, D., Wang, K., & Zhang, D. 2021. Assessing the risks of potential bacterial pathogens attaching to different microplastics during the Summer–Autumn period in a mariculture cage. *Microorganisms*, *9*(9), 1909. https://doi.org/10.3390/microorganisms9091909

Jin, Q., & Kirk, M. F. 2018. pH as a primary control in environmental microbiology: 1. Thermodynamic perspective. *Frontiers in Environmental Science*, *6*, 21 https://doi.org/10.3389/fenvs.2018.00021

Jones, M. K., & Oliver, J. D. 2009. Vibrio vulnificus: Disease and pathogenesis. *Infection and Immunity*, *77*(5), 1723–1733. https://doi.org/10.1128/IAI.01046-08

Krsmanovic, M., Biswas, D., Ali, H., Kumar, A., Ghosh, R., & Dickerson, A. K. 2021. Hydrodynamics and surface properties influence biofilm proliferation. *Advances in Colloid and Interface Science*, *288*, 102336. https://doi.org/10.1016/j.cis.2020.102336

Kulakowski, S., Banerjee, D., Scown, C. D., & Mukhopadhyay, A. 2023. Improving microbial bioproduction under low-oxygen conditions. *Current Opinion in Biotechnology*, *84*, 103016. https://doi.org/10.1016/j.copbio.2023.103016

Li, Y., Li, X., Hao, Y., Liu, Y., Dong, Z., & Li, K. 2021. Biological and physiochemical methods of biofilm adhesion resistance control of medical-context surface. *International Journal of Biological Sciences*, *17*(7), 1769–1781. https://doi.org/10.7150/ijbs.59025

Lories, B., Roberfroid, S., Dieltjens, L., De Coster, D., Foster, K. R., & Steenackers, H. P. 2020. Biofilm bacteria use stress responses to detect and respond to competitors. *Current Biology*, *30*(7), 1231–1244.e4. https://doi.org/10.1016/j.cub.2020.01.065

Lu, J., Hong, Y., Wei, Y., Gu, J.-D., Wu, J., Wang, Y., Ye, F., & Lin, J.-G. 2021. Nitrification mainly driven by ammonia-oxidizing bacteria and nitrite-oxidizing bacteria in an anammox-inoculated wastewater treatment system. *AMB Express*, *11*(1), 158. https://doi.org/10.1186/s13568-021-01321-6

Luo, X., Zhang, B., Lu, Y., Mei, Y., & Shen, L. 2022. Advances in application of ultraviolet irradiation for biofilm control in water and wastewater infrastructure. *Journal of Hazardous Materials*, *421*, 126682. https://doi.org/10.1016/j.jhazmat.2021.126682

Lytle, D. A., Sorg, T. J., Wang, L., Muhlen, C., Rahrig, M., & French, K. 2007. Biological nitrification in a full-scale and pilot-scale iron removal drinking water treatment plant. *Journal of Water Supply: Research and Technology-Aqua*, *56*(2), 125–136. https://doi.org/10.2166/aqua.2007.092

Mahapatra, A. 2015. Study of biofilm in bacteria from water pipelines. *Journal of Clinical and Diagnostic Research,9(3):DC09-DC11.* https://doi.org/10.7860/JCDR/2015/12415.5715

McCormick, A., Hoellein, T. J., Mason, S. A., Schluep, J., & Kelly, J. J. 2014. Microplastic is an abundant and distinct microbial habitat in an Urban River. *Environmental Science & Technology*, *48(20),* 11863–11871. https://doi.org/10.1021/es503610r

Moreira, J. M. R., Simões, M., Melo, L. F., & Mergulhão, F. J. 2015. The combined effects of shear stress and mass transfer on the balance between biofilm and suspended cell dynamics. *Desalination and Water Treatment*, *53*(12), 3348–3354. https://doi.org/10.1080/19443994.2014.933625

Moreno Osorio, J. H., Pollio, A., Frunzo, L., Lens, P. N. L., & Esposito, G. 2021. A review of microalgal biofilm technologies: Definition, applications, settings and analysis. *Frontiers in Chemical Engineering*, *3*:737710. https://doi.org/10.3389/fceng.2021.737710

Moyal, J., Dave, P. H., Wu, M., Karimpour, S., Brar, S. K., Zhong, H., & Kwong, R. W. M. 2023. Impacts of biofilm formation on the physicochemical properties and toxicity of microplastics: A concise review. *Reviews of Environmental Contamination and Toxicology*, *261*(1), 8. https://doi.org/10.1007/s44169-023-00035-z

Muhammad, M. H., Idris, A. L., Fan, X., Guo, Y., Yu, Y., Jin, X., Qiu, J., Guan, X., & Huang, T. 2020. Beyond risk: Bacterial biofilms and their regulating approaches. *Frontiers in Microbiology*, *11*. 928. https://doi.org/10.3389/fmicb.2020.00928

Nadell, C. D., Drescher, K., & Foster, K. R. 2016. Spatial structure, cooperation and competition in biofilms. *Nature Reviews Microbiology*, *14*(9), 589–600. https://doi.org/10.1038/nrmicro.2016.84

Neagu, L., Cirstea, D. M., Curutiu, C., Mitache, M. M., Lazăr, V., & Chifiriuc, M. C. 2017. Microbial biofilms from the aquatic ecosystems and water quality. In *Water Purification* (pp. 621–642). Elsevier. https://doi.org/10.1016/B978-0-12-804300-4.00018-6

Nesse, L. L., Osland, A. M., & Vestby, L. K. 2023. The role of biofilms in the pathogenesis of animal bacterial infections. *Microorganisms*, *11*(3), 608. https://doi.org/10.3390/microorganisms11030608

Ojeda Rodriguez JA, Kahwaji CI. 2023. Vibrio Cholerae Infection. [Updated 2022 Sep 1]. In: *StatPearls [Internet]*. StatPearls Publishing. Available from: www.ncbi.nlm.nih.gov/books/NBK526099/.

Pandey, P. K., Vivekanand, B., & Kundan, K. 2014. Biofilm in aquaculture production. *African Journal of Microbiology Research*, *8*(13), 1434–1443. https://doi.org/10.5897/AJMR2013.6445

Picioreanu, C., Vrouwenvelder, J. S., & van Loosdrecht, M. C. M. 2009. Three-dimensional modeling of biofouling and fluid dynamics in feed spacer channels of membrane devices. *Journal of Membrane Science*, *345*(1–2), 340–354. https://doi.org/10.1016/j.memsci.2009.09.024

Puttamreddy, S., Cornick, N. A., & Minion, F. C. 2010. Genome-wide transposon mutagenesis reveals a role for pO157 genes in biofilm development in *Escherichia coli* O157:H7 EDL933. *Infection and Immunity*, *78*(6), 2377–2384. https://doi.org/10.1128/IAI.00156-10

Quilliam, R. S., Jamieson, J., & Oliver, D. M. 2014. Seaweeds and plastic debris can influence the survival of faecal indicator organisms in beach environments. *Marine Pollution Bulletin*, *84*(1–2), 201–207. https://doi.org/10.1016/j.marpolbul.2014.05.011

Radu, A. I., Vrouwenvelder, J. S., van Loosdrecht, M. C. M., & Picioreanu, C. 2012. Effect of flow velocity, substrate concentration and hydraulic cleaning on biofouling of reverse osmosis feed channels. *Chemical Engineering Journal*, *188*, 30–39. https://doi.org/10.1016/j.cej.2012.01.133

Ramamurthy, T., Chowdhury, G., Pazhani, G. P., & Shinoda, S. 2014. Vibrio fluvialis: An emerging human pathogen. *Frontiers in Microbiology*, *5*. https://doi.org/10.3389/fmicb.2014.00091

Rashid, M. I., Mujawar, L. H., Shahzad, T., Almeelbi, T., Ismail, I. M. I., & Oves, M. 2016. Bacteria and fungi can contribute to nutrients bioavailability and aggregate formation in degraded soils. *Microbiological Research*, *183*, 26–41. https://doi.org/10.1016/j.micres.2015.11.007

Relucenti, M., Familiari, G., Donfrancesco, O., Taurino, M., Li, X., Chen, R., Artini, M., Papa, R., & Selan, L. 2021. Microscopy methods for biofilm imaging: Focus on SEM and VP-SEM pros and cons. *Biology*, *10*(1), 51. https://doi.org/10.3390/biology10010051

Rendueles, O., & Ghigo, J.-M. (2015a). Mechanisms of competition in biofilm communities. *Microbiology Spectrum*, *3*(3). https://doi.org/10.1128/microbiolspec.MB-0009-2014

Rendueles, O., & Ghigo, J.-M. (2015b). Mechanisms of competition in biofilm communities. *Microbiology Spectrum*, *3*(3). https://doi.org/10.1128/microbiolspec.MB-0009-2014

Rodrigues, A., Oliver, D. M., McCarron, A., & Quilliam, R. S. 2019. Colonisation of plastic pellets (nurdles) by E. coli at public bathing beaches. *Marine Pollution Bulletin*, *139*, 376–380. https://doi.org/10.1016/j.marpolbul.2019.01.011

Rusconi, R., Lecuyer, S., Autrusson, N., Guglielmini, L., & Stone, H. A. 2011. Secondary flow as a mechanism for the formation of biofilm streamers. *Biophysical Journal*, *100*(6), 1392–1399. https://doi.org/10.1016/j.bpj.2011.01.065

Samrot, A. V., Abubakar Mohamed, A., Faradjeva, E., Si Jie, L., Hooi Sze, C., Arif, A., Chuan Sean, T., Norbert Michael, E., Yeok Mun, C., Xiao Qi, N., Ling Mok, P., & Kumar, S. S. 2021. Mechanisms and impact of biofilms and targeting of biofilms using bioactive compounds—A review. *Medicina*, *57*(8), 839. https://doi.org/10.3390/medicina57080839

Sauer, K., Camper, A. K., Ehrlich, G. D., Costerton, J. W., & Davies, D. G. 2002. *Pseudomonas aeruginosa* displays multiple phenotypes during development as a biofilm. *Journal of Bacteriology*, *184*(4), 1140–1154. https://doi.org/10.1128/jb.184.4.1140-1154.2002

Schmidt, V. T., Reveillaud, J., Zettler, E., Mincer, T. J., Murphy, L., & Amaral-Zettler, L. A. 2014. Oligotyping reveals community level habitat selection within the genus Vibrio. *Frontiers in Microbiology*, *5*. https://doi.org/10.3389/fmicb.2014.00563

Schoina, E., Doulgeraki, A. I., Miliou, H., & Nychas, G.-J. E. 2022. Dynamics of water and biofilm bacterial community composition in a mediterranean recirculation aquaculture system. *Aquaculture Journal*, *2*(2), 164–179. https://doi.org/10.3390/aquacj2020008

September, S. M., Els, F. A., Venter, S. N., & Brözel, V. S. 2007. Prevalence of bacterial pathogens in biofilms of drinking water distribution systems. *Journal of Water and Health*, *5*(2), 219–227. https://doi.org/10.2166/wh.2007.004b

Silva, M. M., Maldonado, G. C., Castro, R. O., de Sá Felizardo, J., Cardoso, R. P., Anjos, R. M. dos, & Araújo, F. V. de. 2019. Dispersal of potentially pathogenic bacteria by plastic debris in Guanabara Bay, RJ, Brazil. *Marine Pollution Bulletin*, *141*, 561–568. https://doi.org/10.1016/j.marpolbul.2019.02.064

Silva, N. B. S., Marques, L. A., & Röder, D. D. B. 2021. Diagnosis of biofilm infections: Current methods used, challenges and perspectives for the future. *Journal of Applied Microbiology*, *131*(5), 2148–2160. https://doi.org/10.1111/jam.15049

Song, C., & Cui, W. 2020. Review of underwater ship hull cleaning technologies. *Journal of Marine Science and Application*, *19*(3), 415–429. https://doi.org/10.1007/s11804-020-00157-z

Stoodley, P., Dodds, I., Boyle, J. D., & Lappin-Scott, H. M. 1998. Influence of hydrodynamics and nutrients on biofilm structure. *Journal of Applied Microbiology*, *85*(S1), 19S–28S. https://doi.org/10.1111/j.1365-2672.1998.tb05279.x

Strom, M. S., & Paranjpye, R. N. 2000. Epidemiology and pathogenesis of Vibrio vulnificus. *Microbes and Infection*, *2*(2), 177–188. https://doi.org/10.1016/S1286-4579(00)00270-7

Taherzadeh, D., Picioreanu, C., Küttler, U., Simone, A., Wall, W. A., & Horn, H. 2010. Computational study of the drag and oscillatory movement of biofilm streamers in fast flows. *Biotechnology and Bioengineering*, *105*(3), 600–610. https://doi.org/10.1002/bit.22551

Thomen, P., Robert, J., Monmeyran, A., Bitbol, A.-F., Douarche, C., & Henry, N. 2017. Bacterial biofilm under flow: First a physical struggle to stay, then a matter of breathing. *PLOS ONE*, *12*(4), e0175197. https://doi.org/10.1371/journal.pone.0175197

Thompson, F. L., Abreu, P. C., & Wasielesky, W. 2002. Importance of biofilm for water quality and nourishment in intensive shrimp culture. *Aquaculture*, *203*(3–4), 263–278. https://doi.org/10.1016/S0044-8486(01)00642-1

Verdegem, M., Buschmann, A. H., Latt, U. W., Dalsgaard, A. J. T., & Lovatelli, A. 2023. The contribution of aquaculture systems to global aquaculture production. *Journal of the World Aquaculture Society*, *54*(2), 206–250. https://doi.org/10.1111/jwas.12963

Villanueva, V. D., Font, J., Schwartz, T., & Romaní, A. M. 2011. Biofilm formation at warming temperature: acceleration of microbial colonization and microbial interactive effects. *Biofouling*, *27*(1), 59–71. https://doi.org/10.1080/08927014.2010.538841

Wang, C., Jiang, C., Gao, T., Peng, X., Ma, S., Sun, Q., Xia, B., Xie, X., Bai, Z., Xu, S., & Zhuang, X. 2022. Improvement of fish production and water quality in a recirculating aquaculture pond enhanced with bacteria-microalgae association. *Aquaculture*, *547*, 737420. https://doi.org/10.1016/j.aquaculture.2021.737420

Wang, X., Zhang, Y., Qin, G., Luo, W., & Lin, Q. 2016. A novel pathogenic bacteria (*Vibrio fortis*) causing enteritis in cultured seahorses, *Hippocampus erectus* Perry, 1810. *Journal of Fish Diseases*, *39*(6), 765–769. https://doi.org/10.1111/jfd.12411

Wei, Q., & Ma, L. 2013. Biofilm matrix and its regulation in Pseudomonas aeruginosa. *International Journal of Molecular Sciences*, *14*(10), 20983–21005. https://doi.org/10.3390/ijms141020983

Wilson, C., Lukowicz, R., Merchant, S., Valquier-Flynn, H., Caballero, J., Sandoval, J., Okuom, M., Huber, C., Brooks, T. D., Wilson, E., Clement, B., Wentworth, C. D., & Holmes, A. E. 2017. Quantitative and qualitative assessment methods for biofilm growth: A mini-review. *Research & Reviews. Journal of Engineering and Technology*, *6*(4), 1–41.

Yang, H., Qi, S., & Yan, Y. 2021. Influencing factors of biofilm nitrification in long-distance water pipeline. *IOP Conference Series: Earth and Environmental Science*, *702*(1), 012057. https://doi.org/10.1088/1755-1315/702/1/012057

Yeor-Davidi, E., Zverzhinetsky, M., Krivitsky, V., & Patolsky, F. 2020. Real-time monitoring of bacterial biofilms metabolic activity by a redox-reactive nanosensors array. *Journal of Nanobiotechnology*, *18*(1), 81. https://doi.org/10.1186/s12951-020-00637-y

Yin, Z., Bi, X., & Xu, C. 2018. Ammonia-Oxidizing Archaea (AOA) Play with Ammonia-Oxidizing Bacteria (AOB) in nitrogen removal from wastewater. *Archaea*, *2018*, 1–9. https://doi.org/10.1155/2018/8429145

Zhang, Y., Love, N., & Edwards, M. 2009. Nitrification in drinking water systems. *Critical Reviews in Environmental Science and Technology*, *39*(3), 153–208. https://doi.org/10.1080/10643380701631739

Zhao, A., Sun, J., & Liu, Y. (2023a). Understanding bacterial biofilms: From definition to treatment strategies. *Frontiers in Cellular and Infection Microbiology*, *13*. https://doi.org/10.3389/fcimb.2023.1137947

Zhao, A., Sun, J., & Liu, Y. (2023b). Understanding bacterial biofilms: From definition to treatment strategies. *Frontiers in Cellular and Infection Microbiology*, *13,* 1137947. https://doi.org/10.3389/fcimb.2023.1137947

Zheng, S., Bawazir, M., Dhall, A., Kim, H.-E., He, L., Heo, J., & Hwang, G. 2021. Implication of surface properties, bacterial motility, and hydrodynamic conditions on bacterial surface sensing and their initial adhesion. *Frontiers in Bioengineering and Biotechnology*, *9*,643722. https://doi.org/10.3389/fbioe.2021.643722

Zhou, Q., Li, K., Jun, X., & Bo, L. 2009. Role and functions of beneficial microorganisms in sustainable aquaculture. *Bioresource Technology*, *100*(16), 3780–3786. https://doi.org/10.1016/j.biortech.2008.12.037

16 Wetland Ecosystems and Biofilms

Suseela Lanka, Mounika K., Ashok Kota, and Santhilatha Pandrangi

16.1 WETLANDS

Areas of land that are periodically or continuously saturated with water are known as wetlands. They can be found in many different types of environments, including bogs, marshes, and swamps, all throughout the world. Important ecosystems, wetlands offer many advantages to both people and wildlife. In addition, wetlands offer several other crucial ecological services, such as recreation, carbon storage, and flood control.

Areas of land that are consistently or occasionally submerged in water are known as wetlands. They are distinguished by the presence of hydric soils and moisture-adapted vegetation. Wetlands encompass a range of habitats, including bogs, fens, marshes, and swamps, and are found all over the world.

Additionally, there are wetlands made by humans known as "anthropogenic wetlands," which artificially hold water both temporarily and permanently while adjusting to the surrounding environment.

We examine the various kinds of wetlands and the key features of these extremely diverse natural ecosystems in the sections that follow.

16.1.1 Wetland Types

Wetlands come in a variety of forms, each having special qualities and ecological purposes of their own. The following are a few of the primary categories of wetlands.

16.1.1.1 Marshes

Grass and other herbaceous plants predominate in marshes. They are usually found near bodies of water that are still or move slowly, like the banks of lakes and rivers. While groundwater feeds certain marshes, surface water feeds the majority of them. There are plenty of nutrients and typically a neutral pH, which leads to a large number of plants and animals.

16.1.1.2 Swamps

These are marshes where bushes and trees predominate. They are usually found near bodies of water, like floodplains and coastal regions. Saturated soils and stagnant

DOI: 10.1201/9781003487203-16

waters are key swamp characteristics. Water-tolerant plants thrive in the dark, nutrient-rich soils found in swamps due to their high organic content.

16.1.1.3 Bogs

Bogs are types of wetlands that are typified by nutrient-poor, acidic soils that are dominated by plants that love acid, such as mosses. Usually, they are found in places with lots of rain and chilly weather. Rainfall, as opposed to runoff, groundwater, or streams, provides all or most of the water in peatlands.

16.1.1.4 Fens

These are wetlands resemble bogs, but they get their nutrients from other sources apart from precipitation, like groundwater movement and the drainage of nearby mineral soils. Bogs and fens are not the same since fens are higher in nutrients and less acidic. A range of wetland plants, such as ferns, grasses, and sedges, are what define them.

These are some of the types of wetlands. All wetlands serve crucial ecological roles and serve as birth places to a distinct variety of animal and plant species.

16.2 CHARACTERISTICS OF WETLANDS

There are certain features that all wetlands have, even though their physical, chemical, and biological qualities vary based on their hydrology, soil, and flora.
Wetlands are places where water is either always present or just occasionally present. What sets wetlands apart from other kinds of ecosystems is the presence of water. Diverse plant and animal species can be found in wetlands, which are frequently quite varied environments. They provide crucial homes for a variety of migrating birds and other wildlife, including rare and endangered species.

Hydrology, the study of the flow and distribution of water in an environment, has an impact on wetlands. The physical and chemical processes in a wetland are impacted by the presence of water. Hydrophilic vegetation is a defining characteristic of wetlands. If non-hydrophilic plants begin to take the place of hydrophilic vegetation, it could mean that the wetland ecosystem is changing into something else entirely, like a terrestrial setting. Wetlands, often known as mixed ecosystems, are places where aquatic and terrestrial systems gradually transition or alter. Wetlands are intricate ecosystems with a distinct set of chemical, biological, and physical characteristics. Comprehending these attributes is crucial for efficient management and preservation of wetland ecosystems. Figure 16.1 shows wetland types and their characteristics.

16.3 IMPORTANCE OF WETLANDS

Wetlands offer numerous significant ecological and economic advantages. Wetlands are birth places to a great diversity of plant and animal species and many of them are exclusive to these habitats. Indeed, wetlands can provide crucial habitats for threatened or endangered species and can be used in habitat restoration initiatives.

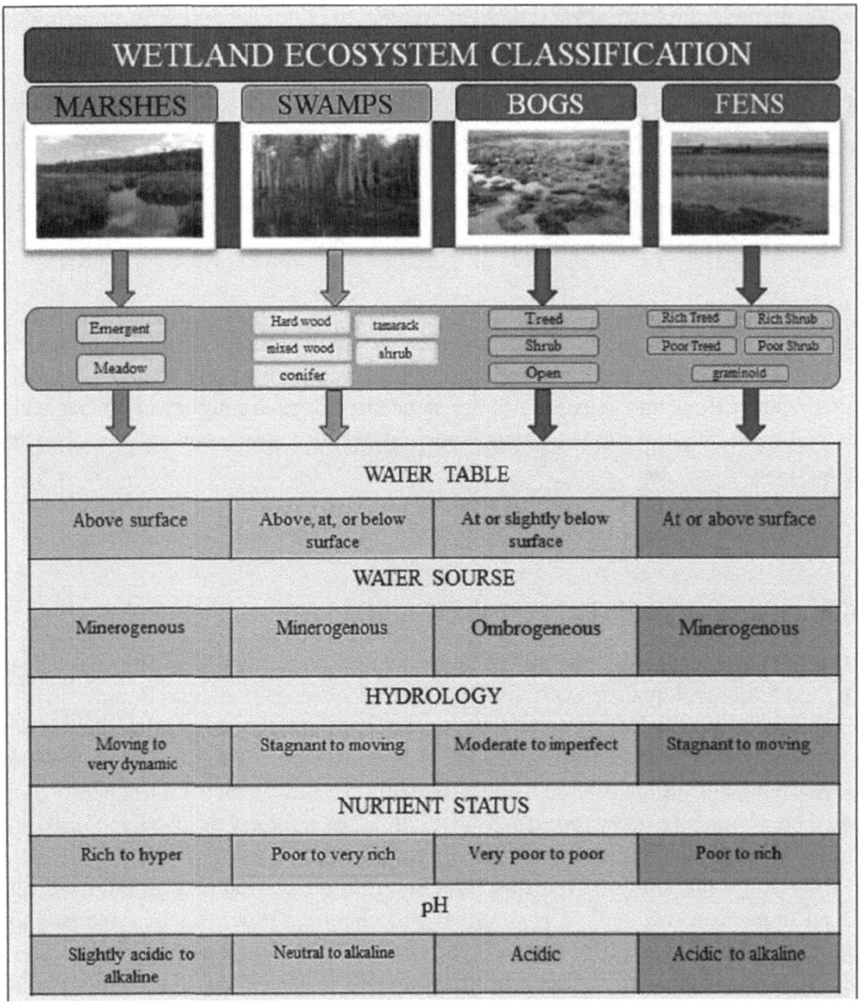

FIGURE 16.1 Classification of wetland ecosystem.

Moreover, wetlands serve as organic filters, drawing impurities and extraneous nutrients out of water before it reaches rivers, streams, and other water bodies. Furthermore, the soils of wetlands hold significant amounts of carbon, which helps to lessen the consequences of climate change.

Lastly, during intense downpours, wetlands absorb and hold onto extra water, reducing the chance of flooding downstream. Additionally, marsh vegetation aids in soil stabilization and reduces erosion, both of which can safeguard nearby infrastructure and towns. Furthermore, wetlands can support the replenishment of groundwater and the maintenance of river and stream water levels, all of which are beneficial to the availability of drinking water.

16.4 INTRODUCTION TO BIOFILMS

According to Vert et al. (2012), biofilms constitute collections of microbes in which the microbial cells are often encased in an auto-produced EPS (extracellular polymeric substance) matrix where they stick to the surface (or) to one another.

Important elements of the wetland ecosystem that are in charge of bioremediating both organic and inorganic contaminants include submerged macrophytes and epiphytic biofilms (Li et al., 2021; Mu et al., 2020). In addition to offering a secure ecological housing for microbes like bacterial species, fungus, algal members, and metazoans that aid in biofilm production, the resident flora supply O_2 for the oxidative destruction of pollution-causing agents (He et al., 2016; Mu et al., 2020). In order to restore the ecology of contaminated sediments/soil, such as that containing PAHs (polycyclic aromatic hydrocarbons), a sustainable natural solution is formed by the networks of epiphytic biofilms and macrophytes (He et al., 2016; Zhao et al., 2018, 2019; Ohore et al., 2021). It is crucial to comprehend the dynamics of microbes and succession of the microbial consortia, especially with regard of epiphytic biofilms and cleanup of PAHs, because the lodged microbial consortia react towards contaminants by switching their pattern of assemblage.

The core and fundamental subject of microbial ecology is the community assemblage mechanics (Rosindell et al., 2011). Microbial community assemblage is mostly determined by stochastic and deterministic mechanisms. The microbial community is driven by biotic interactions and environmental conditions, which are referred to as deterministic processes. Birth, death, speciation, and immigration are considered by stochastic processes to be the chief forces influencing formation of microbial consortia. Regarding assemblage of microbial consortia, especially with regard to epiphytic biofilms of wet land sub-merged macrophytes, not much is known.

Microorganisms in the sediment and aquatic environments differ from those that live in community with regard to diversity and microbe composition, which is determined by community assembly processes. Even during the brief colonization, the epiphytic microbial population underwent succession and turnover due to culture time (Yi et al., 2021). It has been shown that in aquatic environments, local environmental characteristics like pH and DO can cause variations in the microbial composition (Yan et al., 2019; Si et al., 2020). Community assembly mechanisms are responsible for all of these variables influencing the microbe community's changes (Mu et al., 2021; Yi et al., 2021).

Microbial co-existence methods have a notable correlation with community assembly processes. The perspective of stochastic and deterministic processes is significantly influenced by species interactions (Chen et al., 2018).

The community assembly regulates ecological functions, which are closely linked to a variety of microbial communities. Because community assembly affects species identification and abundance, it is essential for creating microbial diversity and establishing functions. The stability of ecological services can be ensured by microbial communities containing functionally redundant species (Allison, 2008). However, in the presence of outside forces (such as nitrification and breakdown), this stability may be lost (Allison and Martiny, 2008).

16.5 SIGNIFICANCE OF BIOFILMS

The the most common microbe on our planet is mainly represented by biofilms, which are colonies encased in a matrix. Biofilms are extremely relevant to all habitats they live in since they are by nature more productive than any comparable planktonic population. Environmental regulators, conservationists, and the general public are still largely unaware of their existence and significance.

The majority of microorganisms found in multicellular organisms, such as people, animals, and plants, exist as real biofilms or structures resembling biofilms, and these structures are essential to the development, physiology, and immunology of these species. In lieu, certain biofilms may adverse to the host's health.

Many terrestrial and marine ecosystems depend on biofilms that primarily grow on non-biological surfaces because they support food webs, maintain nutrient cycling, as well enable bioremediation in natural systems. Ecological biofilms, on the other hand, have the potentiality to seriously impair human well-being and the economy by producing toxic substances, fouling and corroding surfaces in both natural and artificial environments, and encouraging the persistence of human diseases.

Although the significance of biofilms in the epidemiology of infectious diseases that are emerging in wildlife are not well understood, they represent a serious danger to sustainability, biodiversity, and public health. All of these factors have an impact on the form and functioning of biofilms.

16.6 BIOFILM DEVELOPMENT

According to researchers (Costerton et al., 1999; Kraigsley et al., 2002), a biofilm consists of a complex network of microbes colonized either on an inert or on a live surface that are encased in a matrix of EPS auto-produced by them (Davey and O'Toole, 2000).

The intricacy and significance of biofilm in various contexts and disciplines remain unexplained by this extensive and detailed definition. It is now commonly acknowledged that the majority of bacteria that are discovered in industrial, clinical, and ecological environments continue to be associated with various surfaces (Donlan, 2002).

The initiation of biofilm production involves three main processes: (1) Attachment: Planktonic bacteria randomly adhere to fixed surfaces using their polar flagella. At first, attachment is reversible, but it eventually becomes permanent. (2) Maturation: A wide variety of microbial species stick to the initiator biofilm monolayer and produce a 3D biofilm; (3) Replication: Initial colonizers multiply and form a monolayer of microcolonies on a fixed substratum. As per Watnick and Kolter (2000) the various bacteria create exopolysaccharides that enhance and stabilize the biofilm structure as it develops.

16.6.1 Factors Affecting Biofilm Growth

Many factors, such as bacterial species, flow parameters, surface features, and nutrition availability, influence biofilm formation (Watnick and Kolter, 2000). Numerous bacterial species such as *Escherichia* sp. (Lee et al., 2007; Pratt and Kolter, 1998), *Pseudomonas aeruginosa, Vibrio* sp. (Watnick et al., 1999), and *Staphylococcus sp.* (Tormo et al., 2007; Harmsen et al., 2010) have potential to extensively build biofilms.

Gene expression, formation mechanisms, and biofilm morphology were the main topics of those investigations. Research demonstrated that bacteria in biofilm form differ when compared to their planktonic counterparts with regard to gene transcription and rate of growth (Donlan, 2002). Numerous researchers looked into how organic load and nutrients affect single species biofilms (SSBs). For example, it has been demonstrated that P, N, and glucose limitation affect *Citrobacter* sp. biofilm production and cell phenotypic features (Allan et al., 2002).

Concentrations of glucose and as well amino acids influence the morphology of *Serratia marcescens* biofilm (Rice et al., 2005); with reduced carbon or nitrogen, the biofilm consists of microcolonies; in the presence of increased nutrient composition, it seems to be a filamentous type. Multiple species biofilms (MSBs) are also impacted by organic load. Research from the past has demonstrated that elevated organic load increases the biomass and growth rate of complex WWT (wastewater treatment biofilms) (Yung-pin, 2005; Tal et al., 2003). Moreover, a number of investigations have demonstrated the impact of nutrient and organic load on the diversity of microbes (Barton et al., 2004; Hewson et al., 2003).

Weber and Legge (2011) illustrated that variations in the organic load cause variations in the bacterial consortia composition, biofilm morphology, and structural attributes. Thus, the organic load and nutritional concentration have a significant impact on both SSBs and MSBs.

16.7 COMPOSITION AND INTERACTIONS OF MULTI-SPECIES BIOFILM COMMUNITY

Experiments with SSBs are useful for learning the fundamentals of biofilm development, but they do not accurately capture the intricate dynamics of the MSBs that occur in nature. The makeup and function of the biofilm community may be impacted by interspecies relationships such as competition among the microbes according to Omil et al. (1998) and Tsuno et al. (2002). According to Dobretsov et al. (2009), quorum sensing is the key contributor along with the synthesis of distinct chemicals by various other bacterial members (Rao et al., 2005).

As demonstrated by dual-species biofilm formation experiments (Rao et al., 2005) two bacterial species combinations lead to either antagonism or complementation in the biomass. The same finding was reported by Mengying et al. (2007) Multi-species bacterial interactions in complex biofilms can be better understood through experiments with mixed cultures of bacterial isolates (Burmølle et al., 2006; Andersson et al., 2008). Hansen et al. (2007) reported the same. Understanding environmental biofilms may also result from more research on naturally existing biofilms with regard to bacterial community makeup and functioning.

16.8 MANGROVE WETLANDS AND BENTHIC BIOFILMS

According to Lovelock et al. (2015), mangroves constitute intertidal forests found along costal zones of tropics and subtropics that offer various ecosystem benefits, including carbon sequestration, protection of the coastline, and support for fisheries. Mangroves make up to about 0.5% of the world's coast line (~138,000 km2), but are capable of exporting around 10% of the particulate earthly carbon to the marines and contribute 10%–15% (24 Tg C year−1) to the carbon storage of coastal sediments (Alongi, 2014).

Mangroves have been used as suitable disposal areas for refuse produced by humans (Tam, 2006). The mangrove wetland ecosystem serves as an intermediate habitat between terrestrial land and oceans, and its unique ecological functions include the purification of nutrients and pollutants derived from land and the buffering of wave moments to prevent maritime abrasion (Wang and Gu, 2021). Historically, several of the urban and industrial centers were placed near estuaries sequined by mangroves.

Marked differences in microbe diversity and metabolic functioning can be witnessed between varied tidal flats (high-medium-low) in artificial mangrove wetlands and mangrove-associated mudflats (Yin and Yan, 2020; Zhang et al., 2018). The sediment-type biofilms in these wetlands have the capability of resisting the floods by tides and ability to adjust to spatial and temporal modifications.

Geographical heterogeneity of mangrove swamp sediment composition resulting from tidal zonation and the root zone impact was linked to the structure of microeukaryotic communities (Zhu et al., 2018). In mangrove habitats on India's southwest coast, notable seasonal fluctuations in the microphytobenthos community structure were noted (Benny et al., 2021). According to Costa-Boddeker et al. (2017), the two most important factors influencing the composition of surface sediment's diatoms include salinity and nutrients. The associated biofilms of mangrove sediments, notably EPS, are the least studied of the impacts of ecological and spatio-temporal variables aside from microbial composition.

According to Lihua Yang et al. (2022), microbial diversity of sediment-type biofilms in marine environments is associated with season, effects of tide, and related sediment characteristics. Their studies revealed that the most prevalent EPS fraction, B protein, was kept at comparatively constant, in contrast to C & B polysaccharides, mono sugar components, and C protein, and this could be attributed to tidal variations, sampling intervals, and wetland type. While the seaward site contained more central diatoms, the landward site acquired more pennate diatoms. The seaward site had higher monosaccharide component concentrations and lower B polysaccharide concentrations due to the change in diatom composition and possible tidal loss.

16.9 BIOFILMS AND WASTEWATER TREATMENT (WWT) PLANTS

Treatment wetlands (TWs) are a low-cost, low-energy, natural wastewater treatment method (Kadlec and Wallace, 2009). These systems' varied microbial communities' greater rates of biological activity account for a large portion of their complicated wastewater treatment potential (Vymazal, 2011).

The use of artificial wetlands as affordable and sustainable WWT solutions is growing in popularity. While there are naturally occurring wetlands all over the world, the idea of using artificial or manufactured wetlands to improve water quality has just gained traction in the last 50 years (Vymazal, 2011).

The polluted water is channeled through artificial wetlands for treatment by biological (microbial competition, bacterial die-off, microbial decomposition/degradation, nutrient transformations by microbes, uptake from water column/root zone), chemical (precipitation, adsorption, volatilization), and physical (sedimentation, filtration, UV exposure) processes (Vymazal, 2011).

Most of the crucial changes and breakdown of pollutants in wastewater are caused by biofilms adhered to the matrix of the wetland, which consists of solid matter and/or plant roots (Larsen and Greenway, 2004; Faulwetter et al., 2009). Nevertheless, comparatively less is understood about the microbial population that participates actively in biofilm's production (Truu et al., 2009). A range of wetland systems contained bacteria belonging to diverse phyla. Ahn et al. (2007) described microbe consortium in artificial wetlands that were either planted or unplanted and had varying phosphorus loading. α-Proteobacteria (48–60% of clones) dominated the microbial consortium in the wetland microcosm substrate, with Actinobacteria- and Firmicutes- related bacteria coming in second in abundance. Ibekwe et al. (2007) studied the rhizosphere and sediments from a built surface flow wetland model and demonstrated that Proteobacteria members made up the second dominating group, with the bulk of sequences recovered belonging to unclassified species.

In the absence of a biofilm, microorganisms will easily leave the TW system by being carried out with the water flow. In the event that this happens, the microorganism may provide a restricted capacity for water treatment when in contact with the pollutants for the duration that it is present in the TW system, which is limited to a maximum of one hydraulic retention time (HRT) of useful service. The truth is that TWs display a variety of flow distributions that signify distinct flow pathways and clearance kinetics that are incongruent with a pristine, biofilm-free "black box" concept. For TW to operate, biofilms—which are home to microbial populations—are therefore equally as crucial as the microbial communities themselves.

Few studies in the field of TWs have concentrated on characterization of biofilm at the microscale, while many have explored the activity, function, and structural diversity of microorganisms to uncover mechanisms behind their "black box" (Weber, 2016).

Mature microbial biofilms, ranging in complexity from single species to more complex environmental samples, have been demonstrated through imaging techniques to exhibit heterogeneity and complex structures instead of uniform morphological traits (Neu and Lawrence, 1997; Wimpenny et al., 2000). This stands in stark contrast to the conventional TW biofilm model that has been employed in macroscale studies, which was traditionally understood or represented as a homogeneous, uniform entity.

Anbareen J. Farooq et al. (2023) used optical light microscopy, Wet-SEM, and E-SEM to study the morphology of detached, rhizospheric, and gravel-associated biofilms. Images showed that the architecture of TW biofilms is clearly heterogeneous, exhibiting continuous films, aggregate clusters, and a porous matrix with elements like filaments, corrugated surfaces, pores/cavities, and water channels.

In terms of its physical and chemical characteristics, the wetland is considered as extremely dynamic and changing system. Different redox gradients, availability of substrate, and environmental factors like temperature and pH all contribute to this variability. Throughout the wetland ponds, the various gradients produce varying niches where various biochemical activities occur (Scholz and Lee, 2005).

Numerous research findings looked into how the microbial consortium composition of the wetland biofilms was affected by various wetland environments. Truu et al. (2005) evaluated microbial consortia of a planted soil wetland for treating household wastewater and elucidated that microbial consortia was influenced by depth in terms of the communities of bacteria and Archaea, including those that oxidized ammonia. Conversely, the makeup of the wetland microbial community is not significantly impacted by the presence of different substrate materials or plants (Silyn-Roberts & Lewis, 2004; Osem et al., 2006). Furthermore, organic matter (Nguyen, 2000), depth of the wetland (Tietz et al., 2007), in addition to surface characteristics (Ladd et al., 2004), all had an impact on the biomass of the biofilm and as well on activity of various wetland types, indicating the influence of various physical and chemical properties on the assembly of the biofilm and on function.

According to Daims et al. (2016), the majority of functional WWT systems rely on naturally existing microbes for the cycling of nutrients and the breakdown of organic carbon. Now it is known that the majority of the potent bacterial species in WWT plants are found in adherent or flocculent type biofilms, despite the fact that microorganisms engaged in WWT processes are often examined in liquid media (Wuertz et al., 2003). Given the importance of wastewater treatment for both public health and the ecosystem as a whole, research into these biofilms is obviously needed.

By creating high-quality effluent through advanced wastewater treatments, the harm due to discharge of treated wastewater into various water sources can be minimized. Furthermore, in nations like Israel that use treated wastewater for irrigation due to a lack of water resources, the quality of the wastewater treatment is crucial (Smakhtin, 2004, Navon et al., 2011).

16.10 CONCLUSION

Assembly of microbial consortia anchored by EPS commonly known as biofilms occurs in the aquatic systems. Biofilms are being employed in sewage treatment plants to "filter," or remove and break down, organic components from wastewater. Biofilms are essential to the wastewater treatment engineering technologies we use today. In a similar vein, they are crucial to built wetlands (CW), which are human-made systems that mimic natural wetlands and are employed to clean storm water runoff or wastewater. While some artificial wetlands are just gravel "filter" media, the majority are planted with hydrophilic (plants that love water) vegetation. At the water-solid interfaces, biofilms are involved in both. It is now clear from recent studies that planted CW systems, together with the biofilms they support, are far more successful in the initial weeks and months of treating wastewater from agricultural operations. After just three weeks of plant development, the planted system in greenhouse testing was able to remove about 80% of the oxygen requirement from organic loading in effluent from the sugarcane process. About 30% less was removed by the unplanted system. According to UC Davis biological engineer Mark Grismer, "every constructed wetland system has a 'ripening period' during which the biofilm is forming." The term "ripening" refers to the period of time required for bacteria or other microorganisms to establish themselves and start removing organic materials. The ripening period in planted CW systems is shorter, according to the results, and it can even be as short as three weeks if the created wetland contains aquatic plants with a rapid rate of growth. In plant-free settings, it can last for two to three years.

REFERENCES

Ahn C, Gillevet PM, Sikaroodi M (2007) Molecular characterization of microbial communities in treatment microcosm wetlands as influenced by macrophytes and phosphorus loading. *Ecol Indicators* 7: 852–863.

Allan VJM, Callow ME, Macaskie LE, Paterson-Beedle M (2002) Effect of nutrient limitation on biofilm formation and phosphatase activity of a Citrobacter sp. *Microbiology* 148: 277–288.

Allison SD et al. (2008) Resistance, resilience, and redundancy in microbial communities. *Proc Natl Acad Sci USA*. 105:11512–11519.

Alongi DM (2014). Carbon cycling and storage in mangrove forests. *Annu Rev Mar Sci* 6: 195–219.

Andersson S, Rajarao GK, Land CJ, Dalhammar G (2008) Biofilm formation and interactions of bacterial strains found in wastewater treatment systems. *FEMS Microbiol Lett* 283: 83–90.

Barton HA, Taylor MR, Pace NR (2004) Molecular phylogenetic analysis of a bacterial community from an oligotrophic cave environment. *Geomicrobiol J* 21:11–20.

Benny N, Thomas LC, Padmakumar KB (2021) Community structure of microphytobenthos associated with mangrove ecosystems along the southwest coast of India. *Estuar Coast* 44 (5): 1380–1391.

Burmolle M, Webb JS, Rao D, Hansen LH, Sorensen SJ, Kjelleberg S (2006) Enhanced biofilm formation and increased resistance to antimicrobial agents and bacterial invasion are caused by synergistic interactions in multispecies biofilms. *Appl Environ Microbiol* 72: 3916–3923.

Chen S et al. (2018) fastp: An ultra-fast all-in-one FASTQ preprocessor bioinformatics, 34:i884-i890.

Costa-Boddeker S, Thuyen LX, Schwarz A, Huy HD, Schwalb A (2017) Diatom assemblages in surface sediments along nutrient and salinity gradients of thi Vai Estuary and can gio mangrove forest, southern Vietnam. *Estuar. Coast*. 40 (2), 479–492.

Costerton JW, Stewart PS, Greenberg EP (1999) Bacterial bioflms: A common cause of persistent infections. *Science* 284:1318–1322.

Daims H, Lücker S, Wagner M (2016) A New Perspective on Microbes Formerly Known as Nitrite-Oxidizing Bacteria. *Trends Microbiol* 24:699–712.

Davey ME, O'toole GA (2000) Microbial biofilms: From ecology to molecular genetics. *Microbiol Mol Biol Rev* 64: 847–867. https://doi.org/10.1128/MMBR.64.4.847-867.2000

Dobretsov S, Teplitski M, Valerie P (2009) Mini-review: Quorum sensing in the marine environment and its relationship to biofouling. *Biofouling* 25: 413–427.

Donlan RM (2002) Biofilms: Microbial life on surfaces. *Emerg Infect Dis 8(9):881* . www.cdc.gov/ncidod/EID/vol8no9/02-0063.html.

Farooq AJ, Chamberlain M, Poonja A, Mumford KG, Wallace S, Weber KP (2023) Peaks, pores, and dragon eggs: Uncovering and quantifying the heterogeneity of treatment wetland biofilm matrices. *Sci Total Environ* 855: 158857, ISSN 0048-9697. https://doi.org/10.1016/j.scitotenv.2022.158857.

Faulwetter JL, Sundbergc GV, Chazarencd F, Burra MD, Brissonb J, Campera AK, Steina OR (2009) Microbial processes influencing performance of treatment wetlands: A review. *Ecol Eng* 35: 0114–899.

Hansen S, Rainey PB, Haagensen JAJ, Molin S (2007) Evolution of species interactions in a biofilm community. *Nature* 445: 533–536.

Harmsen M, Yang L, Pamp SJ, Tolker-Nielsen T (2010) An update on Pseudomonas aeruginosa biofilm formation, tolerance, and dispersal. *FEMS Immunol Med Microbiol* 59: 253–268.

He Y et al. (2016) Response of bacterial community structure to disappearance of phenanthrene and pyrene from sediment with different submerged macrophytes. *Ecol Eng*. 91, 207–211.

Hewson I, Vargo GA, Fuhrman JA (2003) Bacterial diversity in shallow oligotrophic marine benthos and overlying waters: Effects of virus infection, containment, and nutrient enrichment. *Microb Ecol* 46: 322–336.

Ibekwe AM, Lyon SR, Leddy M, Jacobson-Meyers M (2007) Impact of plant density and microbial composition on water quality from a free water surface constructed wetland. *J Appl Microbiol* 102: 921–36.

Kadlec RH , Wallace S (2009) Treatment Wetlands. 2nd ed, Taylor & Francis, Boca Raton. https://doi.org/10.1201/9781420012514

Kraigsley A, Ronney PD, Finkel SE (2002) Dynamics of self propagating fronts of motile bacteria. https://sites.usc.edu/ronney/research/bacterial-growth-and-electrical-power-production/dynamics-of-self-propagating-fronts-of-motile-bacteria/

Ladd JN, Amato M, van Veen HA (2004) Soil microbial biomass: Its assay and role in turnover of organic matter c and n. *Soil Biol Biochem* 36: 1369–1372.

Larsen E, Greenway M (2004) Quantification of biofilms in a subsurface flow wetland and their role in nutrient removal. *Water Sci Technol* 49: 115–122.

Lee J, Bansal T, Jayaraman A, Bentley WE, Wood TK (2007) Enterohemorrhagic Escherichia coli biofilms are inhibited by 7-hydroxyindole and stimulated by isatin. *Appl Environ Microbiol* 73: 4100–4109.

Li X et al. (2021) Removal of perfluoroalkyl acids (PFAAs) in constructed wetlands: considerable contributions of submerged macrophytes and the microbial community. *Water Res* . 197:117080.

Lovelock CE, Cahoon DR, Friess DA, Guntenspergen GR, Krauss KW, Reef R, et al. (2015). The vulnerability of indo-pacific mangrove forests to sea-level rise. *Nature* 526 (7574): 559–565.

Mengying L, Lu P, Zhang J, Jingliang X, Shunpeng L (2007) Establishment and characterization of dual-species biofilms formed from a 3,5-dinitrobenzoic- degrading strain and bacteria with high biofilm-forming capabilities. *FEMS Microbiol Lett* 278: 15–21.

Mu X et al. (2020) Biofilms attached to Myriophyllum spicatum play a dominant role in nitrogen removal in constructed wetland mesocosms with submersed macrophytes: evidence from 15N tracking, nitrogen budgets and metagenomics analyses. *Environ Pollut*. 266 (2),115203.

Mu X et al. (2021). Water flow and temperature drove epiphytic microbial community shift: Insight into nutrient removal in constructed wetlands from microbial assemblage and co-occurrence patterns. *Bioresour Technol*: 332:125134

Navon R, Hernandez-Ruiz S, Chorover J, Chefetz B (2011) Interactions of carbamazepine in soil: Effects of dissolved organic matter. *J Environ Qual* 40: 942–948.

Neu TR, Lawrence JR (1997) Development and structure of microbial biofilms in river water studied by confocal laser scanning microscopy, *FEMS Microbiol Ecol* 24 1: 11–25. https://doi.org/10.1111/j.1574-6941.1997.tb00419.x

Nguyen LM (2000) Organic matter composition, microbial biomass and microbial activity in gravel-bed constructed wetlands treating farm dairy wastewaters. *Ecol Eng* 16: 199–221.
Ohore OE, Zhang S, Guo S, Manirakiza B, Addo FG, Zhang W (2021) The fate of tetracycline in vegetated mesocosmic wetlands and its impact on the water quality and epiphytic microbes. J. Hazard. Mater. 417, 126148.
Omil F, Lens P, Visser A, Hulshoff Pol LW, Lettinga G (1998). Long-term competition between sulfate-reducing and methanogenic bacteria in UASB reactors treating volatile fatty acids. *Biotechnol Bioeng* 20: 676–85.
Osem Y, Chen Y, Levinson D, Hadar Y (2006) The effects of plant roots on microbial community structure in aerated wastewater-treatment reactors. *Ecol Eng* 29: 133–142.
Pratt LA, Kolter R. (1998) Genetic analysis of Escherichia coli biofilm formation: Defining the roles of flagella, motility, chemotaxis and type I pili. *Mol Microbiol* 30: 285–293.
Rao D, Webb JS, Kjelleberg S (2005) Competitive interactions in mixed-species biofilms containing the marine bacterium Pseudoalteromonas tunicata. *Appl Environ Microbiol* 71: 1729–0935.
Rice SA, Koh KS, Queck SY, Labbate M, Lam KW, Kjelleberg S (2005) Biofilm formation and sloughing in Serratia marcescens are controlled by quorum sensing and nutrient cues. *J Bacteriol* 187: 3477–3485.
Rosindell J et al. (2011) The unified neutral theory of biodiversity and biogeography at age ten. *Trends Ecol Evol.*
Scholz M, Lee BH (2005) Constructed wetlands: A review. *Int J Environ Stud* 62: 421–47.
Sentenac H, Loyau A, Leflaive J, Schmeller DS, The significance of biofilms to human, animal, plant and ecosystem health First published: 12 October 2021. *Br Ecol Soc.* https://doi.org/10.1111/1365-2435.13947
Si T et al. (2020) Bacterial succession in epiphytic biofilms and deciduous layer sediments during Hydrilla verticillata decay: A field investigation. *J Environ Sci.* 93:193–201.
Silyn-Roberts G, Lewis G (2004) Substrata effects on bacterial biofilm development in a subsurface flow dairy waste treatment wetland. *Water Sci Technol* 48: 261–269.
Smakhtin V (2004) A pilot global assessment of environmental water requirements and scarcity. *Int J Water* 29: 307–317.
Tal Y, Watts JEM, Schreier SB, Sowers KR, Schreier HJ (2003). Characterization of the microbial community and nitrogen transformation processes associated with moving bed bioreactors in a closed recirculated mariculture system. Aquaculture 215: 187–202.
Tam NFY (2006). Pollution studies on mangroves in Hong Kong and mainland China, in *Environment in Asia pacific harbours*. Ed. Wolanski,. E. (The Netherlands: Springer): 147–163.
Tietz A, Kirschner A, Langergraber G, Sleytr K, Haberl R (2007) Characterisation of microbial biocoenosis in vertical subsurface flow constructed wetlands. *Sci Total Environ* 380: 163–72.
Tormo MA, Ubeda C, Marti M, Maiques E, Cucarella C, Valle J, et al. (2007) Phase- variable expression of the biofilm-associated protein (Bap) in Staphylococcus aureus. *Microbiology* 153: 1702–1710.
Truu J, Nurk K, Juhanson J, Mander U (2005) Variation of microbiological parameters within planted soil filter for domestic wastewater treatment. *J Environ Sci Health A Tox Hazard Subst Environ Eng* 40: 1191–200.
Truu M, Juhanson J, Truu J (2009) Microbial biomass, activity and community composition in constructed wetlands. *Sci Total Environ* 407: 3958–3971.
Tsuno H, Hidakaa T, Nishimura F (2002) A simple biofilm model of bacterial competition for attached surface. *Water Res* 36: 996–1006.

Vert M, Doi Y, Hellwich KH, Hess M, Hodge P, Kubisa P, et al. (2012) Terminology for biorelated polymers and applications (IUPAC recommendations 2012). *Pure Appl Chem* 84 (2): 377–408.

Vymazal J (2011) Constructed wetlands for wastewater treatment: Five decades of experience. *Environ Sci Technol* 45: 61–69.

Wang YS, Gu JD (2021). Ecological responses, adaptation and mechanisms of mangrove wetland ecosystem to global climate change and anthropogenic activities. *Int Biodeter Biodegr* 162: 105248.

Watnick P, Kolter R. (2000) Biofilm, city of microbes. *J Bacteriol* 182: 2675–2679.

Watnick PI, Fullner KJ, Kolter R. (1999) A role for the mannose-sensitive hemagglutinin in biofilm formation by Vibrio cholerae El Tor. *J Bacteriol* 181: 3606–3609.

Weber KP, Legge RL. (2011) Dynamics in the bacterial community-level physiological profiles and hydrological characteristics of constructed wetland mesocosms during start-up. *Ecol Eng* 37 (5): 666–677, ISSN 0925-8574. https://doi.org/10.1016/j.ecoleng.2010.03.016

Weber, KP (2016) Microbial community assessment in wetlands for water pollution control: Past, present, and future outlook. *Water* 8: 503. https://doi.org/10.3390/w8110503

Wimpenny J, Manz W, Szewzyk U (2000) Heterogeneity in biofilms. *FEMS Microbiol Rev* 24(5), 661–671.

Wuertz S, Bishop PL, Wilderer PA (2003) *Biofilms in Wastewater Treatment: an Interdisciplinary Approach*. IWA Publishing: London, UK.

Yan L et al. (2019) Ammonium loading disturbed the microbial food webs in biofilms attached to submersed macrophyte Vallisneria natans. *Sci Total Environ.*

Yang L, Yang Q, Lin L, Luan T, Yang Q, Li P and Tam NFY (2022) Effects of wetland, tide, and season on benthic biofilms and related sediment properties in mangrove ecosystems. *Front Mar Sci* 9: 1043826. doi: 10.3389/fmars.2022.1043826

Yi Y, Lin CH, Wang W, Song J (2021) Habitat and seasonal variations in bacterial community structure and diversity in sediments of a Shallow lake. *Ecol. Indic.* 120, 106959.

Yin YC, Yan ZZ (2020). Variations of soil bacterial diversity and metabolic function with tidal flat elevation gradient in an artificial mangrove wetland. *Sci Total Environ* 718: 137385.

Yung-Pin T (2005) Simulation of biofilm formation at different assimilable organic carbon concentrations under lower flow velocity condition. *J Basic Microbiol* 45: 475–485.

Zhang XY, Hu BX, Ren HJ, Zhang J. (2018). Composition and functional diversity of microbial community across a mangrove-inhabited mudflat as revealed by 16S rDNA gene sequences. *Sci Total Environ* 633: 518–528.

Zhao Z et al. (2018) Dissipation characteristics of pyrene and ecological contribution of submerged macrophytes and their biofilms-leaves in constructed wetland *Bioresour Technol.* 267:158–166.

Zhao Z et al. The responding and ecological contribution of biofilm-leaves of submerged macrophytes on phenanthrene dissipation in sediments. *Environ Pollut* (2019). 131:104940

Zhu P, Wang YP, Shi TT, Huang GQ, Gong J (2018). Genetic diversity of benthic microbial eukaryotes in response to spatial heterogeneity of sediment geochemistry in a mangrove ecosystem. *Estuar Coast* 41 (3): 751–764.

Index

Note: Emboldened page numbers signify text within tables; italicized page number signify text within figures.

C

D

N

Q

R

S

T